SEED TECHNOLOGY FOR MEDICINAL AND AROMATIC PLANTS

Production, Testing, Storage, and Marketing

SEED TECHNOLOGY FOR MEDICINAL AND AROMATIC PLANTS

Production, Testing, Storage, and Marketing

Priyanka Sharma, PhD
Kishore Kumar Sharma, PhD
Gopal Shukla, PhD
Editors

AAP | APPLE ACADEMIC PRESS

First edition published 2027

Apple Academic Press Inc.
1265 Goldenrod Circle, NE,
Palm Bay, FL 32905 USA

760 Laurentian Drive, Unit 19,
Burlington, ON L7N 0A4, Canada

CRC Press
2385 NW Executive Center Drive,
Suite 320, Boca Raton FL 33431

4 Park Square, Milton Park,
Abingdon, Oxon, OX14 4RN, UK

Library and Archives Canada Cataloguing in Publication

CIP data on file with Canada Library and Archives

Library of Congress Cataloging-in-Publication Data

CIP data on file with US Library of Congress

ISBN: 978-1-77964-118-2 (hbk)
ISBN: 978-1-77964-119-9 (ebk)

DOI: 10.1201/9781779641199

About the Editors

Priyanka Sharma, PhD

Priyanka Sharma, PhD, is an Assistant Professor in Seed Science and Technology, Department of Plant Breeding and Genetics, Assam Agricultural University (AAU), Jorhat, India. She earned her BSc (Agri) and MSc (Agri) in Seed Science and Technology from AAU, Jorhat, and her PhD in Seed Science and Technology from Uttar Banga Krishi Vishwavidyalaya, Cooch Behar, West Bengal, India. She has qualified the ASRB–ICAR–NET examination and has published several research papers, review articles, books, and book chapters related to medicinal plants and other agricultural crops in reputed national and international journals, along with popular articles.

She is the recipient of a Senior Research Fellowship from the Department of Biotechnology, Government of India, New Delhi. She has teaching experience at Beehive College of Management and Technology, Dehradun, Uttarakhand; worked as a Research Associate at the ICAR–Directorate of Rapeseed-Mustard Research, Bharatpur, Rajasthan; and served as Seed Coordinator at Assam Seeds Corporation Limited, Guwahati, Assam.

Currently, she is involved in teaching, research, and conducting farmer training programs related to the production of quality seeds.

Kishore Kumar Sharma, PhD

Kishore Kumar Sharma, PhD, is currently a Professor and Head of the Department of Plant Breeding and Genetics at Assam Agricultural University (AAU), Jorhat, with more than 20 years of teaching and research experience. He holds a master's and a doctoral degree in Plant Breeding and Genetics from AAU, Jorhat. Dr. Sharma has also served as Associate Dean at Lakhimpur College of Veterinary Science, AAU.

He was involved in the development of two medium-duration rice varieties, *Satyaranjan* and *Basundhara*, which were notified by the Central Varietal Release Committee (CVRC) in 2006 for the state of Assam. In addition, he contributed to the development of two boro rice varieties, *Jyotiprasad* and *Bishnuprasad*, which were also notified in 2006. The two submergence-tolerant rice varieties, *Jalashree* and *Jalkuwari*, were recommended for cultivation in the POP meeting in 2009 and notified in March 2017. Furthermore, he developed two semi-deep rice varieties, *Panchanan* and *Padumoni*, which were recommended in the POP meeting held in 2015. *Podumoni* is an aromatic selection of the Thailand variety *KDML105*.

Apart from rice, Dr. Sharma has developed two early-duration mustard varieties, *PM 25* and *PM 26*, which were identified and recommended in the POP meeting in 2015. He has published more than 40 research papers, one book, three newspaper articles, three web articles, four bulletins, and six book chapters in his areas of research.

Dr. Sharma has managed more than nine research projects funded by national and international organizations and agencies. He is currently the Principal Investigator of the Government of India and World Bank–sponsored "National Agricultural Higher Education Project (NAHEP)," with a project cost of Rs. 24.48 crore, implemented in 2018.

Gopal Shukla, PhD

Gopal Shukla, PhD, is a Professor and Head of the Department of Forestry at North-Eastern Hill University (NEHU), Tura Campus, Tura, West Garo Hills, Meghalaya, India. Prior to joining NEHU, he served as a faculty member in the Department of Forestry at Uttar Banga Krishi Viswavidyalaya (UBKV), Cooch Behar, West Bengal, India. He has over 13 years of experience in forestry teaching, research, and extension.

His research expertise lies in forest ecology, with a focus on agroforestry, medicinal plants, conservation, and ecosystem management. Dr. Shukla has been actively involved in several prestigious projects, including the National Agricultural Innovation Project, National Innovations in Climate Resilient Agriculture (ICAR), and a SERB (DST) Young Scientist Project on Carbon Stock of Different Land Uses of the Terai Zone of West Bengal.

Dr. Shukla has made significant contributions to forestry research, authoring over 100 refereed journal articles, 30 book chapters, and ten books. He has successfully led three extramural projects as Principal Investigator and contributed to six projects as Co-PI. He has also played an important role in academic capacity-building by organizing two national seminars, a ten-day training program, and nine farmer training/awareness programs.

As a committed academician, Dr. Shukla has guided four Ph.D. scholars and thirteen master's students in forestry. He has also contributed to academic governance, serving on the Faculty Council, Faculty of Horticulture (UBKV), the Board of Studies in Forestry (Bankura University, Bankura, West Bengal, India), and as a Member of the Research Advisory Committee, ICFRE-Rain Forest Research Institute, Jorhat, Assam, India. He was actively involved in framing undergraduate and postgraduate forestry curricula for Bankura University and Amity University, Uttar Pradesh, India.

In recognition of his contributions, Dr. Shukla has received several awards, including the SERB (DST) Young Scientist Award (2012), the Young Scientist Award (2018), and multiple best oral and poster presentation awards at conferences. He is frequently invited as a speaker and has chaired technical sessions at national and international conferences.

Dr. Shukla earned his M.Sc. (Forestry) and Ph.D. (Forestry) from Uttar Banga Krishi Viswavidyalaya (UBKV), West Bengal, India.

Contents

Contributors

Priya Bansal
Department of Forestry and Natural Resources, Hemvati Nandan Bahuguna Garhwal University, Srinagar Garhwal, Uttarakhand, India

N. Bhola
Odisha University of Agriculture and Technology, Bhubaneswar, Odisha, India

Anchal Bisht
Institute of Wood Science and Technology, Bangalore, Karnataka, India

J. S. Butola
Department of Forestry and Natural Resources, Hemvati Nandan Bahuguna Garhwal University, Srinagar Garhwal, Uttarakhand, India

Rakhi Chandoliya
CSIR–Institute of Himalayan Bioresource Technology (CSIR–IHBT), Palampur, Himachal Pradesh, India; Department of Biotechnology, School of Bioengineering and Biosciences, Lovely Professional University, Phagwara, Punjab, India

Kajal Mog Chaudhuri
Department of Seed Science and Technology, Uttar Banga Krishi Viswavidyalaya, Pundibari, Cooch Behar, West Bengal, India

Amit Chauhan
CSIR–Central Institute of Medicinal and Aromatic Plants (CIMAP), Research Center, Pantnagar, Uttarakhand, India

Deepak Chettri
Department of Botany, Sikkim University, Tadong, Gangtok, Sikkim, India

Aman Dabral
Department of Forest Biology and Tree Improvement, College of Forestry, Ranichauri, VCSG Uttarakhand University of Horticulture and Forestry, Uttarakhand, India

H. K. Das
Odisha University of Agriculture and Technology, Bhubaneswar, Odisha, India

Shreya Das
ICAR–Agricultural Technology Application Research Institute, Kolkata, West Bengal, India

Suddhasuchi Das
Malda Krishi Vigyan Kendra, Uttar Banga Krishi Viswavidyalaya, Malda, West Bengal, India

Sudhir Ch. Das
Indian Forest Service (Retd.), West Bengal, India

Manmohan J. Dobriyal
College of Horticulture and Forestry, Rani Lakshmi Bai Central Agricultural University, Jhansi, Uttar Pradesh, India

Himshikha Gusain
Department of Forestry and Natural Resources, Hemvati Nandan Bahuguna Garhwal University, Srinagar Garhwal, Uttarakhand, India

Suman Gusain
CSIR–Institute of Himalayan Bioresource Technology (CSIR–IHBT), Palampur, Himachal Pradesh, India; Academy of Scientific and Innovative Research (AcSIR), Ghaziabad, Uttar Pradesh, India

Rohit Joshi
CSIR–Institute of Himalayan Bioresource Technology (CSIR–IHBT), Palampur, Himachal Pradesh, India; Academy of Scientific and Innovative Research (AcSIR), Ghaziabad, Uttar Pradesh, India

Shubham Joshi
CSIR–Institute of Himalayan Bioresource Technology (CSIR–IHBT), Palampur, Himachal Pradesh, India; Academy of Scientific and Innovative Research (AcSIR), Ghaziabad, Uttar Pradesh, India

Amir Khan
CSIR–Central Institute of Medicinal and Aromatic Plants (CIMAP), Research Center, Pantnagar, Uttarakhand, India

Sushma Kholiya
CSIR–Central Institute of Medicinal and Aromatic Plants (CIMAP), Research Center, Pantnagar, Uttarakhand, India

Dipender Kumar
Division of Crop Production and Protection, CSIR–Central Institute of Medicinal and Aromatic Plants (CIMAP), Research Center, Pantnagar, Uttarakhand, India

Anita Kumari
CSIR–Institute of Himalayan Bioresource Technology (CSIR–IHBT), Palampur, Himachal Pradesh, India; Academy of Scientific and Innovative Research (AcSIR), Ghaziabad, Uttar Pradesh, India

Khushbu Kumari
CSIR–Institute of Himalayan Bioresource Technology (CSIR–IHBT), Palampur, Himachal Pradesh, India; Academy of Scientific and Innovative Research (AcSIR), Ghaziabad, Uttar Pradesh, India

Nisha Kumari
Department of Biochemistry, Chaudhary Charan Singh Haryana Agricultural University, Hisar, Haryana, India

Nima Tshering Lepcha
Department of Botany, Nar Bahadur Bhandari Degree College, Tadong, Gangtok, Sikkim, India

Patrush Lepcha
Department of Botany, Sikkim University, Tadong, Gangtok, Sikkim, India; Animal Husbandry and Veterinary Services Department, Government of Sikkim, Krishi Bhawan, Tadong, Sikkim, India

Bikram Subba Limboo
Department of Botany, Sikkim University, Tadong, Gangtok, Sikkim, India

Payal Maiti
Department of Post-Harvest Management, Bidhan Chandra Krishi Viswavidyalaya, Mohanpur, Nadia, West Bengal, India

A. R. Malik
Division of Forest Products and Utilization, Faculty of Forestry, Sher-e-Kashmir University of Agricultural Sciences and Technology of Kashmir (SKUAST-K), Benhama, Ganderbal, Jammu and Kashmir, India

Hemanth Kumar Manne
Department of Biochemistry, Chaudhary Charan Singh Haryana Agricultural University, Hisar, Haryana, India

Amit Mishra
Department of Fruit Science, Acharya Narendra Deva University of Agriculture and Technology, Kumarganj, Ayodhya, Uttar Pradesh, India

Manish Kumar Mittal
ICAR–Directorate of Medicinal and Aromatic Plants Research, Boriavi, Anand, Gujarat, India

Rajdeep Mohanta
Department of Agriculture, Brainware University, Barasat, Kolkata, West Bengal, India

Partha Mondal
Department of Agricultural Biochemistry, Bidhan Chandra Krishi Viswavidyalaya, Mohanpur, West Bengal, India

Tanmoy Mondal
Department of Horticulture, MITS Institute of Professional Studies, Rayagada, Odisha, India

Raksha Mukhia
Department of Botany, Sikkim University, Tadong, Gangtok, Sikkim, India

Manas Ranjan Nayak
Odisha University of Agriculture and Technology, Bhubaneswar, Odisha, India

R. C. Padalia
Division of Crop Production and Protection, CSIR–Central Institute of Medicinal and Aromatic Plants (CIMAP), Research Center, Pantnagar, Uttarakhand, India

Pragya Pandey
CSIR–Central Institute of Medicinal and Aromatic Plants (CIMAP), Research Center, Pantnagar, Uttarakhand, India

Reeta Pandey
CSIR–Central Institute of Medicinal and Aromatic Plants (CIMAP), Research Center, Pantnagar, Uttarakhand, India

Akshita Patial
CSIR–Institute of Himalayan Bioresource Technology (CSIR–IHBT), Palampur, Himachal Pradesh, India; Department of Biotechnology, School of Bioengineering and Biosciences, Lovely Professional University, Phagwara, Punjab, India

Meghna Patial
CSIR–Institute of Himalayan Bioresource Technology (CSIR–IHBT), Palampur, Himachal Pradesh, India

Chakradhar Patra
Odisha University of Agriculture and Technology, Bhubaneswar, Odisha, India

Sunayana Rathi
Department of Biochemistry and Agricultural Chemistry, Assam Agricultural University, Jorhat, Assam, India

Meenakshi Rawat
CSIR–Institute of Himalayan Bioresource Technology (CSIR–IHBT), Palampur, Himachal Pradesh, India

Bidhan Roy
Department of Seed Science and Technology, Uttar Banga Krishi Viswavidyalaya, Pundibari, Cooch Behar, West Bengal, India

Monish Roy
Department of Seed Science and Technology, Uttar Banga Krishi Viswavidyalaya, Pundibari, Cooch Behar, West Bengal, India

Biplov Ch. Sarkar
Faculty of Agricultural Sciences, GLA University, Mathura, Chaumuhan, Uttar Pradesh, India

Kishore Kumar Sharma
Department of Plant Breeding and Genetics, Faculty of Agriculture, Assam Agricultural University, Jorhat, Assam, India

Pankaj Sharma
Himachal Pradesh Biodiversity Board, Paryavaran Bhawan, Shimla, Himachal Pradesh, India

Priyanka Sharma
Department of Plant Breeding and Genetics, AICRP on Crops, Faculty of Agriculture, Assam Agricultural University, Jorhat, Assam, India

Vikas Sharma
Department of Biotechnology, School of Bioengineering and Biosciences, Lovely Professional University, Phagwara, Punjab, India

Gopal Shukla
Department of Forestry, North-Eastern Hill University, Tura Campus, Tura, Meghalaya, India

Anasuya Sil
Department of Plantation Spices, Medicinal and Aromatic Crops, Bidhan Chandra Krishi Viswavidyalaya, Mohanpur, Nadia, West Bengal, India

Naresh Singh
Department of Forest Biology and Tree Improvement, College of Forestry, Ranichauri, VCSG Uttarakhand University of Horticulture and Forestry, Uttarakhand, India

Sonveer Singh
Division of Crop Production and Protection, CSIR–Central Institute of Medicinal and Aromatic Plants, Research Center, Pantnagar, Uttarakhand, India

Kumari Punam Sinha
Department of Seed Science and Technology, Uttar Banga Krishi Viswavidyalaya, Cooch Behar, West Bengal, India

Yenkokpam Supriya
Department of Agricultural Biochemistry, Bidhan Chandra Krishi Viswavidyalaya, Mohanpur, West Bengal, India

C. P. Suresh
Department of Forestry, North Eastern Hill University, Tura Campus, West Garo Hills District, Chasingre, Tura, Meghalaya, India

Manish Kumar Sutha
ICAR–Directorate of Medicinal and Aromatic Plants Research, Boriavi, Anand, Gujarat, India

Ameda Swarnalatha
Department of Plantation Spices, Medicinal and Aromatic Crops, Bidhan Chandra Krishi Viswavidyalaya, Mohanpur, Nadia, West Bengal, India

T. Tamrakar
Odisha University of Agriculture and Technology, Bhubaneswar, Odisha, India

Kodidhala Vaishnavi
Department of Plant Pathology, Uttar Banga Krishi Viswavidyalaya, Cooch Behar, West Bengal, India

K. T. Venkatesha
Division of Crop Production and Protection,
CSIR–Central Institute of Medicinal and Aromatic Plants, Research Center, Pantnagar, Uttarakhand, India

Prawal Pratap Singh Verma
Division of Crop Production and Protection, CSIR–Central Institute of Medicinal and Aromatic Plants (CIMAP), Research Center, Pantnagar, Uttarakhand, India

S. K. Verma
Department of Fruit Science, Acharya Narendra Deva University of Agriculture and Technology, Kumarganj, Ayodhya, Uttar Pradesh, India

Abbreviations

ABA	abscisic acid
AM	arbuscular mycorrhizal
BGCI	Botanic Garden Conservation International
CCRAS	Central Council for Research in Ayurvedic Sciences
CCRH	Central Council for Research in Homoeopathy
CCRS	Central Council for Research in Siddha
CCRUM	Central Council for Research in Unani Medicine
CIMAP	Central Institute of Medicinal and Aromatic Plants
CCP	composted coir pith
CSIR	Council of Scientific and Industrial Research
DBT	Department of Biotechnology
EC	electrical conductivity
EME	emergence in a greenhouse
ELISA	enzyme-linked immunosorbent assay
FYM	farmyard manure
FID	flame ionization detector
FFA	free fatty acid
GC	gas chromatography
GC-IMS	gas chromatography-ion mobility spectrometry
GHS	gas hermetic fumigation
GPI	Germination Potential Index
GBH	Global Biodiversity Hotspots
HS	hermetic storage
HEPA	high-efficiency particulate air
IFC	immunofluorescence colony-staining
IF	immunofluorescence microscopy
IMS	immunomagnetic separation
ICFRE	Indian Council of Forestry Research and Education
IHR	Indian Himalayan Region
ITK	Indigenous Traditional Knowledge
IF	infrared fluorescence

CIAT	International Center for Tropical Agriculture
IFOAM	International Federation of Organic Agriculture Movements
IRRI	International Rice Research Institute
ITC	International Trade Centre
IUCN	International Union for Conservation of Nature
KI	Kováts Index
LC	Least Concern
LMV	lettuce mosaic virus
LRI	Linear Retention Index
LAMP	loop-mediated isothermal amplification
MCH	magnetic-capture hybridization
MSD	mass spectrometry detector
MS	mass spectrometry
MGT	mean germination time
MAP	medicinal and aromatic plants
MPCP	Medicinal Plants Conservation Park
MSB	millennium seed bank
MSI	multispectral imaging
NAGS	National Active Germplasm Sites
NAEB	National Afforestation and Eco-Development Board
NBPGR	National Bureau of Plants Genetic Resources
NGB	National Gene Bank
NSC	National Seeds Corporation
NAD	Nicotinamide Adenine Dinucleotide
NTP	non-thermal plasma
NMR	nuclear magnetic resonance
NASH	nucleic acid spot hybridization
NSCA	nutrient-starch cycloheximide agar
OHS	organic hermetic storage
OPPP	oxidative pentose phosphate pathway
PGR	plant genetic resources
PGPB	plant growth-promoting bacteria
PM	plasma membrane
PEG	polyethylene glycol
PCR	polymerase chain reaction
PPO	polyphenol oxidase

PUFA	polyunsaturated fatty acids
PTRI	Programmed-Temperature Retention Index
PSV	protein storage vacuole
PEF	pulse electric field
PP-IP	pyrophosphate containing inositol phosphates
ROS	reactive oxygen species
RRL	Regional Research Laboratory
RBG Kew	Royal Botanic Gardens, Kew
COU	seed screening and counting
SSB	Soil Seed Bank
TGBRI	Tropical Botanical Gardens and Research Institute
VHF	Vacuum Hermetic Fumigation
VOC	volatile organic compounds
WHO	World Health Organization

Preface

The most successful and efficient method of crop propagation is through seeds. The production and preservation of medicinal plant genetic resources depend on high-quality seeds. For medicinal and aromatic plants, the forest environment constitutes a distinctive ecosystem and offers hospitable conditions for growth and development. The Eastern Sub-Himalayan regions of India have suffered because of the overuse of these therapeutic plants. Important plant species are at risk of going extinct in their specific habitats. In forests, seed is the primary source of regeneration, and it requires particular growing conditions. Thus, preserving trees and their resources is closely related to understanding seeds.

After harvesting and storing the seeds of medicinal plants for a few months, they quickly deteriorate and lose viability. It is necessary to maintain seed viability for a long time in order to preserve the genetic integrity of stored samples. Farmers must store these seeds properly because they will be used in the following growing season. Proper storage is necessary to prevent low germination due to variations in the storage environment. Using improper storage conditions can result in low germination, seed degradation, and loss of viability.

Medicinal plants have been used for therapeutic purposes since the dawn of humankind and are essential for human well-being, as they play a significant role in human welfare. Worldwide, about 350,000 plant species are used as dietary supplements and medicines by more than 80% of the population. The high cost of allopathic treatments and their negative side effects have made the use of plant-based medicines increasingly important.

The needs of various employees working in the seed industry, as well as the seed technology requirements for students at agricultural universities and institutions, have been taken into consideration in writing this book. Topics include: general and specific seed production technology of medicinal and aromatic plants; seed production planning and economics; organic seed production; detection of volatile organic compounds emitted by seeds; seed storage (open, bulk, controlled, hermetic, germplasm, cryopreservation); safe seed storage, including ecofriendly techniques; estimation of carbohydrates, proteins, fats, enzyme activities, respiration rate, and nucleic acids in fresh and aged seeds; the effects of drying temperature and duration on seed germination and storability, with particular reference to medicinal plant seeds; the concept and principles of seed processing; packaging principles, practices, and materials; bagging and labeling; detection of seed-borne fungi, bacteria, and viruses; identification of storage fungi; and biotechnological interventions for producing quality planting materials of medicinal and aromatic plants, along with the various challenges and opportunities in the seed industry for these species.

The content of this book is comprehensive enough to meet the needs of graduate and postgraduate students and faculty in agriculture, horticulture, and forestry; training personnel; field and laboratory officials; policy-makers; seed quality control officials; and those interested in the welfare of farmers. In addition, it will be highly useful for faculty members in State Agricultural Universities and Central Agricultural Universities for improving the quality of teaching.

—Editors

Introduction

This book will cover topics in detail related to seed production, processing, storage, and seed quality testing techniques. The chapters are divided into several parts, each addressing a specific aspect of the subject matter.

India, in addition to having a long history of traditional medical practices such as Ayurveda, Siddha, and Unani that rely heavily on medicinal plant species, is renowned for its rich biodiversity. The pharmaceutical, nutraceutical, and herbal sectors all depend greatly on medicinal plants because of their valuable therapeutic properties. Due to growing public awareness of the potential health benefits of medicinal plants and the trend toward using natural and traditional remedies instead of pharmaceuticals, there has been an increased demand for these plants and their derivatives in recent years.

This rising demand has had both positive and negative impacts on the seed industry in India. To address these challenges and maximize the benefits, it is crucial to encourage sustainable cultivation practices, implement appropriate legislation, support research and development, conserve medicinal plant species, and promote fair trade practices.

The selection of seeds is a key factor in determining how well a species of medicinal or aromatic plant will grow in the field. The germination rate, vigor, and overall performance of the plants are all directly affected by seed quality.

To ensure high-quality seed production, the cultivation of medicinal and aromatic plant species requires careful planning and consideration of numerous factors. The economics of seed production for these plants is influenced by several elements, including market demand, cultivation requirements, seed yield, and processing costs. For the preservation and future use of medicinal plants, it is essential to store the seeds of medicinal and aromatic species from tropical and subtropical climate zones. Proper seed storage preserves the viability and genetic integrity of these plants, ensuring their availability for research, cultivation, and biodiversity conservation activities.

For medicinal and aromatic plant species, enhancing seed quality is essential for increasing germination rates, seedling vigor, and overall crop productivity. It is important to note that specific methods may vary depending on the species, the characteristics of their seeds, and regional requirements.

Processing and packaging are critical in the medicinal and aromatic plant industries for maintaining product quality, safety, and shelf life. To ensure the end product's safety, efficacy, and marketability, compliance with regulatory requirements, Good Manufacturing Practices (GMP), and quality control procedures is essential throughout the processing and packaging stages. For preserving the long-term viability and quality of stored seeds, effective disease and pest management is also necessary. Practices should be tailored to the types of medicinal and aromatic plants involved, as well as to the prevalent diseases and pests in a given region.

This book also addresses the use of biotechnological tools and techniques for producing quality planting materials of endangered medicinal plant species, as well as the various challenges and opportunities in the medicinal and aromatic plants seed industry. For certain plant species, seeking advice from regional horticultural or agricultural specialists, seed banks, or research organizations can be invaluable in managing diseases and pests in seed storage.

CHAPTER 1

Principles and Practices of Seed Storage for Maintaining Seed Quality of Medicinal and Aromatic Plants

AMIR KHAN,[1] SUSHMA KHOLIYA,[1] PRAGYA PANDEY,[1] REETA PANDEY,[1] DIPENDER KUMAR,[2,*] R. C. PADALIA,[2] AMIT CHAUHAN,[1] and PRAWAL PRATAP SINGH VERMA[2]

[1]*CSIR–Central Institute of Medicinal and Aromatic Plants (CIMAP), Research Center, Pantnagar, Uttarakhand, India*

[2]*Division of Crop Production and Protection, CSIR–Central Institute of Medicinal and Aromatic Plants (CIMAP), Research Center, Pantnagar, Uttarakhand, India*

*Corresponding author

ABSTRACT

Seeds serve as the cornerstone of plant propagation, agriculture, biodiversity preservation, and scientific exploration. Seeds are viable, but they are susceptible to influence from elements like temperature, humidity, light, and the onslaught of pathogens. If seeds are subjected to suboptimal conditions, these seeds can forfeit their capacity to germinate. Contrary to other crops that have seed dormancy and seed inhibitors, which result in lower seed quality, medicinal and aromatic crops depend heavily on seed quality. In this situation, “seed enhancement” through a variety of procedures and storage techniques has a considerable potential in achieving full seed germination and optimal field establishment, thereby improving the quality of the product and production of these commercially significant crops. Therefore, the

Seed Technology for Medicinal and Aromatic Plants: Production, Testing, Storage, and Marketing. Priyanka Sharma, Kishore Kumar Sharma, and Gopal Shukla (Eds.)

DOI: 10.1201/9781779641199-1

importance of seed preservation emerges—an endeavor that safeguards the viability of seeds for posterity and genetic stability. The process of seed storage commences even before harvest, ensuring that seeds attain the pinnacle of their vitality within the parent plant. A practice steeped in history; it has evolved from rudimentary techniques to modern marvels like cryogenic tank storage and gene repositories. Within this intricate narrative, an array of methods and techniques come into play: open and bag storage, cryopreservation, hermetic storage, germplasm repositories, and soil seed banks (SSB)—each technique holds a unique key to prolonged viability. These methods, whether time-honored or cutting-edge, share a common purpose: to preserve the legacy of life contained within these seeds. Ultimately, the manuscript at hand delves deeply into the significance and methodologies of seed preservation. It intricately explores the intricate measures required to ensure the integrity of stored seeds. These precautions stand as a testament to humanity's commitment to nurturing life's potential, a commitment spanning generations and transcending time.

1.1 INTRODUCTION

A major portion of the plant world is made up of medicinal and aromatic plants (MAP), which are important suppliers of raw materials for the pharmaceutical, cosmetics, and medicine industries. Due to their long-term advantages and few side effects, plant-based herbal medicines are becoming more and more popular as the price and side effects of allopathic medications rise. Many medicinal plant species have been brought dangerously close to extinction by a number of circumstances, including overexploitation, rising human activity, commercial collection, poor seed viability, poor management practices, and unrestrained ayurvedic practices (Nishteswar, 2014). These plants are rapidly becoming extinct due to insufficient cultivation and conservation efforts, as well as the global demand for herbal items made from them. Therefore, it is crucial to understand that after being harvested and stored, the seeds of medicinal plants quickly deteriorate and lose their vitality. It is essential to maintain seed viability for long periods of time in order to protect the genetic integrity of preserved samples. Furthermore, it is critical to maintain a steady supply of novel, underutilized, or threatened medicinal plant species if we are to preserve our system of traditional medicine. Resource conservation is inextricably tied to understanding the science of seed storage. Given that the interval between the initial harvest and the later planting can be as long as 100 to 200 days or more, seed storage is an essential step in ensuring effective germination for subsequent planting (Dadlani et al., 2023). Under off-field conditions, seeds are extremely brittle and vulnerable to fungal and insect attacks. Depending on the type of seed and the region, different storage conditions apply. Seeds can be kept in ambient conditions for extended durations in temperate climates, whereas it is necessary to adjust the temperature and moisture content in tropical and subtropical climates in order to maintain the vigor of the seeds. It is crucial to preserve the seeds' original quality, which includes elements like germination rate, physical purity, and vigor. For the purpose of storing seeds for later use, a variety of *in situ* and *ex situ* storage methods are used. The current chapter, which offers helpful advice for seed storage and preservation practices, concentrates on methods and methodologies for the conservation of medicinal seed resources.

1.2 IMPORTANCE OF SEED AND SEED STORAGE

In agriculture, the significance of aromatic and medicinal plant seeds transcends multiple domains, impacting cultivation, culinary applications, cosmetics, fragrances, and pharmaceutical advancements. These seeds serve as the foundation for human sustenance while harboring the genetic potential of crops. Moreover, they serve as a vital source of various raw materials, including edible oils such as soybean, mustard, groundnut, and coconut oil, constituting a substantial portion of human diet. Additionally, the seeds of coffee contribute to the manufacturing of diverse beauty products. For many individuals who predominantly follow vegetarian diets, seeds are the cornerstone of their nutritional intake. However, it's crucial to recognize that medicinal plant seeds have a limited shelf life, losing viability within months of harvest and storage. Preserving the genetic integrity of these stored samples becomes paramount, necessitating the maintenance of seed viability over extended periods. Numerous factors, both abiotic (such as temperature, moisture content, and oxygen) and biotic (including insects, rodents, mites, bacteria, and common fungi like *Aspergillus, Penicillium, Alternaria,* and *Rhizopus*), pose threats to the genetic potential and quality of seeds, causing deterioration (De Vitis et al., 2020; Savvides et al., 2016). Hence, effective seed storage is imperative. The primary objective of storing seeds of economic plants is to safeguard planting materials from one season to the next. This is especially crucial since seed demand can fluctuate significantly from year to year, and suppliers may not consistently meet market needs due to seasonal constraints (Nagaraju et al., 2018). Therefore, seed storage, achieved through domestication and multiplication of superior varieties, plays a pivotal role in addressing these challenges. Therefore, proper seed storage is crucial to retain viability of seed for longer duration. Seed banks or storage facilities emerge as valuable assets in the preservation of seed viability, enabling seeds to remain viable for decades or even centuries.

1.3 PRINCIPLES OF SEED STORAGE

The primary goal of storing seed is to preserve it in optimal, physiologically active, and viable conditions until it is used for planting, and so forth. Sometimes, seed producers have unsold seeds from one planting season, and such seeds are often carried over into storage for future marketing. In certain cases, seeds are preserved for extended periods to enhance germination rates by allowing them to naturally come out of dormancy; this kind of seed needs proper storage. However, it is important to note that seed stored at ambient or natural temperature is sensitive to temperature fluctuations, exposure to oxygen, and changes in relative humidity, causing seed deterioration. To safeguard seeds during the storage period, specific principles and practices must be followed in storage facilities. Principles and practices of seed storage are as follows (Justice and Bass, 1978; Parimala et al., 2013).

1.3.1 CONTROLLED ENVIRONMENTAL CONDITIONS

Managing seeds under controlled conditions helps to reduce their metabolism, slowing down the aging process and extending their longevity. Maintaining a cool and dry environment is crucial for most seeds. According to Harrington's "Thumb Rules," the effects of both temperature and moisture are additive and independent, halving the seed life span for every 5°C increase in seed storage temperature and 1% increase in seed moisture content. Harrington also issued a warning, saying that the rule is not applicable in temperatures below zero or above 50 (Harrington and Kozlowski, 1972).

1.3.2 SEED PREFERENCE

To ensure optimal seed storage, it is important to select seeds that are genetically pure, physiologically active, and exhibit high viability. For orthodox seeds, which are more common, cooler, and drier conditions contribute to greater longevity.

- **Pest Control:** Effective measures should be in place to control pests during storage, preventing damage to the seeds.
- **Cleanliness and Hygiene:** Seed storing facilities should be kept clean and hygienic to avoid contamination or spoilage.
- **Moisture Management:** There are numerous methods for seed storage which require different moisture levels; therefore, before storing seeds, they should be properly dried to reach optimal moisture levels suitable for the chosen storage method. According to Harrington's thumb rules, reducing moisture in seeds up to a certain limit can increase their lifespan.
- **High-Quality Seeds:** Only viable seeds of high quality should be stored, meaning seeds which are well-cleaned, treated, and showing strong germination and vigor.
- **Adapted Storage Conditions:** Storage requirements should be determined based on the intended storage duration and local climate throughout the storage period. Seed stored for the long term demands a more precise environment than short-term storage, and regions with lower relative humidity require less advanced storage facilities compared to high-humidity areas.

1.4 FACTORS AFFECTING SEEDS LONGEVITY

The significance of preserving seeds has been acknowledged, dating back to the early days of plant domestication by humans. It has been recognized that the vitality of a medicinal plant's seed can vary among genotypes and is also influenced by various abiotic factors, that is, temperature, initial seed quality, relative humidity, moisture content, gaseous exchange, storage structures, physio-chemical composition of seeds, and packaging materials (Kuntal, 2014). In addition, Roberts (1972) has also reported that seed deterioration during the storing period is primarily due to damage to seed membranes, proteins, enzymes, and nucleic acids. Over time, these successive degenerative changes lead to the disorganization of membranes and cell organelles, ultimately resulting in the death of the seed and loss of its germination potential. Considering the delicacy of MAPs, which exhibit poor seed germination, prolonged seed dormancy, and challenges in acclimatization, it is imperative to store them under optimal conditions to preserve their genetic integrity. These plants are highly vulnerable to various abiotic and biotic threats due to the poor acclimatization and amelioration process. Understanding and effectively addressing these factors is essential to ensure the long-term preservation of seeds for future use. Therefore, it becomes crucial to consider the factors influencing the longevity of seeds during storage, some of which are outlined in the following section.

1.4.1 TEMPERATURE

Temperature is a pivotal factor for the storage of seeds. It significantly affects the preservation of seeds. It is reported that increasing temperature leads to the infestation risk by insects and molds, as well as the metabolism of seeds, causing potential deterioration during storage. Moreover, seeds having higher moisture content are particularly susceptible to temperature-induced deterioration. Consequently, seeds stored at elevated temperatures may experience a more rapid decline in viability, whereas lower temperatures generally slow down the aging process and prolong the longevity of seeds. While the ideal storage temperatures can vary for different species, they typically fall within the range of –20°C to 5°C for long-term storage (Pradhan and Badola, 2012).

1.4.2 MOISTURE CONTENT

The moisture content within seeds is the most critical factor determining seed viability throughout the storing period. Seeds achieve a specific moisture content under exposure to a particular level of atmospheric humidity, referred to as equilibrium moisture content. The establishment of moisture-equilibrium within seeds is a time-dependent process because initially seeds contain a high level of moisture content, but over time, towards its maturity, its moisture declines and reaches up to a certain limit, which is crucially required for seed viability. The duration required for moisture-equilibrium establishment varies with seed type, initial moisture content, average relative humidity, and temperature. Consequently, maintaining the moisture-equilibrium of seeds during storage is influenced primarily by relative humidity, with temperature playing a lesser role. At equilibrium moisture content, there is no net gain or loss of moisture in seeds. In sealed storage conditions, seed moisture content determines the relative humidity inside the containers. Seeds with high moisture content are most vulnerable to deterioration because there is a greater likelihood of microbial attack in moist environments. Therefore, adequately dried seeds, with moisture content reduced to appropriate levels, tend to have a longer shelf life. Conversely, if seeds are stored with high moisture content, losses can occur rapidly due to mold growth. While excessively low moisture content, that is, below 4%, can damage seeds through desiccation. Typically, seed drying up to 10% moisture content is recognized as an appropriate level. In contrast, for storing orthodox seeds in sealed containers, drying up to 5% to 8% moisture content, depending on the specific seed type, is necessary (Pradhan and Badola, 2012).

1.4.3 RELATIVE HUMIDITY

Relative humidity represents the amount of water vapor present in the air at a specific temperature relative to its maximum moisture-holding capacity. Considering its significance, maintaining the appropriate relative humidity is vital for ensuring the long-term viability of seeds. Seeds stored in high humid environments are susceptible to mold growth and fungal infections, which can reduce their viability. Conversely, excessively low humidity levels can result in seed desiccation, also posing a threat to their viability (Suma et al., 2013).

1.4.4 OXYGEN LEVELS

Oxygen is essential for the respiration of living seeds. However, researches done so far into the impact of gaseous environments on seed viability has revealed that elevated oxygen levels can promote oxidation processes. These processes generate free radicals and reactive oxygen species (ROS) within the seeds, which are highly reactive compounds that can damage cellular components, including DNA, proteins, and lipids, ultimately leading to seed deterioration. To extend seed storage life, it is advisable to employ vacuum sealing or store seeds in airtight containers with reduced oxygen levels (Groot et al., 2012).

1.4.5 SEED QUALITY

Seed quality encompasses the overall condition and health of seeds, including factors such as physiological status, genetic purity, devoid of genetic mutations or impurities, and absence of any contaminants. Seeds with genetic abnormalities may exhibit reduced viability and may produce offspring with undesirable traits. Barton (1941) observed that seeds with high initial viability exhibit greater resistance to unfavorable storage conditions than seeds with low viability. When seeds start to deteriorate, the process usually picks up speed over time because initial deterioration triggers a chain reaction of chemical and biological processes that lead to a quicker degradation. Mechanically damaged seeds, in particular, suffer substantial viability and vigor losses rapidly. In general, small seeds tend to escape injury, while large seeds are more susceptible to extensive damage. This underscores the importance of carrying over only high-quality seeds for storage.

1.4.6 TYPE/VARIETY OF SEED

Seeds belonging to different species or varieties exhibit diverse levels of seed longevity, and this characteristic is primarily determined by their genetic composition. Certain seeds possess inherent durability, enabling them to endure even under suboptimal storage conditions and thus have extended storage lifespans over less healthy seeds. In contrast, some seeds are more perishable and tend to deteriorate more rapidly. It is essential to gain insight into the specific characteristics of each type or variety of seed for effective long-term storage.

Seeds are categorized based on their genetic makeup into three groups:

1. **Microbiotic Seeds (Short-Lived):** Seeds falling into this category have a short storage life of not more than 3 years. They are least resilient and tend to deteriorate quickly under storage conditions that are not ideal. Their examples include Soybean, Onion, River Maple, and wild rice.

2. **Mesobiotic Seeds (Medium-Lived):** These seeds offer relatively a moderate storage life; their life span ranges from 3–15 years, for example, Barley, Flax, Tall fescue, and Carrot seeds. They can maintain their viability for a reasonable duration, demonstrating a balanced level of resilience.

3. **Macrobiotic Seeds (Long-Lived):** Macrobiotic seeds exhibit the ability to endure suboptimal storage conditions for an extended period. Their life span ranges from 15 to more than 100 years, for example, Lupin seeds, Indian lotus, Silk tree seeds, Trifolium seeds. They possess a high level of resilience and can maintain viability over a long storage life.

Understanding the genetic characteristics and storage requirements of specific seed types or varieties is essential for effective seed preservation and management (Bewley and Black, 1982).

1.4.7 BIOTIC FACTORS

Various organisms, including bacteria, fungi, mites, insects, rodents, and birds, are associated with seeds in storage and influence seed viability (Parimala et al., 2013).

i. **Bacteria and Fungi:** The control of seed microflora, including bacteria and fungi, is greatly influenced by the relative humidity. The population and action of microflora are regulated by factors such as moisture content and temperature. Treated seeds with antibacterial and fungicidal agents can be stored for longer periods, and fumigation can help control insects, leading to extended storage durations. Fumigants, such as methyl bromide, ethylene dichloride, carbon tetrachloride, carbon disulfide, and naphthalene, as well as aluminum phosphine, are used to control storage fungi and insects. The presence of fungi in seeds often exhibits an exponential relationship with relative humidity. Research indicates that storage fungi are entirely inactive below 62% relative humidity, and their activity is limited below approximately 75% relative humidity. Storage bacteria require at least 90% relative humidity for growth and become significant only under conditions where fungi are already active. High temperatures can accelerate the growth of microflora, but deep freezing (–20°C or lower) is the most practical method of controlling their activity by temperature alone.

ii. **Insects and Mites**: Insects are not active below 8% seed moisture content. However, in the case of infected grains, insect activity can be seen up to 15% moisture content. Storage insects are most active in the range of 28 to 38°C. Temperatures below 17 to 22°C are considered unsafe for insect activity. While it is preferable to control insect and mite activity by manipulating the seed environment, chemical control methods, however, some chemicals, for example, fipronil and imidacloprid, may have adverse effects on seed viability and pose handling risks.

iii. **Rodents and Birds**: Birds can pose a constant threat to seed storage if even small openings exist in storage facilities. These openings should be sealed or screened, as needed for ventilation. Rodents, on the other hand, can result in a complete loss of seeds. To control rodents, measures may include building the storage facility or store with the floor 90 cm above ground level at entrances, creating a 15 cm lip around the building at the 90 cm level of the floor, and providing a removable deck at the entrance for use only during seed loading and unloading.

1.4.8 EFFECT OF PROVENANCE

As previously mentioned, various factors, both before and during harvest, can impact seed viability. Interestingly, samples of seeds acquired from different sources may exhibit variations in their viability behavior. The geographical origin of the seed crop greatly influences its storability. For instance, red clover seeds grown in Canada maintained 80% germination after 4 years of storage, whereas seeds grown in England and New Zealand retained the same germination rate but for only 3 years. This disparity can be attributed to the distinct climatic conditions and soil types prevalent in different geographic locations (Dadlani et al., 2023).

1.4.9 SEED PACKAGING

Selecting the appropriate packaging material and conditions is of paramount importance. Seeds are typically stored in sealed containers like glass jars, foil pouches, or moisture-resistant bags. The packaging must be airtight, impervious to moisture, and capable of withstanding temperature fluctuations. Proper storage practices are essential not only in warehouses but also during transit, at the retailer's store, and on the user's farm. Neglecting adequate storage precautions at any of these points can compromise seed viability (Tiwari et al., 2022).

1.4.10 STORAGE DURATION

The duration of storage plays a crucial role in retaining the vitality and viability of seeds. It is observed that the longer the seeds are stored, the greater the likelihood of viability deterioration. Different plant species have varying storage potentials, with some having longer viability retention than others. It is advisable to regularly inspect and replace old seeds with fresh ones to maintain genetic diversity and uphold seed quality.

1.4.11 OTHER FACTORS

Besides the factors discussed above, several other elements can influence seed storage life. These include exposure to direct sunlight, the frequency and type of fumigation, the impact of seed treatments, and various other factors that can either directly or indirectly affect seed viability. Understanding, considering, and providing the appropriate response to such factors can extend the shelf life of seeds during storage.

1.5 CLASSIFICATION OF SEED BASED ON MOISTURE CONTENT AND STORABILITY

The quality of the seeds determines whether or not a crop of plants can be produced in a sufficient quantity under a variety of soil and climatic conditions. One of the more popular approaches for dividing seeds into groups is based on their susceptibility to desiccation. Three categories can be used to categorize seeds: orthodox or desiccation-tolerant seeds, recalcitrant or desiccation-intolerant seeds, and intermediate seeds.

1.5.1 ORTHODOX SEEDS (DESICCATION-TOLERANT)

Orthodox seeds are a type of seed that can be dried to a very low moisture content level without losing their viability (Wyse and Dickie, 2016). The recommended conditions for the extended preservation of orthodox seeds are storing them at a temperature of –20 ± 4°C and a relative humidity of 15±3%. The maturation of orthodox seeds may tolerate desiccation with minimal moisture levels (at least 2 to 6%) depending upon their species, and they are normally easy to store if basic processing and preservation facilities are offered (Fetene et al., 2012). In late embryogenesis, orthodox seeds produce an abundance of proteins which includes small hydrophilic proteins, heat shock proteins, and sugars, all of which contribute to the seed's propensity to lose water. These characteristics allow the orthodox seeds to enter a vitrified state, a quiescent stage in which metabolic activity is reduced due to the minimum moisture level. Storage of orthodox seeds for the long duration in seed banks is often done in conventional or walk-in freezers at –18°C (Pritchard and Nadarajan, 2008). The majority of the economically significant species of the Northern Temperate Zone's tree genera, such as *Acer saccharinum* L, Alnus Mill. (alder), Betula L. (birch), *Fraxinus excelsior* L. (common ash), *Prunus armeniaca* L. (apricot), *Prunus avium* L. (sweet cherry), *Pinus resinosa* (red pine) and so forth, are classified as having true orthodox seeds and stored for relatively long periods at subfreezing temperatures with 5 to 10% moisture content. Numerous valuable tropical and subtropical genera, including *Acacia* L. (wattles), *Albizia Durz.* (siris), numerous Fabaceae, *Eucalyptus* L' Her. (gum tree), and *Tectona* L. f. (teak), also produce orthodox seeds. Sub-orthodox seeds can be stored for a substantially shorter amount of time under the same circumstances as genuine orthodox seeds. This category also includes *Oryza sativa* L. (rice), *Zea mays* L. (maize), *Triticum aestivum* L. (wheat), *Glycine max* L. (soybean), and *Hordeum vulgare* L. (barley).

1.5.2 RECALCITRANT SEEDS (DESICCATION-INTOLERANT)

The term "recalcitrant" was first used by Eric Roberts in 1973 to describe seeds that cannot be normally preserved. Many species with tropical or subtropical origins have recalcitrant seeds that are delicate to dryness and chiling and are unable to be stored in conventional gene banks. They can tolerate a negligible amount of post-shedding desiccation and are frequently sensitive to chiling. They have a very short lifespan and cannot be dried to moisture content below 20–30% (Pradhan and Badola, 2012). Typical examples include avocado, cocoa, coconut, jackfruit, lychee, mango, rubber, a few horticulture plants, and a variety of medicinal herbs.

Since it is shed at a range of species-related higher water content and cannot sustain dehydration to levels suitable for low relative humidity/sub-zero temperature storage, recalcitrant seeds cannot be conserved using conventional GenBank techniques. Recalcitrant seeds can be stored for a short duration of time, which is a few to several months under the best conditions feasible, but those conditions must prevent moisture loss. Hydrated storage in environments with high relative humidity (RH) is necessary for staying functional in short to medium storage (Al-Zoubi and Normah, 2015). The long-term cryopreservation of the germplasm of recalcitrant-seeded species is considered to be the only feasible method. However, dehydration and ice crystallization are two additional key risks connected with cryopreservation. To prevent ice crystallization, cryopreservation techniques employ vitrification (rapid chiling and rewarming) (Engelmann, 2011). This method uses a rapid cooling process to lower the substance's temperature in order to prevent the formation of crystalline. Because of this, the substance transforms into an amorphous solid with glass-like qualities (Berjak and Pammenter, 2013). Avocado, coffee, mango, lychee, cocoa, and rubber tree seeds are examples of recalcitrant seed species that have historically been preserved in field gene banks. Depending on the environmental conditions of the geographical area, there are two different types of recalcitrant seeds, such as temperate and tropical kinds.

i. **Temperate-Recalcitrant Seeds:** This variety of seeds cannot be desiccated, but they thrive at temperatures near or below freezing. These species can be kept for three to five years if maintained at –3 to +4°C temperature and a moisture content of 30 to 50%. There are several genera with seeds that are resistant to the temperate climate, including *Acer saccharinum L.* (silver maple), *Castanea ozarkensis* (Ozark Chinquapin), *Quercus robur* L. (European oak), *Cocos nucifera* L. (coconut), and red bay (*Persea* Mill.).

ii. **Tropical-Recalcitrant Seeds:** They share the same sensitivity to desiccation and low temperatures as seeds that are resistant to temperate climates. The viability of these seeds has been shown to diminish even after a brief exposure to temperatures below 10 to 15°C. This group of species includes *Araucaria angustifolia* (Paraná pine), *Hopea odorata* Roxb. (*White thingan*), *Shorea tumbuggaia* Roxb. (Green dammar tree), and *Theobroma cacao* L. (cocoa tree) (Berjak and Pammenter, 2013).

1.5.2.1 INTERMEDIATE SEEDS

These seeds, known as intermediate seeds, share functional traits with both categories and were proposed by Ellis (1981) as a middle category between the orthodox and recalcitrant seed groups. The seeds are believed to shed according to intermediate storage behavior at relatively high-water concentrations, despite the fact that they can withstand significant dehydration, just not to the same extent as orthodox seeds (Ballesteros et al., 2021). These seeds, like recalcitrant seeds, may resist significant dehydration and cannot be preserved under conventional GenBank conditions owing to their sensitive nature to cold. These seeds could only be able to tolerate between 12 and 15% dryness, although they are vulnerable to freezing temperatures. Every seed is susceptible to damage from further dehydration below a certain critical moisture level (Walters et al., 2013). At room temperature, it is acceptable to dry conventional seeds to water concentrations of 0.03 to 0.08 g of overall mass (3 to 7%), however, it is unsafe to dry recalcitrant seeds to water levels of more than 0.20 g of whole mass (> 20%). Intermediate seeds fall in the middle of these two ranges. *Citrus aurantifolia* (Lime), *C. madurensis* (Golden lime), *C. reticulate* (Tangerine), and *Coffea liberica* (Liberian Coffee) are a few examples that fit the intermediate group.

1.6 TYPES OF STORAGE

There are four basic kinds of storage techniques used to preserve the quality of seeds:

1. **Storage of Commercial Seeds:** Commercial seed storage refers to the controlled preservation of plant seeds for agricultural enterprises, horticultural, and other industrial purposes. The largest storage requirement, accounting for 75 to 80% of the total, pertains to preserving seeds from the time of harvest until the following year's planting season. This storage duration can vary from a few days to as long as eight or nine months. It involves storing seeds under specific conditions to maintain their viability and quality (V et al., 2022). This typically includes controlling temperature, humidity, and airflow to prevent degradation (Li et al., 2017). Such commercial seed storage systems use hermetic storage techniques,

which involve sealing seeds in airtight containers or bags to create a controlled atmosphere, reducing oxygen levels and preventing moisture ingress. These storage facilities can handle a wide range of seeds, including those for food crops (e.g., grains, vegetables), cash crops (e.g., cotton, tobacco), forage crops (e.g., Alfalfa), and ornamental plants (e.g., lavender). Commercial seed storage of cereals should be dried until their moisture content falls below 14%, while oil seeds should be dried to a moisture content of less than 11%.

2. **Carryover Seeds:** Carryover seeds encompass approximately 20–25% of the overall seed stock that has been preserved for the upcoming planting season. These seeds can be stored for a potential duration of 1 to 1½ years. To safeguard carryover seeds from moisture and air exposure, it is advisable to store them in hermetic containers like glass jars or resealable plastic bags (Onyekwelua and Fayose, 2007). However, in order to be successful, it necessitates close consideration of quality, potential cross-pollination problems, legal challenges, and disease/pest management (Bureau of Land Management, 2016).

3. **Foundation Stock and Enforcement Seed Sample:** The terms "foundation stock" and "enforcement seed samples" are frequently used in relation to seed storage initiatives (seed banks and crop preservation). In terms of plant breeding, farming, and intellectual property protection, enforcement seed samples are utilized to preserve the integrity of particular plant kinds, whereas foundation stock serves as the starting genetic pool for breeding operations. These phrases refer to particular steps and methods intended to safeguard and preserve the genetic diversity of plant species (Griffth et al., 2015).

   i. **Foundation Stock:** When building a seed bank or germplasm repository, foundation stock refers to the initial stock of seeds or plant material from a certain crop or plant species. Idealistically, this initial collection should display a variety of genetic variation within the species (Yang et al., 2005). To ensure that they accurately reflect the genetic variation that occurs in the plant species, whether in the wild or in cultivated forms, these seeds are carefully selected and gathered (Faisal and Alatar, 2019). The purpose of foundation stock is to protect a species' genetic variety, particularly if that species is threatened or endangered. In order to preserve genetic diversity, these seeds are meticulously recorded, categorized, and kept in a safe and controlled environment (Bettoniet al., 2019).

ii. **Enforcement Seed Sample:** In order to ensure the integrity of the foundation stock, it is common practice to remove a smaller subset of seeds or plant material for use as an enforcement seed sample, sometimes called authentication, or reference seed sample. This subset is utilized for routine testing and identification of the stored germplasm's identity and purity (Faisal and Alatar, 2019). These samples are necessary to guarantee the precision and consistency of the seed bank's collections throughout time. To reduce the chance of genetic drift or contamination, the enforcement seed sample is kept under stricter control and monitoring than the larger foundation stock collection.

4. **Germplasm Seed Storage:** Germplasm seed storage, commonly referred to as a seed bank or gene bank, involves the preservation and protection of the genetic variety within plant species by carefully storing seeds in controlled environments. This crucial undertaking is vital for the preservation of biodiversity and the safeguarding of genetic materials for the benefit of future generations. It plays a critical role in the context of challenges like environmental shifts, diseases, and other perils faced by plant species (Faisal and Alatar, 2019). Germplasm conservation encompasses the collection, protection, and maintenance of seeds from diverse plant varieties, encompassing crops, wild flora, and endangered species. Notable instances of germplasm preservation centers include the Svalbard Global Seed Vault in Norway, the International Center for Tropical Agriculture (CIAT) in Colombia, and the rice GenBank at the International Rice Research Institute (IRRI). These facilities play a crucial role in guaranteeing the enduring survival and availability of a broad spectrum of plant genetic resources (PGR). Seed banks are strategically designed to shield the stored seeds from pests, diseases, and environmental stress factors that might compromise their integrity, thereby ensuring the long-term viability of the genetic material contained

within the seeds (van Treuren et al., 2018). It serves as a valuable asset for agricultural research and breeding initiatives aimed at creating novel crop varieties exhibiting enhanced characteristics, including increased yields, resistance to diseases, or adaptability to challenging environmental conditions. Cryopreservation stands as an economical and enduring technique for preserving germplasm, wherein biological materials are stored at an extremely low temperature of approximately –196°C. At this frigid temperature, cellular metabolic processes come to a halt, enabling the prolonged preservation of these materials without any changes or modifications (Dagnachew et al., 2023).

Germplasm preservation can occur in two distinct ways: either in *ex situ* locations, such as seedbanks and genetic repositories, or within *in vitro* environments, such as farms, natural parks, and reserves. *Ex situ* conservation represents the primary approach for safeguarding germplasm acquired from both cultivated and wild plant sources. Genetic materials, in the form of seeds or those maintained through *in vitro* cultures (comprising plant cells, tissues, or organs), can be stored for extended periods under suitable conditions in gene banks. In contrast, *in vitro* techniques employing shoots, meristems, and embryos are particularly well-suited for safeguarding germplasm from plants that are propagated vegetatively. Additionally, this *in vitro* method can also be applied to the preservation of plants with recalcitrant seeds and those containing genetically engineered materials.

1.7 SEED STORAGE METHODS, THEIR ADVANTAGES AND LIMITATIONS

A robust storage facility for seeds is essential, as it serves as the critical genetic bridge between two consecutive generations of a plant species. To ensure the preservation of their germination potential and vitality, seeds should be stored meticulously. Unfortunately, approximately 30% of seeds are lost during storage due to the presence of insects, rodents, and microorganisms. The storage period commences from the moment seeds reach physiological maturity in the field and extends until they are planted in the subsequent season. It is imperative to safeguard the seeds' ability to germinate, their viability, and their overall vigor. In this section, we will explore the following seed storage methods along with their advantages and disadvantages.

1.7.1 OPEN STORAGE

Open seed storage refers to a method of storing seeds in a well-ventilated and open environment, typically in bins, trays, or containers that allows air circulation. This approach is often used for short-term seed storage, such as seeds that will be planted in the next growing season (McCormack, 2004).

- **Advantages:** Open seed storage is relatively low-cost compared to more specialized storage methods like cold storage or vacuum packaging. Seeds stored openly are easily accessible, making it convenient to select and retrieve seeds when needed for planting. Moreover, adequate air circulation helps prevent the buildup of moisture, reducing the risk of fungal and microbial growth that can damage seeds. In addition, in an open storage, it is easier to monitor the condition of seeds as it can be visually inspected and quickly detect any signs of deterioration or pests.
- **Disadvantages:** Seeds stored in an open environment are more susceptible to fluctuations in temperature and humidity, which shorten their viability and germination rate. In an open storage system, seeds are exposed to the risk of insect infestation, rodents, and other pests, which lead to significant seed loss if not properly managed. This storage system may require more space compared to other storage methods. Moreover, openly stored seeds may be at risk of cross-contamination, compromising their genetic purity.

1.7.2 BAG STORAGE

The "bag seed storage method" is not a widely recognized term in the field of seed storage, because it is sometimes practiced informally, particularly in small-scale or traditional agricultural settings. This method involves storing seeds in bags or pouches made of various materials like cloth, paper, or plastic. Here are the advantages and disadvantages of this approach (Hare et al., 2008).

- **Advantages:** Bag seed storage is cost-effective because it does not require specialized infrastructure or equipment, and bags are readily available and affordable. It is a simple method that is easy to implement, making it accessible to small-scale farmers and seed growers with limited resources and technical knowledge.

Some bag materials, like cloth, can provide limited air circulation, which may help prevent moisture buildup and fungal growth. Bag storage is flexible and adaptable to different seed quantities and types. It can accommodate various seed sizes and shapes. Bag stored seeds are easily accessible, allowing for quick retrieval and planting when needed.

- **Disadvantages:** It can be challenging to monitor and maintain the conditions inside a bag, because it does not provide temperature and humidity control, which can be crucial for preserving seed viability over extended periods. It offers limited protection against pests, rodents, and moisture, which lead to seed damage. Moreover, compared to more controlled seed storage methods, seeds stored in bags may have a shorter shelf life due to exposure to environmental fluctuations. If bags are not properly sealed or stored in a clean environment, there is a risk of contamination, which affects seed quality.

1.7.3 CONTROLLED SEED STORAGE

It is a method of preserving seeds under carefully regulated environmental conditions to maximize their viability and longevity. This approach is commonly used in seed banks, gene banks, and research institutions to maintain the genetic diversity of plant species (McCormack, 2004).

- **Advantages:** Controlled storage facilities are designed to be pest-proof and resistant to microbial contamination, reducing the risk of seed damage and loss due to insects, rodents, and pathogens. Apart from this, by controlling temperature, humidity, and other environmental factors, controlled seed storage helps maintain the genetic diversity and viability of seeds, allowing them to remain viable for many years, even decades, in some cases. Properly managed controlled storage facilities maintain detailed records of each seed accession, including its origin, characteristics, and storage conditions. This information is essential for cataloging and managing seed collections. Controlled storage can accommodate a wide range of seed types, from orthodox seeds (those that can be dried for storage) to recalcitrant seeds (those that cannot be dried) and seeds with varying temperature and humidity requirements. This method is crucial for conserving rare and endangered plant species like *Crocus sativus* (Saffron), *Rauwolfia serpentine* (Indian Snakeroot), and so forth.

- **Disadvantages**: Establishing and maintaining controlled seed storage facilities can be expensive and need technical expertise. Not all organizations or regions have access to controlled storage facilities, limiting the availability of this method for seed preservation. Controlled storage facilities depend on a continuous supply of electricity and other infrastructure, making them vulnerable to power outages and equipment failures. Large collections of seeds require significant storage space, which can be a challenge for institutions with limited space availability. Maintaining controlled storage facilities may have environmental impacts due to energy consumption and refrigerant use.

1.7.4 HERMETIC SEED STORAGE

Hermetic storage (HS) technology is a storage system that does not rely on chemicals. Its fundamental concept involves creating an atmosphere within an airtight storage structure that is depleted of oxygen and enriched with carbon dioxide and/or nitrogen. This can occur naturally through the respiration of living organisms in the stored bulk or can be artificially accelerated using specific methods (Jonfia et al., 2010). There are three primary types of hermetic storage: Organic hermetic storage (OHS), Vacuum hermetic fumigation (VHF), and Gas hermetic fumigation (GHS) (Okolo et al., 2017).

- **Advantages:** Hermetic storage helps minimize physical losses of stored crops. Hermetic storage can significantly extend the viability of seeds, allowing them to remain viable for longer periods, often decades or even centuries. Hermetic storage helps preserve the quality of seeds by preventing the entry of moisture, oxygen, and pests that can lead to deterioration and loss of viability. Unlike controlled seed storage, hermetic storage does not require extensive infrastructure, making it accessible to a wider range of organizations and regions. Once seeds are placed in hermetic containers, ongoing costs are relatively low, as hermetic storage does not rely on continuous energy consumption or complex monitoring equipment. Hermetic containers create a barrier against insect infestations, rodents, and microbial contamination, reducing the risk of seed damage and loss. Some hermetic storage methods can be used for seeds with higher moisture content, reducing the need for extensive drying processes (Somvat et al., 2016).

- **Disadvantages:** Acquiring hermetic containers or equipment for large-scale storage can require an initial investment, which may be a barrier for some organizations. Access to hermetic storage solutions may be limited in certain regions, particularly in developing countries, due to the unavailability of suitable containers and equipment. Also, it may not be suitable for all seed types. While hermetic containers are space-efficient, large collections of seeds can still require significant storage space. Moreover, monitoring the condition of seeds within hermetic containers can be challenging, as they are sealed and not easily accessible for routine inspection. In OHS, it is crucial to avoid trapping a significant volume of air, as it could sustain pests and microorganisms, potentially causing damage to stored crops. Vacuum and GHS (by Sulfuryl fluoride) are more technically complex methods, requiring expertise and precise equipment for oxygen level measurement. Hermetic containers are typically made of plastic, which may have environmental implications if not properly managed at the end of their life cycle. Hermetic storage structures have high thermal conductivity, which can lead to localized or internal heating of grains. This can result in browning and mustiness of grains, loss of moisture content, and moisture condensation inside the storage structures (Okolo et al., 2017).

1.7.5 GERMPLASM

When plants, seeds, or cultures are preserved for the purpose of studying, controlling, or harnessing their genetic information, they are considered germplasm. Among these forms, seeds are the most frequently stored type of germplasm. Germplasm seed storage is a method used to preserve the genetic diversity of plant species by storing seeds representing various genetic traits and characteristics. This method plays a crucial role in maintaining biodiversity and ensuring the availability of genetic resources for breeding programs and future agricultural needs.

Two primary approaches for conserving germplasm are recognized, namely *ex situ* and *in situ* conservation (Azough et al., 2019). *Ex situ* conservation is essential to safeguard PGR from their natural surroundings. These resources can be preserved as living plant specimens within field gene banks, botanic gardens, or arboreta. Alternatively, they can be maintained as seeds, pollen, explants, or DNA in specialized controlled environments (Nwachukwu et al., 2016). Conversely, *in situ* conservation pertains to the preservation of species within their natural habitats, promoting the sustainability of both the environment and the ecosystem. In the case of cultivated species, this approach involves storing them within the ecosystem where their distinct characteristics evolved (Maxted, 2013). In short, germplasm seed storage is a valuable method for preserving genetic diversity and supporting agricultural and research efforts.

- **Advantages:** Germplasm seed storage allows for the long-term preservation of genetic diversity within plant species. Therefore, breeders can access a wide range of traits and characteristics to improve crops. Germplasm storage is essential for the conservation of endangered or rare plant species. It provides a backup repository in case of over exploited species. Germplasm collections may contain seeds with natural resistance to pests and diseases, which can be valuable in developing pest-resistant crop varieties.
- **Disadvantages:** Maintaining germplasm collections requires significant resources, including funding, personnel, and infrastructure for seed storage and management. Over time, the viability of stored seeds can decline, even under optimal storage conditions. Germplasm collections can take up a substantial amount of space, which can be a challenge for institutions with limited storage capacity. There is a risk of disease or contamination within the stored seeds, which can spread to other stored seeds if not properly managed. Access to germplasm collections can be a subject of debate, as it involves issues of ownership, intellectual property rights, and benefit-sharing, particularly when collections are accessed for commercial breeding purposes. Maintaining detailed

records of the stored germplasm, including its characteristics and origins, requires a complex and organized management system.

1.7.6 CRYOPRESERVATION

Cryopreservation, a valuable technique for germplasm storage, has been developed in recent years and is now in use within contemporary germplasm banks. This method involves preserving the genetic material of seeds in extremely low temperatures ranging from −135°C to −196°C using liquid nitrogen (LN), thereby enabling seeds to retain their viability and genetic integrity. This is accomplished by cryopreservation, which effectively stops cellular aging, as the metabolic, physiological, and biochemical processes are nearly completely immobilized at this stage (Acosta et al., 2020; Agarwal et al., 2019). This technique is particularly valuable for conserving seeds of rare, endangered, or difficult-to-preserve plant species.

- **Advantages:** Seeds stored through cryopreservation retain their genetic stability and viability, potentially for many decades or even centuries. Cryopreservation can be applied to a wide range of plant species, including those with recalcitrant seeds (seeds that do not tolerate drying well) and orthodox seeds (seeds that can tolerate drying). Seeds stored at ultra-low temperatures are effectively protected from pests, diseases, and microbial contamination. This ensures that the stored seeds can be used for breeding programs, research, and restoration efforts without significant loss of genetic diversity.
- **Disadvantages:** Cryopreservation requires specialized equipment and expertise to establish and maintain, compared to traditional seed storage methods. Maintaining liquid nitrogen and the infrastructure for cryopreservation is expensive. While cryopreservation can preserve genetic integrity effectively, there is still a minimal risk of genetic changes over extremely long storage periods. Preparing seeds for cryopreservation often involves complex procedures, including dehydration and cryoprotectant treatments, which can be labor-intensive and may not be suitable for all seed types. Accessing and utilizing seeds stored through

cryopreservation may be more challenging than seeds stored using traditional methods, as it requires specific equipment and expertise.

1.7.7 SOIL SEED BANK (SSB)

A natural bank of healthy seeds that can be found in or on the soil is known as the soil seed bank (SSB). Despite the fact that the SSB has been the topic of substantial research across the globe, it has rarely been visualized or subjected to quantitative analysis (Shi et al., 2020). The seed bank is often evaluated using two traditional techniques: seedling emergence and seed screening (achieved by washing, sieving, and counting the seeds in the soil). There are three different types of SSB that is, seedling emergence in a greenhouse (EME), seedling emergence under submersion (SUB), and seed screening and counting (COU). Seedling EME offers a primary advantage by facilitating species identification. However, it should be noted that it may not unveil the presence of all species within the soil due to seed dormancy (Capon, 2007). Also, for wetlands, it is required to flood part of the soil samples to evaluate the aquatic community (Bao et al., 2018). Seedling EME shows the community of terrestrial and amphibious plants, which are better to identify. This method allows the creation of a database of images of plants in different stages of life. Submerged seedling emergence under submersion shows the aquatic community. It allows the creation of a database of images of aquatic plants in different stages of life. COU reflects the entire plant community (aquatic, terrestrial, and amphibious plants). It allows the creation of a database of images of seeds and makes it possible to evaluate seed predation in the seed bank (Bao et al., 2020).

- **Advantages:** The SSB acts as a pool of seeds that, in the right circumstances, can sprout and develop into new plants. This is essential for plant populations that need to recover from disturbances caused by humans, fires, or floods. The genetic variety of plant species within an environment can be preserved by SSBs. The durability and long-term adaptability of plant populations to shifting environmental conditions depend on this diversity. Many times, a wide variety of plant species, both widespread and unusual, can be found in SSBs. The resilience and general health of the ecosystem are aided by this diversity.

- **Disadvantages:** Seedling EME does not show all seedlings; some late or dormant seeds remain in the soil without germinating. Seedling emergence under submersion (SUB) is a complementary method; it does not work alone in temporary ponds, and it does not show terrestrial plants. The biggest disadvantage of the seed screening method is that it does not show seeds' viability. COU spends so much time and is detailed. Many seeds remain unidentified, and the seeds are too small for *tetrazolium testing* (Bao et al., 2020).

1.7.8 CONTAINERIZED SEED BANK

A containerized seed bank refers to a specialized storage facility or system designed for the preservation and conservation of plant seeds. Unlike traditional seed banks that may use various storage methods, containerized seed banks typically involve the use of sealed, airtight containers to protect seeds from moisture, oxygen, pests, and other environmental factors.

- **Advantages:** Containerized seed banks are well-suited for long-term conservation efforts, especially for rare and endangered plant species. It offers excellent protection for seeds, preserving their viability and genetic integrity for extended periods, often decades or longer. These seed banks are typically designed to maximize storage space, making them suitable for large seed collections without the need for extensive facilities. Sealed containers provide a barrier against pests, rodents, and microbial contamination, reducing the risk of seed damage and loss. In containerized seed banks, there is precise control over temperature, humidity, and other environmental factors, ensuring optimal conditions for seed storage. Containers are labeled and organized, allowing for easy access to specific seed accessions.
- **Disadvantages:** Setting up a containerized seed bank can require a significant initial investment in specialized containers, climate control systems, and monitoring equipment. Its operation and maintenance can be costly and require technical expertise.

Accessing seeds stored in containers may require special protocols and can be less accessible for research compared to seeds stored using other methods. Containerized seed banks are dependent on a continuous supply of electricity and functioning equipment, making them vulnerable to power outages and equipment failures. The energy consumption of climate control systems and the materials used in container construction may have environmental impacts.

1.7.9 OTHER SEED STORAGE METHODS

In addition to the methods discussed earlier, there are several other methods and techniques for seed storage, each with its own advantages and disadvantages. These methods are often chosen based on the specific requirements of the seeds being stored and the resources available. Here are some other methods for seed storage:

- **Traditional Seed Storage:** In many traditional agricultural societies, seeds are stored using age-old techniques such as storing them in dry, cool, and well-ventilated locations. These methods are often based on local knowledge and may not involve advanced technology.
- **Seed Libraries:** Some communities and organizations establish seed libraries or seed-sharing programs to promote the exchange and conservation of local and heirloom seed varieties. The examples of seed libraries include Hudson Valley Seed Library, Hampshire College's seed library, Bay Area Seed Interchange Library, and so forth.
- **Freeze Drying:** Freeze drying is a method that involves freezing seeds (–20°C or lower) and then removing water by sublimation, converting ice directly into vapor. This process helps preserve the seeds by reducing their moisture content to very low levels.
- **Desiccation:** Desiccation involves drying seeds to a low moisture content, typically by exposing them to low humidity conditions. This method is suitable for seeds that can tolerate drying.
- **Passive Storage:** In some cases, seeds may be stored passively in their natural environment, relying on the SSB or natural seed dispersal mechanisms for preservation.

- **Field Gene Banks:** Some institutions maintain living collections of plants in the field, known as field gene banks. These collections consist of entire plants or plots where seeds are continually produced and harvested.
- **Drying Beads or Silica Gel:** Drying beads or silica gel can be placed in seed storage containers to help maintain low humidity levels and prevent moisture-related damage.
- **Seed Vaults:** Similar to seed banks, seed vaults are facilities designed to store large quantities of seeds in cold, dry, and secure conditions. The Svalbard Global Seed Vault in Norway is one such example, known as the "Doomsday Vault." Svalbard Global Seed Vault is not a gene bank, but rather a safe-keeping facility for duplicate seed collections on behalf of gene banks. The Seed Vault will only be accessible if the original seed collections are lost for whatever reason. The depositors will maintain ownership of the seeds. The Seed Vault can hold up to 4.5 million distinct seed samples. Each sample contains 500 seeds on average, allowing for a total of 2.25 billion seeds to be kept in the Seed Vault.

1.8 CONCLUSION

Recognizing the profound importance of medicinal seeds and the multifaceted challenges that confront them—from issues such as overexploitation, escalating human activities, widespread commercial harvesting, low seed viability, inadequate management practices, and unregulated ayurvedic approaches—underscores the compelling need for the immediate implementation of robust and effective seed storage strategies. The chapter extensively explores a range of seed storage methods, meticulously detailing their individual merits and limitations. The overarching message is that, through the cautious application of suitable storage techniques in optimal environmental conditions, the viability of medicinal seeds can be vigilantly preserved, ensuring their vigorous germination over prolonged durations. Nonetheless, this critical analysis underscores an urgent and imperative demand for further research, aimed at identifying the most advantageous materials and optimal conditions for both short-term and long-term storage of medicinal seeds. This pursuit holds the potential to offer a promising and decisive solution to address the prevailing crisis.

ACKNOWLEDGMENT

The authors extend their heartfelt gratitude to the Director, CSIR-CIMAP, and the Scientist in Charge, Research Centre Pantnagar, for their invaluable support in generously providing the essential infrastructure necessary for the comprehensive process of the preparation of this manuscript.

KEYWORDS

- **cryopreservation**
- ***ex situ* and *in situ* conservation**
- **germplasm conservation**
- **medicinal seeds**
- **seed storage**
- **soil seed bank**

REFERENCES

Acosta, Y., Pérez, L., Linares, C., Hernández, L., Escalante, D., Pérez, A., & Lorenzo, J. C. (2020). Effects of *Teramnus labialis* (Lf) Spreng seed cryopreservation on subsequent seed and seedling growth and biochemistry. *Acta Physiologiae Plantarum, 42,* 1–7.

Agrawal, A., Singh, S., Malhotra, E. V., Meena, D. P. S., & Tyagi, R. K. (2019). *In vitro* conservation and cryopreservation of clonally propagated horticultural species. *Conservation and Utilization of Horticultural Genetic Resources,* 529–578.

Al-Zoubi, O. M., & Normah, M. N. (2015). Ultrastructural response of embryonic axes of *Fortunella polyandra* to dehydration and cryopreservation. *Cryo Letters, 36,* 379–391.

Anju, M., Warrier, R. R., & Kunhikannan, C. (2022). Significance of soil seed bank in forest vegetation—A review. *Seeds, 1*(3), 181–197.

Azough, Z., Kehel, Z., Benomar, A., Bellafkih, M., & Amri, A. (2019). Predictive characterization of ICARDA genebank barley accessions using FIGS and machine learning. In *Intelligent Environments (Workshops)* (pp. 121–129).

Ballesteros, D., Fanega-Sleziak, N., & Davies, R. M. (2021). Cryopreservation of seeds and seed embryos in orthodox-, intermediate-, and recalcitrant-seeded species. In *Cryopreservation and Freeze-Drying Protocols* (pp. 663–682).

Bao, F., Assis, M. A., & Pott, A. (2020). Applicability of seed bank assessment methods in wetlands: Advantages and disadvantages. *Oecologia Australis, 25*(1), 22–33.

Bao, F., Elsey-Quirk, T., de Assis, M. A., & Pott, A. (2018). Seed bank of seasonally flooded grassland: Experimental simulation of flood and post-flood. *Aquatic Ecology, 52,* 93–105.

Barton, L. V. (1941). Relation of certain air temperatures and humidities to viability of seeds. *Contribution from Boyce Thompson Institute, 12*(2), 85–102.

Berjak, P., & Pammenter, N. W. (2013). Implications of the lack of desiccation tolerance in recalcitrant seeds. *Frontiers in Plant Science, 4,* 478.

Bettoni, J. C., Bonnart, R., Shepherd, A., Kretzschmar, A. A., & Volk, G. M. (2019). Cryopreservation of grapevine (*Vitis* spp.) shoot tips from growth chamber-sourced plants and histological observations. *Vitis, 58*(2).

Bewley, J. D., & Black, M. (1982). Viability and longevity. In *Physiology and Biochemistry of Seeds in Relation to Germination.* Springer, Berlin, Heidelberg.

Bureau of Land Management. (2016). *Technical Protocol for the Collection, Study, and Conservation of Seeds from Native Plant Species for Seeds of Success.*

Capon, S. J. (2007). Effects of flooding on seedling emergence from the soil seed bank of a large desert floodplain. *Wetlands, 27*(4), 904–914.

Dadlani, M., Gupta, A., Sinha, S. N., & Kavali, R. (2023). Seed storage and packaging. In M. Dadlani & D. K. Yadava (Eds.), *Seed Science and Technology* (pp. 239–266). Springer Nature.

Dagnachew, S., Teketay, D., Demissew, S., & Awas, T. (2023). Evaluation of germination under different storage conditions of four endemic plant species from Ethiopia: Implications for *ex situ* conservation in seed banks. *Seeds, 2*(1), 45–59.

De Vitis, M., Hay, F. R., Dickie, J. B., Trivedi, C., Choi, J., & Fiegener, R. (2020). Seed storage: Maintaining seed viability and vigor for restoration use. *Restoration Ecology, 28,* S249–S255.

Ellis, R. A. (1981). The quantification of agent and survival in orthodox seeds. *Seed Science and Technology, 9,* 373–409.

Engelmann, F. (2011). Cryopreservation of embryos: An overview. In T. A. Thorpe & E. C. Yeung (Eds.), *Plant Embryo Culture: Methods and Protocols* (pp. 155–184). Totowa: Humana Press.

Faisal, M., & Alatar, A. A. (2019). Synthetic seeds: Germplasm regeneration, preservation and prospects. Springer Nature.

Fetene, A., Yeshitela, K., & Desta, H. (2012). Approaches to conservation and sustainable use of biodiversity—A review. *Nature & Science, 10*(12), 51–62.

Griffith, M. P., Calonje, M., Meerow, A. W., Tut, F., Kramer, A. T., Hird, A., & Husby, C. E. (2015). Can a botanic garden cycad collection capture the genetic diversity in a wild population? *International Journal of Plant Sciences, 176*(1), 1–10.

Groot, S. P. C., Surki, A. A., De Vos, R. C. H., & Kodde, J. (2012). Seed storage at elevated partial pressure of oxygen, a fast method for analyzing seed ageing under dry conditions. *Annals of Botany, 110*(6), 1149–1159.

Hare, M. D., Tatsapong, P., & Phengphet, S. (2008). Effect of storage duration, storage room and bag type on seed germination of brachiaria hybrid cv. Mulato. *Tropical Grasslands, 42*(4), 224.

Harrington, J. F., & Kozlowski, T. T. (1972). Seed storage and longevity. *Seed Biology, 3,* 145–245.

Jonfia-Essien, W., Varro, S., & Villers, P. (2010). Hermetic storage: A novel approach to the protection of cocoa beans. *African Crop Science Journal, 18*(2).

Justice, O. L., & Bass, L. N. (1978). *Principles and Practices of Seed Storage* (No. 506). U.S. Department of Agriculture.

Kuntal, D. (2014). Impact of differential storage conditions on seed germination and viability of some medicinal plants. *African Journal of Agricultural Research, 9*(20), 1578–1585.

Li, C., Xiao, B., Wang, Q., Zheng, R., & Wu, J. (2017). Responses of soil seed bank and vegetation to the increasing intensity of human disturbance in a semi-arid region of northern China. *Sustainability, 9*(10), 1837.

Maxted, N. (2013). *In situ, ex situ* conservation. In *Encyclopedia of Biodiversity: Second Edition* (pp. 313–323). Elsevier.

McCormack, J. H. (2004). *Seed Processing and Storage: Principles and Practices of Seed Harvesting, Processing, and Storage: An Organic Seed Production Manual for Seed Growers in the Mid-Atlantic and Southern US* (p. 28).

Nagaraju, K. S., Medar, V. S., & Aruna, K. (2018). Seed quality enhancement techniques in medicinal and aromatic crops. *Journal of Pharmacognosy and Phytochemistry, 7*(3S), 104–109.

Nishteswar, K. (2014). Depleting medicinal plant resources: A threat for survival of Ayurveda. *Ayu, 35*(4), 349.

Nwachukwu, I., Ndukwe, O. K., & Ekaiko, M. U. (2016). Germplasm preservation and propagation; The foundation of agricultural development–A review. *IOSR, 11,* 70–73.

Okolo, C. A., Chukwu, O., Adejumo, B. A., & Haruna, S. A. (2017). Hermetic storage technology: The way forward in solving numerous cereal grains storage challenges in developing countries. *International Journal of Engineering Research & Technology, 6.*

Onyekwelu, J. C., & Fayose, O. J. (2007). Effect of storage methods on the germination and proximate composition of *Treculia africana* seeds. In *Proceedings of the International Conference on Agricultural Research for Development.*

Parimala, K., Subramanian, K., Mahalinga Kannan, S., & Vijayalakshmi, K. (2013). *Seed Storage Techniques–A Primer.* CIKS, Chennai, India.

Pradhan, B. K., & Badola, H. K. (2012). Effect of storage conditions and storage periods on seed germination in eleven populations of *Swertia chirayita*: A critically endangered medicinal herb in Himalaya. *The Scientific World Journal.* https://doi.org/10.1100/2012/128105 (accessed on 12th August 2025).

Pritchard, H. W., & Nadarajan, J. (2008). Cryopreservation of orthodox (desiccation tolerant) seeds. In B. M. Reed (Ed.), *Plant Cryopreservation: A Practical Guide* (pp. 485–501). Springer, New York, USA.

Roberts, E. H. (1972). Storage environment and the control of viability. In *Viability of Seeds* (pp. 14–58). Dordrecht: Springer Netherlands.

Savvides, A., Ali, S., Tester, M., & Fotopoulos, V. (2016). Chemical priming of plants against multiple abiotic stresses: Mission possible? *Trends in Plant Science, 21*(4), 329–340.

Shi, Z., Zhang, J., & Wei, H. (2020). Research progress on soil seed bank: A bibliometrics analysis. *Sustainability, 12*(12), 4888.

Somavat, P., Huang, H., Kumar, S., Garg, M. K., Danao, M. G. C., Singh, V., & Rausch, K. D. (2017). Comparison of hermetic storage of wheat with traditional storage methods in India. *Applied Engineering in Agriculture, 33*(1), 121.

Suma, A., Sreenivasan, K., Singh, A. K., & Radhamani, J. (2013). Role of relative humidity in processing and storage of seeds and assessment of variability in storage behavior in *Brica* spp. and *Eruca sativa. The Scientific World Journal,* 504141. https://doi.org/10.1155/2013/504141 (accessed on 12th August 2025).

Tiwari, R. S., Chandra, K. K., Dubey, S., & Tripathi, S. (2022). Influence of packaging materials and storage conditions on seed germination ability and biochemical changes in some medicinal plants of Indian forests. *Frontiers in Forests and Global Change, 5,* 868237.

van Treuren, R., Bas, N., Kodde, J., Groot, S. P., & Kik, C. (2018). Rapid loss of seed viability in *ex situ* conserved wheat and barley at 4°C as compared to −20°C storage. *Conservation Physiology, 6*(1), coy033.

Walters, C., Berjak, P., Pammenter, N., Kennedy, K., & Raven, P. (2013). Preservation of recalcitrant seeds. *Science, 339*(6122), 915–916.

Wyse, S. V., & Dickie, J. B. (2017). Predicting the global incidence of seed desiccation sensitivity. *Journal of Ecology, 105*(4), 1082–1093.

Yang, Q. H., Ye, X. W. H., Deng, H. L., Cao, Y., Zhang, & Xu, K. Y. (2005). Seed germination ecophysiology of *Mikania micrantha* H.B.K. *Botanical Bulletin of Academia Sinica, 46*(4), 293–299.

CHAPTER 2

Organic Farming and High-Quality Seed Production of Medicinal and Aromatic Plants

KUMARI PUNAM SINHA,[1] KODIDHALA VAISHNAVI,[2] HEMANTH KUMAR MANNE,[3,*] PARTHA MONDAL,[4] NISHA KUMARI,[3] and SUNAYANA RATHI[5]

[1]*Department of Seed Science and Technology, Uttar Banga Krishi Viswavidyalaya, Cooch Behar, West Bengal, India*

[2]*Department of Plant Pathology, Uttar Banga Krishi Viswavidyalaya, Cooch Behar, West Bengal, India*

[3]*Department of Biochemistry, Chaudhary Charan Singh Haryana Agricultural University, Hisar, Haryana, India*

[4]*Department of Agricultural Biochemistry, Bidhan Chandra Krishi Viswavidyalaya, Mohanpur, West Bengal, India*

[5]*Department of Biochemistry and Agricultural Chemistry, Assam Agricultural University, Jorhat, Assam, India*

*Corresponding author

ABSTRACT

For decades, the people of India have been engaging in the practice of Ayurvedic medicine, which involves the utilization of medicinal and fragrant crops. These plants have many applications in the production of cosmetics, pharmaceuticals, health foods, and other natural health products. Additionally, they may be used for their fragrant, medicinal,

Seed Technology for Medicinal and Aromatic Plants: Production, Testing, Storage, and Marketing.
Priyanka Sharma, Kishore Kumar Sharma, and Gopal Shukla (Eds.)

DOI: 10.1201/9781779641199-2

and gastronomic properties. It is crucial to enhance the production of biomass for aromatic and medicinal plants in many industries without the use of harmful chemicals. The association between microorganisms and aromatic and medicinal plants, together with the use of organic fertilizers and organic farming methods, may enhance both the quantity and the quality of the crops. Studies on plants with medicinal and aromatic properties indicate that the use of biodegradable fertilizers and the implementation of organic fertilizers might result in optimal crop productivity and superior quality. Some of the things that can be used are beginning to grow organically, making a management, and looking at scented and healing plants all over the world. Biological and natural fertilizers have not been extensively explored on aromatic and medicinal plant growth and production. To achieve sustainable development and agricultural policy goals, organic agriculture must use biologically organic solutions that meet food demands.

2.1 INTRODUCTION

The words "organic" and "organic farming" were first proposed by North Bourne in the 1940s. Organic farming is the practice of growing food without the use of artificial fertilizers, pesticides, or other synthetic materials. To help promote sustainable agricultural practices, use organic fertilizers, natural fertilizers, biological pest control using materials from plants and animals, and other approaches. The main goals are to restore ecological balance and fix the damage that the Green Revolution caused. All forms of organic farming adhere to the same standards. As part of soil management, natural fertilizers are used to replenish soil nutrients after each growing season. To control weeds, it is best to use natural techniques such as mulching and cutting rather than chemical herbicides. Environmentally friendly farming practices also back biological pest control measures. Organic farming has several benefits, one of which is that it yields the healthiest and most nourishing foods. Increasing agricultural yields in this way is an environmentally friendly option. The fundamental components of agricultural systems include crop rotation, the use of crop residues, animal manure, green manure, biofertilizers, mechanical cultivation, rocks that contain minerals, biological control mechanisms, and organic waste from sources outside of farms. Soil fertility and production are enhanced by organic agricultural practices, which also manage pests like weeds and insects.

As a means of combating worldwide health crises, the use of aromatic and medicinal plants in healthcare systems is gaining prominence. Current research on drug discovery from medicinal plants uses a flexible method integrating botanical, phytochemical, biological, and molecular techniques (Sastry et al., 2015). Essential oils, which are a byproduct of many medicinal plants, are also employed in the perfume industry (El-Hennawy, 2018). Improving the quality of aromatic and medicinal plants may be made easier with an organic crop management system, which can be implemented globally and last for a long time. Aromatic and therapeutic plants grew better and produced more when fertilized with organic manures. In various medical systems, including Allopathy, Ayurveda, Homeopathy, Siddha, Unani, Folklore, and so forth, medicinal plants are used to treat illness. These plants can be used in their whole or in part, including roots, stems, leaves, bark, flowers, fruits, seeds, and so forth, or for their chemical composition. Medicinal plants may be used directly in a variety of ways, including powder, extracts, medicinal chemicals (such as alkaloids and glycosides), plant remedies, and plant parts (such as roots, stems, leaves, seeds, and bark). They were, however, unaffected by the quality of the essential oil. In addition to organic manures, other methods for increasing yield and minimizing nutrient consumption include crop rotations, phosphorus-solubilizing bacteria, organic mulches, arbuscular mycorrhizal (AM) fungi, and biofertilizers like *Azotobacter* and *Azospirillum.*

Sustainable crop growth requires a number of inputs, the most important of which is seed. It is believed that seed quality accounts for 20–25% of output. In order to maintain physical and genetic purity, crops are grown in a systematic way with the express purpose of producing seeds. The National Seeds Corporation (NSC) came into being about 1963. Its founders set out to create a system that would ensure the scientific processing, storage, and selling of seeds while simultaneously fostering the healthy expansion of the seed market in India. India has the perfect combination of cheap labor costs, a climate that supports the growth of robust, high-quality seeds, and a wide variety of crops that are exported and consumed domestically. The rural poor will get the skills they need to start their own businesses and generate revenue, and we will save money in the process. Making and selling premium seeds is the best course of action.

The seed business is expected to remain busy and vigorous in the future, thanks to the creation of hybrid and open-pollinated cultivars for local home markets and commercial farmers. Vegetable seed and various crops, including aromatic and medicinal varieties, are essential to economic growth and will never go out of style.

2.2 WHAT IS ORGANIC FARMING AND WHY?

Look to the Land, written by Lord Northbourne, is credited as the origin of the phrase "organic farming." Sir Albert Howard, a botanist from Britain, is widely considered the father of modern organic farming. Organic farming helps keep ecosystems, humans, and soil healthy, says the International Federation of Organic Agriculture Movements (IFOAM). Rather than relying on inputs that cause harm, it is dependent on localized biological processes, biodiversity, and cycles. By combining tradition, creativity, and science, organic farming seeks to enhance the common environment, promote equitable relationships, and raise the standard of living for everyone. According to the FAO, organic farming is a specific kind of production management that prioritizes the health of the agro-ecosystem, which includes soil biological activity, biological cycles, and biodiversity, by removing all synthetic off-farm inputs and using on-farm agronomic, biological, and mechanical methods. The absence of chemical inputs and the preservation of natural spaces within and surrounding organic fields provide ideal homes for wildlife at the ecosystem level, bringing in migratory and permanent species to the organic zone, including beneficial creatures (such as pest predators and pollinators) and wild flora and animals (like birds). In addition to reducing or eliminating the need for pesticides, it provides food, shelter, and a home for new species. In organic farming, plants are grown utilizing several crop rotations and extensive ploughing of the soil, all without the use of synthetic fertilizers or pesticides (Woese et al., 1997).

Organic farming is a way of managing a farm that benefits both humans and the environment in the long run. It is both a cost-effective cultural approach and an investment strategy with a good return. Highly adaptable microbes increase resistance to pests and excitation assaults. The plant's ability to absorb and use nutrients is also enhanced by this. Meeting India's growing demand and preserving future generations are both aided by the achievement of sustainable agricultural practices. Organic agricultural practices aim to minimize or eliminate the use of synthetic fertilizers, herbicides, growth regulators, GMOs, and feed additives (Woese et al., 1997).

The green revolution and modern agriculture have had an effect on human health, natural resources, and agriculture itself in the last several decades. Customer preferences for food quality have changed due to the growing public awareness of the importance of environmental protection and the health risks linked to agrochemicals. Customers are increasingly drawn to foods that are both safe and organic. People and the planet both benefit from organic farming in the long term. One of the main tenets of organic farming is the progressive improvement of soil fertility.

- Avoid hurting the environment while keeping illnesses and pests at bay.
- Nutritious food: free of toxic additives.
- Resolve farming-related environmental concerns.

2.3 MEDICINAL AND AROMATIC PLANTS AND THEIR USES

Medicinal plants refer to complete plants or specific plant components, such as roots, stems, leaves, bark, flowers, fruits, seeds, and so forth, or the chemical compounds obtained from these components. These plants and compounds are utilized in different medical systems (Allopathy, Ayurveda, Homeopathy, Siddha, Unani, Herbomineral, Folklore, etc.) to treat diseases. Medicinal plants may be used in many forms, including powder, extracts, medicinal components such as alkaloids and glycosides, plant pharmaceuticals, and different plant parts such as the root, stem, leaf, seed, and bark.

Aromatic plants are characterized by their ability to produce aromatic essential oils, which are extracted and used for various applications such as fragrance, cosmetics, flavoring, medicine, and more. Aromatic plants may be defined as those that contain volatile compounds with a strong scent, which can be extracted from various portions of the plant to produce essential oils, gum exudates, balsam, and similar substances. At normal ambient conditions, the bulk of them exhibit volatility due to their composition as essential oils. Aromatic products include essential oils (thymol, citronellol, citronellal, terpinen-4-ol), fragrance compounds, absolutes, balsams, concentrates, gums, hydrosols, mosses, scented water, pomades, resins, and resinoids. Provide a comprehensive inventory of notable aromatic and medicinal plants, together with their respective uses and bioactive compounds (Ganaie, 2021; Lubbe and Verporte, 2011).

2.4 THE CORRELATION BETWEEN ORGANIC AGRICULTURAL PRACTICES AND MEDICINAL AND AROMATIC PLANTS

The symbiotic relationship between the earth and living organisms has been enhanced as a result of organic farming. The primary determinant of soil fertility is the quantity of organic matter present in the soil. The organic husbandry approach emphasizes the use of organic resources to maintain soil health, enhance the development of beneficial microbes, and minimize health hazards associated with food production. In our contemporary society, there is a significant need for both sugary and therapeutic crops, as well as vivid natural products for the terminally ill. It has gained global acclaim and is highly sought after by pharmaceutical companies, as well as scent and taste investigators across the globe. The emulsion obtained from organically cultivated medicinal and sweet crop plants has superior physical and chemical qualities in comparison to conventional systems. Integrating the notions of productivity and sustainability into an organic husbandry system for these crops is a challenge. Organic farmers must consider how the many elements of their system, including irrigation, weed and pest control, and soil fertility, will contribute to the long-term productivity and financial viability of their operation.

Crop rotation is an essential component of a successful periodic organic farming system; however, the specific techniques may vary across farms and regions.

- Enhance soil conservation and augment its organic matter concentration.
- Offer services for the management of weeds, nuisances, and nonentities.
- Safeguard and maintain the purity of water.
- Encourage the development of habitats for different species of animals and enhance the variety of natural ecosystems.
- Ensure the husbandry system is financially lucrative.

An effectively designed rotation system is not only the combination of its individual components, but rather a comprehensive approach that considers the interconnections between all elements. This system serves as the main operational instrument for managing all aspects of agricultural production, including weeds, pests, insects, soils, and crop products. An economically viable and environmentally sustainable organic system relies on achieving a favorable rotation and maintaining fertile soil.

2.5 THE IMPORTANCE OF USING ORGANIC FARMING METHODS IN THE CULTIVATION OF MEDICINAL AND AROMATIC PLANTS

India is one of the leading countries in the world when it comes to the cultivation of aromatic and medicinal plants. It was recommended by both the Food and Agriculture Organization and the National Medicinal Plants Board that all aromatic and medicinal plants be farmed in an organic manner. The medicinal and aromatic products that are cultivated organically attract greater prices than those that are grown conventionally. This is in addition to the fact that organically grown products are readily accepted on foreign markets. For example, in March of 2006, the rate of Psyllium (*Plantago ovata*) husk that was grown organically was more than six times higher than the rate of conventional or dirty product. On the other hand, the leaf rate of senna (*Cassia angustifolia*) plants that were cultivated organically was almost twice as high as the leaf rate of plants that were produced conventionally. Senna (*Cassia angustifolia*) pods that were cultivated organically scored around 40% higher than those that were produced using conventional methods. As a result, there is a potentially lucrative worldwide market for products that are produced organically in the years to come. India has to act swiftly in order to make the transition from chemical or conventional farming to organic farming in order to convert the medicinal and aromatic plant (MAP) sectors. Given the ecological significance of aromatic and medicinal plants, it is clear that their needs for soil, water, and temperature are quite specific. It is important to ensure that the soil in an area has a significant quantity of organic matter before it is introduced to new plants. The plants will be able to handle the little concentrations of individual ions that are present in the soil and irrigation water without experiencing any adverse effects on their development as a result of this buffering effect. Many aromatic and medicinal plants display chlorosis when there is a deficiency of organic matter in the soil and a high residual sodium carbonate concentration in the soil or irrigation water. This condition is characterized by the presence of chlorosis (Aishwath, 2004, 2005).

2.6 IMPORTANT ASPECTS OF ORGANIC FARMING AS IT RELATES TO AROMATIC AND MEDICINAL PLANTS

Concentrated manures (oil cakes, meat and blood, fish, horn, and hoof meal, etc.), microbial fertilizers, crop management and rotation, green manure and farmyard manure (FYM), composted urban and agricultural waste, vermi-compost and vermi-wash, crop residue management, cover crops, and mulching are the main components. Everything listed above is a soil input since it contains organic matter, which is the main component of humus after decomposition. Soil physical, chemical, and biological properties are altered by decomposition of organic materials (Aishwath et al., 2003). As a substrate for microbes and an improved home for macrofauna (including worms, spiders, and millipedes), it aids in soil aeration. Because of the improved physical conditions, the inoculum's potential (density and capacity) is altered. Enhanced physicochemical properties promote root growth and nutrient absorption in agricultural plants (Chandra et al., 2003). These factors contributed to the plant's resilience, which allowed it to either endure or escape the illness. Enhanced production ultimately masks the effects of disease. Microbiological infections, crop revolutions, cornucopias, hoof muck, concentrated coprolites (oil galettes, meat and blood, fish, cornucopia, etc.), and crop residue management are the main causes. Green sewage, farming waste, vermi-compost, and vermi-marshland are other major contributors, as are compost from megacities and ranches. Everything on this list is an input that drains organic matter from the soil after corruption.

An experiment was conducted in Madhya Pradesh, India, using composts and vermicompost's. Nine different types of medicinal plants were examined in a pot trial using degraded soil. *Plantago ovata, Andrographis paniculata, Plumbago zeylanica, Abelmoschus moschatus, Plantago somnifera, Hyoscyamus niger, Ruta graveolens, Plantanum xanthocarpum,* and *Solanum xanthocarpum* were among the plants that were included. A total of nine medicinal plant species were subjected to the trial's growing conditions, which included cadaverous (demoralized) soil (Singh et al., 1998). The findings indicated that, before planting, the field soil should be amended with compost in a 12-part ratio. 20 t FYM ha^{-1}, 50:50:50 kg NPK ha^{-1}, NPK alone, a control, composted coir pith (CCP) at 29 and 39 t ha^{-1}, vermicompost (VC) at 21 and 28 t ha^{-1}, and coconuts were interplanted with these different materials (Aguilera et al., 2000). In Kharagpur, India, aloe vera performances were polluted by pollutants from both organic and inorganic sources (Saha et al., 2005). In comparison to non-treated plants, plants treated with the toxin show a considerable improvement in plant height, leaf density, natural and gel yields, and chlorophyll content. Adding gel, humidity, gel ash, and aloin to vermicompost and vermin-marshland made them just as potent as inorganic toxin sources. Vermicomposting increased the fresh and dry weight of turmeric (*Curcuma longa*) rhizomes when used in combination with N, P, and K toxins instead of insulating (Vishwanath et al., 2004).

2.7 REACTIVE ACTIONS OF BIOFERTILIZERS

Some of the microbes are beneficial as natural nitrogen supplements and function as biofertilizers. They may be divided into two groups. Examples of free-living nitrogen fixers include *Azolla*, blue-green algae, *Clostridium, Azotobacter,* and *Azospirillum*. Tricalcium phosphate is insoluble in water, but phosphorus-solubilizing bacteria can dissolve it. Symbiotic nitrogen fixers, such as *Rhizobium* spp., also contribute to improving phosphorus availability. The breakdown of silicate minerals by *Bacillus siliceus* allows plants to absorb K. Some bacteria improve crop nutrient availability by changing the solubility of minerals, including iron, sulfur, and boron. In the rhizosphere soil of 25 medicinal plants and 21 aromatic plants, the populations of Azotobacter were lowest in *Plantago ovata* and highest in *Rauvolfia serpentina,* whereas palmarosa had the highest overall population. *Datura stramonium*, when combined with *Azotobacter, Azospirillum,* and AM fungi, enhanced the alkaloid and dry matter contents. By increasing root weight and development, *Azospirillum* seed treatment enhanced senna (*Cassia angustifolia*) production of dry leaves, pods, and total dry matter.

After looking at 20 aromatic plants and 25 medicinal plants, the ones with the highest *Azotobacter* populations were palmarosa rhizosphere soils; the ones with the lowest numbers were *Rauvolfia tetraphylla* and *Plantago ovafa* (Rao et al., 1987). Both rainfed palmarosa (Maheshwari et al., 1998) and vetiver (Sinkanagam et al., 1999) yields were significantly increased by the use of *Azotobacter*, with or without mineral fertilizers. The oil quality was unaffected. The growth, yield, and oil content of palmarosa were enhanced by dual inoculation with *Glomus aggregatum* and *Azospirillum brasilense* compared with those in uninoculated plants or plants inoculated with vesicular-arbuscular mycorrhiza (VAM) or *Azospirillum* alone (Neelima and Janardhanan, 1996). For bergamot mint (*Mentha citrata*) (Madhav, 2004), the effects of *Azotobacter* alone on plant height, oil content, and oil yield were similar to those of the application of the prescribed quantities of NPK. However, regarding herb yield and N and K uptake, the latter was more effective. The combination of 75% recommended NPK, castor cake, and Azotobacter was determined by Madhav (2004) to produce the best results across all parameters. According to Saleh et al. (1998), inoculating *Datura stramonium* plants with *Azotobacter, Azospirillum*, and VAM fungus significantly increased the alkaloid and dry matter contents. Further studies are needed to understand the response of aromatic and medicinal plants to biofertilizers.

Kapoor (2004) made the observation that the use of mycorrhiza and organic phosphate was increased when fennel essential oil was utilized. In comparison to chemical fertilizers, they claimed that the bio-fertilizer treatments offered a greater number of benefits. According to their argument, the microbiological activity of the soil generates favorable circumstances for the macro- and microminerals, which in turn improve the quality of the essential oil. A research study was carried out by Leithy et al. (2006) to investigate the impact that *Azotobacter* biofertilizer had on the growth of *Rosmarinus officinalis* L., which is a particular kind of rosemary that is often used in the manufacture of oil. According to the findings of the research, cultivating marjoram herbs with bacteria that are capable of nitrogen fixation and phosphate solubilization results in an increase in the quantity of essential oil that is produced (Gharib et al., 2008). During the process of inoculation, the oil content of these fertilizers is enhanced as a consequence of an increase in the formation of monoterpenes and gland secretion. According to the findings of Smith and Read (1997), an increase in the availability of phosphorus in the soil may lead to a large rise in the concentration of chlorophyll A (28%) and the total chlorophyll concentration (19%). In 2007, Abdul-Jaleel performed a research study in which he discovered that subjecting Vinca (*Catharanthus roseus*) plants to stressful conditions and inoculating them with *Pseudomonas fluorescence* bacteria led to an increase in the alkaloid content of the plants as well as an increase in the amount of biomass that they produced. According to Bachman (1998), the presence of organic matter, improved soil water retention, enhanced plant hormone-like effects, increased nutrient absorption by plants, and overall enhancement of the chemical and physical composition of plant debris are some of the factors that have been reported to contribute to the increase in the yield of organic fertilizer. According to Khalid et al. (2006), basil that was cultivated using organic ways produced more than twice as much as basil that was grown using conventional chemical fertilizers.

2.8 GREEN MANURE AND FARMYARD MANURES

The presence of sufficient water is essential for the establishment of green manuring crops. In general, leguminous green manuring crops aid in the restoration of soil nitrogen, the utilization of atmospheric nitrogen, and the enhancement of nutrient availability. The crop also absorbs and utilizes the nutrients that are located under the topsoil. Most organic coprolites investigations incorporate pharmacy and confectionery-driven integrated nutrition operation. Safed musli (*Chlorophytum borivilianum*) had a positive response in terms of root output when subjected to a single application of 10 to 15 tons per hectare of FYM. The operation of the 10–15 t FYM resulted in enhanced absorption of macro and micronutrients by safed musli and soil parcels. Increased cardamom output was obtained by using agricultural fertilizer alone or in combination with inorganic disease control measures. In the study conducted by Rajeswara Rao et al. (2004), it was shown that although adding 10 tons of FYM increased the shoot production of the ginseng crop, the treatment including the application of NPK with 20 tons of FYM resulted in the highest yield. According to Prakash Rao et al. (2003), the application of FYM in Bacopa increased the yield of the first crop. Additionally, when FYM and NPK were combined, it resulted in the greatest potential total herb output. Periwinkle did not respond to the addition of FYM, as stated by Prakash Rao et al. (1988). The use of FYM did not affect the quality of the oils obtained from palmarosa (Rajeswara Rao, 2001), jamrosa (Chand et al., 1996), davana (Prakash Rao et al., 1997).

2.9 PEST AND DISEASE MANAGEMENT

Most medicinal plants are either ingested or used in different therapeutic methods in traditional medical systems (such as Ayurveda, Siddha, and Unani). However, some of the active components are extracted or preserved for use in allopathic medicine. Therefore, it is crucial to address the management of pests and environmental conditions in medicinal and sweet plants in order to mitigate the adverse consequences of fungicide residues or other pesticides. Botanical species with sweet and therapeutic properties often exhibit remarkable resilience to various environmental conditions and pests. Organic husbandry is a beneficial strategy for cultivating therapeutic and pleasant plants, as it provides them with the natural resilience to resist, tolerate, and repel diseases and pests. However, to avoid the disadvantages indicated earlier, traditional farming should use organic farming principles in the control of these pests and situations (Upadhyay and Mishra, 1999).

2.10 UTILIZING MEDICINAL AND AROMATIC PLANTS AS BIO-PESTICIDES FOR OTHER CROPS

Some fragrant and therapeutic plants that have antibacterial and pesticidal properties can be used as biopesticides in organic farming. For example, *Catharanthus* sp., pink and white periwinkle, controls nematodes throughout the plant. Additionally, this particular crop is employed as a trap crop to mitigate okra infection and manage root knot nematodes (*Meloidogyne incognita* and *javanica*) in different crops (Patel et al., 1991). Pyrethrum leaves contain pyrethrins, which possess intrinsic insecticidal properties (Head, 1967). Neem oil and cake are widely acknowledged for their insecticidal properties, while pyrethrum flowers are utilized as repellents against mosquitoes. *Tebernamotena corneria* and *Calotropis* spp. possess potent insecticidal properties that aid in the control of pests and diseases in many crops. Specific types of essential oils, such as citronella and lemon-scented gum, have the ability to repel mosquitoes. Farmers utilize various substances such as custard apple seed and leaf extracts, *Vitex negundo* leaf extract, garlic and chili extracts, fragrant crop distillation water, and plant extracts as insect repellents and pest controllers. However, it is necessary to scientifically validate these methods (Rajeswara Rao and Rajput, 2005). Rao et al. (1996) found that the combination of *Calotropis procera* leaf and AM fungus had a significant impact on reducing *Meloidogyne incognita* infection and enhancing the growth of tomato seedlings. Extracts from the shoots and roots of *Cannabis sativa*, *Chenopodium ambrosioides*, and *Solanum hispidum* were found to be lethal to nematodes such as *Haplolimus indicus* and *Rotylenchulus reniformis* (Haseeb et al., 1978). Deshpande et al. (1990) demonstrated the efficacy of a mixture of acetone extracts from *Glycosmis pentaphyllum*, *Catharanthus roseus*, *Salvodora oleododesh*, and *Breneya* species in deterring oviposition by *Phthorimaea operculella*. The aqueous extract inhibited 78.6% of aflatoxin production by *Aspergillus flavus*. Additionally, crude extracts of *Andrographis paniculata* and its andrographolide showed anti-feeding efficacy against the rice pest *Nephotettix cincticeps* (Kumar and Prasad, 1992; Wadiarta et al., 1997). The growth of rice tango virus was suppressed by leaf extracts obtained from periwinkle, *Tribulus terrestris*, and aloe vera, as demonstrated by Selvaraj and Swamy (1991). The essential oil of *cymbopogon* species (*C. Martinii, C. flexuousus,* and *C. winterianus*) was found to be toxic to nematode species such as *Anguina tritici, Meloidogyne javanica*, and *Heterodera avenae*. This toxicity was attributed to the presence of the main constituents of the essential oil, namely geraniol, citron, and citronellol (Sangwan et al., 1985). The *cymbopogon* essential oil was subjected to testing against the activity of 18 bacteria (Pattnaik et al., 1996). The efficacy of lemongrass oil (*Cymbopogon citratus*) in killing eggs and larvae of the cotton leaf worm (*Spodoptera exigua*) has been demonstrated by Sharaby (1988). The researchers Tiyagi et al. (1990) effectively controlled root knots induced by *Meloidogyne incognita* on tomatoes by immersing tomato roots in a water extract of lemongrass throughout the transplantation process.

Pyrethrins, which are natural insecticidal substances, are present in the flowers of the Pyrethrum plant, which is recognized for its pesticidal properties (Head, 1967). The flowers are often used in domestic pesticides and treatments that deter insects. Neem cake and neem oil, derived from the *Azadirachta indica* plant, are both acknowledged as biopesticides. Most therapeutic plant essential oils and extracts are pesticidal and antibacterial. Certain essential oils, including citronella and gum with a lemon flavor, might deter insects. In order to avoid or control pests on various crops, farmers are using the following herbal medicines:

- Extracts or decoctions made from the seeds and leaves of custard apple.
- The foliage of *Vitex negundo.*
- The extract derived from garlic and chilies.
- The distillation fluids of aromatic crops.

Botanical and herbal insecticides show promise as a viable option. Several studies have shown that herbal extracts may be effective insecticides, but a viable technology has prevented their commercialization. It is necessary to thoroughly evaluate the durability of these botanical extracts.

2.11 ARBUSCULAR MYCORRHIZAL (AM) FUNGI'S ACTION

The association between AM fungus and plants has been seen not only in field crops but also in almost all aromatic and medicinal plants (Gautam and Sharma, 1996; Rao et al., 1987, 1989; Sharma and Roy, 1991, Venkateswara Rao et al., 2000). In order to accurately assess their actual influence on crops, more field research is required, since the bulk of the trials were carried out in controlled pot environments. The *Datura stramonium* (Saleh et al., 1998) and four other medicinal plants (Camprubi et al., 1990) showed increased growth yield and nutrient absorption when they were infected with *Glomus mosseae.* Isabgol (*Plantago ovata*) had enhanced growth, productivity, and seed weight when it was subjected to inoculation with AM-*Glomus fasciculatum, G. Macrocarpus*, and *G. mosseae*. However, *G. mosseae* demonstrated superior performance compared to the other inoculums, as reported by Bloss (1982). The vesicular-arbuscular mycorrhizal fungus *Glomus mosseae*, which infects paprika (*Capsicum frutescens* cv. Passion), exhibited the highest absorption of soil phosphorus at a pH of 6.8, while simultaneously benefitting from nitrate feeding. The increased bio-availability of phosphorus (P) in the soil environment around the roots may explain the 53% improvement in P uptake seen in paprika plants (Jiahnyao and Hongchi, 2000).

2.12 HERBICIDE MANAGEMENT IN MEDICINAL AND AROMATIC CROPS

Cultural practices such as mulching, hand weeding, manual weeding, and intercropping are sometimes used to manage weeds. According to Singh et al. (1986), the most economically efficient strategies for managing medicinal yam (*Dioscorea floribunda*) include intercropping with elephant yam and black gram, as well as using menthol mint waste, palmarosa spent grass, and organic mulch. Hand weeding, mulching with waste grass, and tight spacing resulted in a much higher yield of citronella compared to the control group. The intercropping of citronella resulted in an increase in the quantity of essential oil in green gram, while maintaining its quality, as reported by Rajeswara Rao et al. (1993). The use of citronella waste grass resulted in enhanced biomass and oil output of citronella, palmarosa, and lemongrass, leading to better menthol mint crop production (Singh et al., 1991, 1997). Around 40% of weeds were stopped from growing by the intercropping of rose-scented geranium and menthol mint, without inflicting any damage. In order to minimize the use of herbicides and effectively control weeds, *Mucuna pruriens* may be employed as a living mulch and cover crop. Weeds may use allelochemicals produced by the crop as a means of control (Fujii et al., 2000). Specific species of *Cymbopogon* were used to assess the efficiency and cost-effectiveness of organic mulch compared to herbicides. The findings indicated that organic mulch had superior performance. However, palmarosa and lemongrass showed better weed-suppression ability than *Citronella java* (Singh et al., 1985). *Withania somnifera* and other crops act as trap crops for broomrape parasite of opium poppy (*Papaver somniferum*) by employing allelochemicals to stimulate the parasite's germination and then destroying it (Ramanathan, 1985).

2.13 THE USE OF MEDICINAL AND AROMATIC PLANTS IN ORGANIC FARMING IS BENEFICIAL FOR ENHANCING SOIL FERTILITY AND ADDRESSING ENVIRONMENTAL CONCERNS

Spent grass and wash from fragrant crops are great organic fertilizers for other crops. The leftover byproduct may also be utilized as mulch for moisture conservation and weed control. Prakash Rao et al. (2001) reported that rainfed palmarosa produced more biomass and oil and that weed biomass was decreased when palmarosa distilled grass was added at a rate of 5 t ha^{-1} in conjunction with N. The application of 6 t ha^{-1} of citronella waste grass, combined with nitrogen, resulted in a crop yield equivalent to 450 kg N ha^{-1}. This approach conserved 150 kg of nitrogen without affecting the quality of the plant's oil (Puttanna and Rao, 1997).

Both pongamia cake (Sahrawat, 1981) and neem cake (Prasad et al., 1971) exhibit nitrification inhibitory properties, which effectively limit the leaching of nitrogen from the soil. Compared to neem cake, menthol mint and pyrethrum flower distilled waste inhibits nitrification (Ram et al., 1993). These organic products and by-products may be used in integrated agricultural nutrient management to enhance the efficiency of nitrogen utilization. Nitrification inhibition enhances nitrogen availability for long-duration crops by reducing nitrate leaching and preventing subsurface water contamination, while minimizing nitrogen loss. According to Ram and Kumar (1997), the yield of the main and ratoon mint harvests was enhanced by applying 180 kg of nitrogen, using citronella wasted grass and menthol mint as mulch, after the plant's rhizomes had sprouted. Adding 20 t ha^{-1} of *Pogoma glabra* leaf mulch to citronella crops resulted in a 72% reduction in weed biomass, maintained 45% of the soil's moisture, and enhanced the crop's oil and herb output. These effects were similar to adding 200 kg of nitrogen (Prakasa Rao and Puttanna, 2000).

2.14 THE IMPORTANCE OF INTERCROPPING IN ORGANIC MEDICINAL AND AROMATIC PLANT PRODUCTION

Time and space may increase plant interactions in the agroecosystem, stabilizing yields and reducing pests and illnesses (Malezieux et al., 2009). Growing two or more plant species in the same plot is called multiple cropping, and "intercropping" means growing one crop between rows of another. Intercropping, which enhances plant variety in the agro-ecosystem, reduces diseases and pests easily. Many countries use this method (Ma et al., 2006). Intercropping increases crop yields by reducing weed and pest abundance (Hatcher and Melander, 2003; Poggio, 2005) and better using environmental resources (compared to continuous cropping).

Scientific data suggests this method may increase biodiversity (Huang et al., 2002). When brown mustard was intercropped with *Artemisia annua L., Coriandrum sativum L.,* and *Anethum sowa Kurtz.*, the brown mustard aphid infestation was reduced. On the other hand, when mustard was intercropped with coriander, the number of plant aphids was highest, and when mustard was intercropped with fennel, the number of aphids was lowest. Arecont plantations have the potential to increase their productivity and net revenue per unit area via the cultivation of aromatic and medicinal plants by intercropping, as stated by Sujatha et al. (2011). Different aromatic and medicinal plants had different kernel equivalent yields, with *Cymbopogon flexuosus* having 1218 kg ha^{-1} and *Piper longrum* having 272 kg ha^{-1}. When compared to all other MAP, the *Asparagus racemosus* plant was responsible for the highest kernel equivalent output, which was 1524 kg ha^{-1}. Additionally, it generated a fresh root yield of 10.666 kg ha^{-1} from the production of arecanuts. Intercropping benefits the economy. The plants listed by Sujatha et al. (2011) as having the greatest net return per rupee investment were *Cymbopogon flexuosus* (₹4.25), *Bacopa monnieri* (₹3.64), and *Artemisia pallens* (₹3.12). Maffei and Mucciarelli (2003) found that intercropping peppermint with soya beans boosts essential oils.

2.15 QUALITY SEED PRODUCTION

Quality seed production is crucial for the successful cultivation of medicinal and aromatic crops. Reports by various researchers have indicated that medicinal and aromatic crops are typically harvested once their seeds have reached market maturity, as determined by specific harvest indices based on the crop.

2.15.1 EXTRACTING SEEDS

Fruits, pods, and spikes are harvested when they reach full maturity. Shade-drying follows harvesting and gathering fruits, spikes, or pods. Seeds and pods split when pressed against a hard surface. The procedure is typically finished by late morning. Everything undergoes a meticulous cleaning process, followed by thorough threshing, and finally, the seeds are carefully separated. As much as feasible, the crops and cultivars were maintained separately. After being dried in partial shade until the moisture level was 8% or below, high-quality ripe seeds were kept, and any immature seeds were separated. Here is the data on seed production per acre: 30 kg of Ashwagandha; Senna weighs 20 kg (Rao et al., 2012); Velvet Bean (30 kg); Kalmegh – 5 kg; Isabgol weighs 240 kg (Meena et al., 2015); 30 km of palmarosa; a batch of sweet basil weighing 25 kg; a batch of 10 kg of basil (Jithendra Kumar, 2014) with special significance.

2.15.2 ANALYZING AND PURIFYING

After arriving at the processing shed, the seed will undergo cleaning and treatment to ensure its quality. It is planned to collect a sample in order to assess the purity of the genetic, germination, and physical characteristics. All remaining seeds will be incinerated. Using the seed that had been cleansed, Padmavati et al. (2012) conducted an analysis to determine whether or not the seed had been cleaned, upgraded, treated without mixing, or damaged the seed lot. Seed processing improves physical seed quality by removing extraneous material and effectively removes broken and undersized seed. The primary steps in this procedure involve preconditioning, grading, drying, and basic cleaning.

2.15.3 SEED DRYING

Seed drying is a crucial factor that directly impacts the longevity and germination rate of seeds. There are two commonly used techniques: sun drying and forced-air drying.

2.15.4 SEED CLEANING

Seed cleaning is an essential process in the field of biology. Preconditioning and precleaning practices are implemented in compliance with CAFT 2017–18,308: Progress in High-Quality Vegetable Crop Seed Production. Using simple seed cleaning techniques to remove admixtures, dust, and sand improves the quality of cleaned seed.

2.16 SEED TREATMENT IS AN IMPORTANT ASPECT OF PLANT CULTIVATION

Seed treatment involves subjecting seeds to various substances that help protect them from pests and enhance their overall health during the growth process. A 3 g/kg seed treatment called Dithane M-45 was applied to a variety of plants, including ashwagandha, sweet basil, and sacred basil, prior to storage (Jithendra Kumar, 2014; Meena et al., 2015; Rao et al., 2012).

2.17 SEED PACKAGING OR LABELING

Before marketing or storage, seed packing, also known as bagging, is essentially the last stage. After processing and preparation, seeds are placed into net-weight containers (Vanangamudi, 2014). These procedures are involved in the packaging process: A specified weight is poured into seed bags or containers, placing leaflets on improved growing methods in the containers or seed packets, and sealing the seed bags or containers after attaching certification tags and labels.

2.18 VARIOUS TYPES OF PACKAGING MATERIAL

Container with the ability to allow moisture vapor to pass through: There are various options available for bags, including non-woven bags, jute bags, cloth bags, paper bags, and multiwall paper bags.

Containers that are resistant to moisture vapor can be made from materials such as polythene bags (200–300 gauge) or jute bags coated with a thin polythene film. These are just a couple of examples of how they can be utilized (Vanangamudi, 2014). Containers that are effective at preventing moisture vapor: for example, tin cans, polythene bags with a gauge of over 700, aluminum foil pouches, and glass bottles were used. Seed bags and containers contained detailed information about the crop variety. Seed variety address of the producer tags are affixed to bags to indicate the specific type of seed contained within. Including information about the physical purity percentage on the seed tag is essential. Percentage of genetic material purity, percentage of moisture, and germination rate should be written on the bag.

2.19 SEED STORAGE

Seed storage involves preserving seeds with minimal loss in viability and vigor until they are needed for sowing. India has the advantage of low-cost production for various crops, thanks to its abundance of affordable labor and a favorable environment that promotes the growth of robust, bold, and top-notch seeds for both local consumption and international trade. In addition to generating foreign exchange, this initiative will also contribute to its conservation and empower the rural underprivileged by equipping them with the necessary skills to generate employment and income. Producing and exporting high-quality seeds is the optimal choice. With a deep understanding of the intricacies of plant genetics and the needs of both domestic markets and commercial farmers, the seed industry is poised to thrive and remain dynamic in the coming years.

2.20 METHODS TO ENCOURAGE ORGANIC FARMING

Research should be conducted to enhance knowledge and relevant methods that would increase productivity in the Ayurvedic pharmaceutical industry.

- It is essential to establish standardized guidelines for organic farming in Ayurvedic medicine.
- It is crucial to enhance exports and support local markets.
- Local Ayurvedic pharmacies should focus on marketing and processing organic items domestically.
- Supporting the conservation of medicinal plants, particularly endangered species, is of utmost importance.
- Offering training courses on therapeutic plant production would be highly beneficial.

2.21 CONCLUSION

Presently, the chemical toxin's imbalanced activity leads to a decline in product quality that is both substandard and, via lingering impacts, infiltrates the food chain, posing a risk to the well-being of humans and other animals. Nevertheless, the organic coprolites progressively enhance soil fertility while also promoting higher crop yields, suppressing weed growth, and supplying the soil with organic material and essential minerals. However, considering the Indian viewpoint, it is not practical to go from contemporary to organic farming. The symbiotic relationship between mycorrhizal fungi and MAP, together with the use of organic manures and biofertilizers, has a positive impact on their development. Mulches and biopesticides (neem, garlic, onion, and citronella) are effective methods for controlling weeds and insect pests, respectively. Intercropping and crop rotation are two methods that may help manage weeds, pests, and diseases, in addition to enhancing the value of the land per unit area. Organic manures and mulches not only boost productivity and inhibit weed development, but they also provide essential minerals and organic matter to the soil, therefore improving soil health over time. Nevertheless, transitioning from conventional farming to organic farming is not feasible from an Indian perspective. Many studies employed lab-studied organic or biopesticides and pot plant biofertilizers. In a natural environment, field verification is necessary for the findings to be comparable. However, no thorough investigation has been conducted to assess the amount and quality of food produced by organic manures vs. fertilizers. Micronutrient use as chemical fertilizer boosted production in many studies without compromising quality. According to specific findings, the application of FYM to the actual roots of *Withania somnifera* has an adverse impact on their quality. This phenomenon may be ascribed to the process of root branching and the growth of fibers in the plant. Chemical fertilizers used to *Plantago ovata* have no effect on the plant's quality but instead enhance the overall yield of husks. Some quickly eaten or processed plant material may be useful for organic farming. That being the case, generalizations about aromatic and therapeutic plants drawn from other crops may not hold. Results from soil studies are not always reliable for predicting how aromatic and medicinal plants will respond to fertilizers and manures. Consequently, the outcomes obtained from various sources cannot be compared, and at times they may provide inaccurate data about the suggested amounts of manures. This issue has to be resolved by providing corroborating proof.

KEYWORDS

- **cryopreservation**
- **medicinal and aromatic crops**
- **medicinal seeds**
- **organic agriculture**
- **organic fertilizers**
- **seed production**

REFERENCES

Abdul-Jaleel, C., Manivannan, P., Sankar, B., Kishorekumar, A., Gopi, R., Somasundaram, R., & Panneerselvam, R. (2007). *Pseudomonas fluorescens* enhances biomass yield and ajmalicine production in *Catharanthus roseus* under water deficit stress. *Colloids and Surfaces B: Bio Interfaces, 60,* 7–11.

Aguilera, D. B., Souza, J. R. P., & De Miglioranza, E. (2000). Effect of controlled release fertilizer and vermicompost on chamomile (*Matricaria chamomilla* L.). *Revista Brasileira de Plantas Medicinais, 3,* 61–65.

Aishwath, O. P. (2004). Growth, yield and mineral composition of *Withania somnifera* grown on saline black soil. *Indian Journal of Agricultural Chemistry, 37,* 1–8.

Aishwath, O. P. (2005). Appraisal of underground water in four consecutive years at NRC MAP, Boriavi (Gujarat) and impact of its irrigations on medicinal and aromatic plants. *Ecology, Environment and Conservation, 11,* 373–380.

Aishwath, O. P., Chandra, R., Kumar, D., & Jha, B. K. (2003). Influence of farmyard manure on yield, nutrient content and uptake by *Chlorophytum borivilianum* (Safed musli). In *Proceedings of the National Seminar on Developments in Soil Science* (p. 135). C.S.A.U. Agriculture and Technology, Kanpur.

Bachman, G. (1998). The use of vermicompost as a media amendment for the growth of tomatoes. *Plant Nutrition,* 1107–1123.

Bloss, H. E. (1982). Formation of mycorrhizae in *Plantago ovata. Mycologic, 74,* 505–508.

Camprubi, A., Estaun, V., Calvet, C., & Pera, J. (1990). Infectivity and effectivity of *Glomus mosseae* mycorrhizae in four different species of medicinal plants. *Symbiosis, 9,* 305–307.

Chand, S., Kaul, P. N., Bhattacharya, A. K., Singh, K., Singh, C. P., & Rao, R. B. R. (1996). Response of Jamarosa (*Cymbopogon* species) to spacings, FYM, nitrogen, phosphorus and zinc. *Indian Perfumer, 40,* 41–46.

Chandra, R., Kumar, D., Aishwath, O. P., & Jha, B. K. (2003). Response of *Chlorophytum borivilianum* to organic manure. *Journal of Tropical Medicinal Plants, 4,* 261–264.

Deshpandey, S. G., Nagasampangi, B. A., & Sharma, R. N. (1990). Synergistic oviposition deterrence activity of extracts of *Glycosmis pentaphyllum* (Rutaceae) and other plants for *Phthorimaea operculella* (Zell) control. *Current Science, 59,* 932–933.

El-Hennawy, H. M. (2018). Medicinal, aromatic and toxic plants in Arab countries. *FAO Document Repository, 7*(2), 2116–2119.

Fujii, Y., Oono, K., Vaughan, D., Tomooka, N., Kaga, A., & Miyazaki, S. (1999). Allelopathy in *Vicia, Mucuna* and related leguminous species. In *International Workshop on Genetic Resources,* Ibaraki, Japan, 13–15 October, Part I Wild Legumes (2000, pp. 191–204).

Ganaie, H. A. (2021). Review of the active principles of medicinal and aromatic plants and their disease-fighting properties. *Medicinal and Aromatic Plants,* 1–36.

Gautam, S. P., & Sharma, R. B. (1996). Improvement in medicinal plants through VAM fungi. *Journal of Phytological Research, 9,* 36, 37.

Gharib, F. A., Moussa, L. A., & Massoud, O. N. (2008). Effect of compost and bio-fertilizers on growth, yield and essential oil of sweet marjoram (*Majorana hortensis*) plant. *International Journal of Agriculture & Biology, 10*(4), 381–387.

Haseeb, A., Singh, B., Khan, A. M., & Saxena, S. K. (1978). Evaluation of nematicidal properties in certain alkaloid-bearing plants. *Geobios, 5,* 116–118.

Hatcher, P. E., & Melander, B. (2003). Combining physical, cultural and biological methods: Prospects for integrated non-chemical weed management strategies. *Weed Research, 43*(5), 303–322.

Head, S. W. (1967). A study of the insecticidal constituents of *Chrysanthemum cinerariaefolium* 3. Their composition in different clones. *Pyrethrum Post, 9,* 3–7.

Huang, W., Luukkanen, O., Johanson, S., Kaarakka, V., Räisänen, S., & Vihemäki, H. (2002). Agroforestry for biodiversity conservation of nature reserves: Functional group identification and analysis. *Agroforestry Systems, 55*(1), 65–72.

Jiahnyao, W., & Hongchi, L. (2002). Effect of soil pH, nitrogen form and VA-mycorrhiza infection on the acquisition of soil phosphorus by the paprika plant. *Food Science and Agricultural Chemistry, 2,* 169–173.

Jitendra, K. (2014). *Cultivation of Ocimum* (Technical Report). Medicinal and Aromatic Plants Research Directorate, ICAR.

Kapoor, R., Giri, B., & Mukerji, K. G. (2004). Improved growth and essential oil yield and quality in *Foeniculum vulgare* Mill. on mycorrhizal inoculation supplemented with P-fertilizer. *Bioresource Technology, 93,* 307–311.

Khalid, K. A., Hendawy, S. F., & El-Gezawy, E. (2006). *Ocimum basilicum* L. production under organic farming. *Research Journal of Agriculture and Biological Sciences, 2*(1), 25–32.

Kumar, S., & Prasad, G. (1992). Efficacy of medicinal plants (*Andrographis paniculata*) extract on aflatoxin production and growth of *Aspergillus flavus. Letters in Applied Microbiology, 15,* 131–134.

Leithy, S., El-Meseiry, T. A., & Abdallah, E. F. (2006). Effect of biofertilizers, cell stabilizer and irrigation regime on rosemary herbage oil yield and quality. *Journal of Applied Research, 2,* 773–779.

Lubbe, A., & Verpoorte, R. (2011). Cultivation of medicinal and aromatic plants for specialty industrial materials. *Industrial Crops and Products, 34,* 785–801.

Ma, X. M., Liu, X. X., Zhang, Q. W., Zhao, J. Z., Cai, Q. N., Ma, Y. A., & Chen, D. M. (2006). Assessment of cotton aphids, *Aphis gossypii,* and their natural enemies on aphid-resistant and aphid-susceptible wheat varieties in a wheat–cotton relay intercropping system. *Entomologia Experimentalis et Applicata, 121*(3), 235–241.

Maffei, M., & Mucciarelli, M. (2003). Essential oil yield in peppermint/soybean strip intercropping. *Field Crops Research, 84*(3), 229–240.

Maheshwari, S. K., Sharma, R. K., & Gangrade, S. K. (1998). Response of palmarosa (*Cymbopogon martinii* var. motia) to biofertilizers, nitrogen and phosphorus in a shallow black soil under rainfed condition. *Indian Journal of Agronomy, 43,* 175–178.

Malezieux, E., Crozat, Y., Dupraz, C., Laurans, M., Makowski, D., Ozier-Lafontaine, H., & Rapidel, B., (2009). Mixing plant species in cropping systems: Concepts, tools and models. A review. *Agronomy for Sustainable Development, 29,* 43–62.

Meena, V. D., Dotania, M. L., Meena, B. P., & Das, H., & Beena. (2015). Isabgol (*Plantago ovata* Forsk.)–Medicinal herb: Good source of income generation in water scarcity regions. *Popular Kheti, 3*(4), 2–5.

Neelima, R., & Janardhanan, K. K. (1996). Response of dual inoculation with VAM and *Azospirillum* on the yield and oil content of palmarosa (*Cymbopogon martinii* var. motia). *Microbiological Research, 151,* 325–328.

Padmavati, S., Prakash, M., Ezhel, S., & Kamraj, S. (2012). *A Textbook of Seed Science and Technology* (pp. 254–262). New India Publication Agency (NIPA), New Delhi.

Patel, H. R., Patel, D. J., Patel, C. C., & Thaker, N. A. (1991). Management of root-knot nematodes by periwinkle. *Nematology Mediterranea, 19*(1), 65–66.

Pattnaik, S., Subramanyam, V. R., Kole, C. R., & Sahoo, S. (1996). Antibacterial activity of essential oils from *Cymbopogon*: Inter- and intra-specific differences. *Microbes, 84,* 239–245.

Poggio, S. L. (2005). Structure of weed communities occurring in monoculture and intercropping of field pea and barley. *Agriculture, Ecosystems & Environment, 109*(1), 48–58.

Prakasa Rao, E. V. S., & Puttanna, K. (2007). Role of leaf mulch and farmyard manure in economy of Java citronella (*Cymbopogon winterianus*). In *Proceedings of the National Seminar on Development in Soil Science* (p. 97). Bangalore.

Prakasa Rao, E. V. S., Ganesha Rao, R. S., & Puttanna, K. (2001). Studies on *in situ* soil moisture conservation and additions of phosphorus and potassium in rainfed palmarosa (*Cymbopogon martinii* var. motia) in a semi-arid tropical region of India. *European Journal of Agronomy, 14,* 167–172.

Prakasa Rao, E. V. S., Narayana, M. R., & Rajeshwara Rao, B. R. (1997). The effect of nitrogen and farmyard manure on yield and nutrient uptake in davana (*Artemisia pallens* Wall. ex DC). *Journal of Herbs, Spices & Medicinal Plants, 5,* 39–48.

Prakasa Rao, E. V. S., Puttanna, K., & Singh, M. (1988). Effect of nitrogen, phosphorus and farmyard manure on yield and alkaloid concentration in *Catharanthus roseus* (L). *Journal of Agricultural Science, Cambridge, 110,* 205–206.

Prakasa Rao, E. V. S., Puttanna, K., Ganesha Rao, R. S., & Chandravamshi, P. (2003). Response of *Bacopa monnieri* to organic and inorganic sources of nutrients. In *Proceedings of the National Seminar on New Perspectives in Spices, Medicinal and Aromatic Plants* (pp. 166–168). Calicut, Kerala.

Prasad, R., Rajale, G. B., & Lakhadive, B. A. (1971). Nitrification retarders and slow-release nitrogen fertilizers. *Advances in Agronomy, 23,* 337–383.

Puttanna, K., & Prakasa Rao, E. V. S. (1997). Effect of incorporation of citronella spent grass in soil on nitrogen economy and yield of citronella (*Cymbopogon winterianus* Jowitt.). *Journal of the Indian Society of Soil Science, 45,* 835–837.

Rajeswara Rao, B. R. (2001). Biomass and essential oil yields of rainfed palmarosa (*Cymbopogon martinii* Roxb. Wats. var. motia) supplied with different levels of organic manure and fertilizer nitrogen in semi-arid tropical climate. *Industrial Crops and Products, 14,* 171–178.

Rajeswara Rao, B. R., & Rajput, D. K. (2005). Organic farming in medicinal and aromatic plants. In *Compilation Notes of Winter School on Organic Farming–A Step Towards Eco-Farming for Sustainable Agriculture* (pp. 66–77). Department of Agronomy, College of Agriculture, ANGRAU, Hyderabad, India.

Rajeswara Rao, B. R., Bhattacharya, A. K., Chand, S., & Kaul, P. N. (1993). Cultural and chemical methods for weed control for citronella java (*Cymbopogon winterianus* Jowitt). *Indian Perfumer, 37,* 270–274.

Rajeswara Rao, B. R., Rajput, D. K., Shastri, K. P., Kothari, S. K., Singh, K., Singh, C. P., & Bhattacharya, A. K. (2004). Integrated nutrient management in king of bitters. In *Proceedings of the Symposium on Organic Farming–Prospects and Challenges in the Millennium* (pp. 149–152). Society of Agronomists, Acharya N. G. Ranga Agricultural University, Hyderabad.

Rajeswara Rao, B. R., Rakut, Nagaraju, G., & Adinarayanam, G. (2012). Opportunities and challenges in the cultivation of ashwagandha (*Withania somnifera* (L.) Dunal). *Journal of Pharmacognosy, 3*(2), 88–91.

Ram, M., & Kumar, S. (1997). Yield improvement in regenerated and transplanted mint (*Mentha arvensis*) by recycling the organic wastes and manures. *Bioresource Technology, 59,* 141–149.

Ram, M., Patra, D. D., Subrahmanyam, K., & Singh, D. V. (1993). Nitrification inhibitory properties in mint-spent and pyrethrum flowers. *Journal of the Indian Society of Soil Science, 41,* 176–177.

Ramanathan, V. S. (1985). Studies on control of broomrape (*Orobanche papaveris*), a root parasite menace on *Papaver somniferum* L. in India. *Indian Journal of Agricultural Research, 19,* 161–171.

Rao, M. S., Reddy, P. P., & Das, S. M. (1996). Effect of integration of *Calotropis procera* leaf and *Glomus fasciculatum* on the management of *Meloidogyne incognita* infesting tomato. *Nematologia Mediterranea, 24,* 59–61.

Rao, Y. S. G., Suresh, C. K., Suresh, N. S., & Mallikarjunaiah, R. R. (1987). Rhizosphere populations of *Azotobacter* in aromatic and medicinal plants. *Current Research, 16,* 144–145.

Rao, Y. S. G., Suresh, C. K., Suresh, N. S., Mallikarjunaiah, R. R., & Bagyaraj, D. J. (1989). Vesicular–arbuscular mycorrhizae in medicinal plants. *Indian Phytopathology, 42,* 476–478.

Saha, A. K. (1988). Note on response of turmeric to manures and source of nitrogen and phosphorus under terrace conditions of mid-altitude Mizoram. *Indian Journal of Horticulture, 45,* 139–140.

Sahrawat, K. L. (1981). Karanja (*Pongamia glabra* Vent) as a source of nitrification inhibitors. *Fertilizer News, 26,* 29–33.

Saleh, E. A., Nokhal, T. H., El Borollosy, M. A., Fendrik, I., Saraf, M. S., & El-Sawy, M. (1998). Effectiveness of dual inoculation with diazotrophs and vesicular–arbuscular mycorrhizae on growth and medicinal compounds of *Datura stramonium. Arab University Journal of Agricultural Science, 6,* 343–355.

Sangwan, N. K., Verma, K. K., Mali, B. S., & Dhindsa, K. S. (1985). Nematocidal activity of essential oils of *Cymbopogon* grasses. *Nematologica, 31,* 93–99.

Sastry, K. P., Kotesh, K. J., & Srinivas, K. S. (2015). Promotion of cultivation and processing of medicinal and aromatic plants (MAPs) and phytochemical exploration of MAPs for future drug discovery. *Central Institute of Medicinal and Aromatic Plants–Research Center,* Boduppal, Hyderabad, India, *3*(5), 76–79.

Selvaraj, C., & Narayana Swamy, P. (1991). Effects of plant extracts in controlling rice tungro. *International Rice Research Newsletter, 16,* 21–22.

Sharaby, A. K. (1988). Anti-insect properties of the essential oil of lemongrass, *Cymbopogon citratus,* against the cotton leaf worm *Spodoptera exigua* (Hbn). *Insect Science and Its Application, 9,* 77–80.

Sharma, M., & Roy, A. K. (1991). Vesicular–arbuscular mycorrhizae in hypoglycemic plant. *National Academy of Science Letters, 14,* 467–468.

Singh, A. K., Bisen, S. S., Singh, R. B., & Biswas, S. C. (1998). Effectiveness of compost towards increasing productivity of some medicinal plants in skeletal soil. *Advances in Forestry Research in India, 18,* 64–83.

Singh, A., Singh, K., & Singh, D. V. (1991). Suitability of organic mulch (distillation waste) and herbicides for weed management in perennial aromatic grasses. *Tropical Pest Management, 37,* 162–165.

Singh, A., Singh, M., & Singh, D. V. (1985). Suitability of organic mulch (distillation waste) and herbicides for weed management in perennial aromatic grasses. *Tropical Pest Management, 31,* 162–165.

Singh, A., Singh, M., & Singh, D. V. (1986). The successful use of intercropping for weed management in medicinal yam (*Dioscorea floribunda* Mart. and Gal.). *Tropical Pest Management, 32,* 105–107.

Singh, A., Singh, M., & Singh, D. V. (1987). Continuous use of organic mulch (distillation waste) and herbicides in palmarosa (*Cymbopogon martinii* var. motia) II. Physico-chemical properties of soil and residual effects on succeeding Japanese mint (*Mentha arvensis*) crop. *Indian Perfumer, 41,* 91–97.

Sinkanagam, B., Siripin, S., Tecratorn, A., & Oates, C. G. (1999). Evaluation of plant growth regulators and effect of nitrogen-fixing bacteria on growth of vetiver grass. In *The 37th Kasetsart University Annual Conference,* Bangkok, Thailand (pp. 40–44).

Smith, S. E., & Read, D. J. (1997). *Mycorrhizal Symbiosis.* Academic Press, 2281–2283.

Sujatha, S., Bhat, R., Kannan, C., & Balasimha, D. (2011). Impact of intercropping of medicinal and aromatic plants with organic farming approach on resource use efficiency in arecanut (*Areca catechu* L.) plantation in India. *Industrial Crops and Products, 33,* 78–83.

Tiyagi, S. A., Ahamed, A., & Alam, M. M. (1990). Control of root-knot, reniform and stunt nematodes by root dip in leaf extract of lemongrass. *International Pest Control, 32,* 70–71.

Upadhyay, S., & Mishra, R. C. (1999). Efficacy and economics of insecticides and neem (*Azadirachta indica*)-based products on incidence of aphid (*Aphis gossypii*) on isabgol (*Plantago ovata*). *Indian Journal of Agricultural Science, 69,* 161–162.

Vanangamudi, K. (2014). *Seed Science and Technology* (pp. 357–366). New India Publication Agency (NIPA), New Delhi.

Venkateswara Rao, G., Manoharachari, C., & Kunwar, I. K., & Rajeswara Rao, B. R. (2000). Arbuscular mycorrhizal fungi associated with some economically important spices and aromatic plants. *Philippine Journal of Science, 129,* 51–55.

Venu Madhav, N. (2004). Integrated nutrient management with castor cake, *Azotobacter* and nitrogenous fertilizer on growth, herbage and oil yield and nutrient uptake in bergamot mint (*Mentha citrata* Ehrh.). (Master's thesis, Acharya N. G. Ranga Agricultural University, Hyderabad).

Vishwanath, Y. C., Hanamashetti, S. I., Patil, P. L., & Hagde, N. K. (2004). Effect of organic and inorganic manures on growth, dry matter accumulation and yield of turmeric. In R. K. Pathak, R. Kishun, R. M. Khan, & R. A. Ram (Eds.), *Organic Farming in Horticulture* (pp. 247–249). CISH, Lucknow.

Wadiarta, I. N., Harmawan, W., Oya, S., Nakajima, S., & Nakasuji, F. (1997). Antifeedant activity of constituents of *Andrographis paniculata* (Acanthaceae) against the green rice leaf hopper, *Nephotettix cincticeps* (Hemiptera: Cicadellidae). *Applied Entomology and Zoology, 32,* 561–566.

Woese, W., Lange, D., Boess, C., & Bogl, K. L. (1997). Comparison of organically and conventionally grown foods—Results of a review of the relevant literature. *Journal of the Science of Food and Agriculture, 74,* 281–293.

CHAPTER 3

Detection of Volatile Organic Compounds Emitted by Seeds

HEMANTH KUMAR MANNE,[1,*] KODIDHALA VAISHNAVI,[2] KUMARI PUNAM SINHA,[3] PARTHA MONDAL,[4] NISHA KUMARI,[1] and SUNAYANA RATHI[5]

[1]*Department of Biochemistry, Chaudhary Charan Singh Haryana Agricultural University, Hisar, Haryana, India*

[2]*Department of Plant Pathology, Uttar Banga Krishi Viswavidyalaya, Cooch Behar, West Bengal, India*

[3]*Department of Seed Science and Technology, Uttar Banga Krishi Viswavidyalaya, Cooch Behar, West Bengal, India*

[4]*Department of Agricultural Biochemistry, Bidhan Chandra Krishi Viswavidyalaya, Mohanpur, West Bengal, India*

[5]*Department of Biochemistry and Agricultural Chemistry, Assam Agricultural University, Jorhat, Assam, India*

*Corresponding author

ABSTRACT

Understanding the function of volatile organic compounds (VOCs) in insect interactions with seeds, plants, and other creatures is crucial for gaining a comprehensive understanding of their ecological significance. This knowledge can then be used to develop innovative tactics for controlling pests and disease vectors. This study examines some of the common strategies used in this discipline, with a specific emphasis on the technology they use. Conventional detection techniques are sometimes costly,

Seed Technology for Medicinal and Aromatic Plants: Production, Testing, Storage, and Marketing.
Priyanka Sharma, Kishore Kumar Sharma, and Gopal Shukla (Eds.)

DOI: 10.1201/9781779641199-3

cumbersome, and need specialized knowledge and training. Thus, there is a need for cost-effective, portable, and user-friendly techniques. The detection of VOCs is achieved through the utilization of gas-sensing technologies such as gas chromatography (GC), nuclear magnetic resonance (NMR), and mass spectrometry (MS). Additionally, GC-ion mobility spectrometry (GC-IMS) is employed to provide essential information for identifying the chemical composition of volatile metabolites. These methods frequently fulfill these criteria, but they have their own distinct positive and negative aspects, including the multitude of variables that influence the composition of the VOCs emitted, such as susceptibility to climatic factors and the presence of soil minerals or water, as well as sensor drift. This chapter specifically examines the use of several gas-sensing technologies during the last 10 years to improve the detection of volatile organic chemicals emitted by seeds.

3.1 INTRODUCTION

Seeds are a prominent category of organisms that have the ability to withstand desiccation and thrive without water. The metabolism of these organisms undergoes significant alterations when water is extracted, and cells with less than 15% water exhibit minimal detectable responses (Walters et al., 2005). Although the chemical activity is severely limited, it is evident that some processes take place in organisms that lack moisture, since they degrade over time and ultimately perish. The consequences of this decline, or aging, are often recognized as diminished growth capacity in rehydrated organisms, and the early phase is very difficult to identify. Ultimately, a point of criticality is reached, and a precipitous decline follows. To comprehend the process of aging and anticipate the point at which significant illness will arise, it is essential to get a more comprehensive grasp of the characteristics and dynamics of chemical activity in dehydrated organisms. Currently, there are several types of chemical reactions that are known to occur in dry conditions. These include fermentation, glycation (such as the Maillard reaction), oxidation, and peroxidation. The latter reactions are triggered or enhanced by activated oxygen molecules or free radicals (Bailly, 2004; Colville et al., 2012; Job et al., 2005; Kranner et al., 2010; Labuza, 1980; Mira et al., 2010; Walters et al., 2010). The reactions can use various substrates, resulting in a broad spectrum of products that can be categorized into two main groups: (i) low molecular weight carbonyl compounds that are released as volatile molecules into the air, and (ii) advanced glycation end-products, where molecules become interconnected (Chan, 1987; Frankel, 1983; Grosch, 1987; Halliwell and Gutteridge, 1999; Knutson et al., 2000). These reactions often exhibit modest rates when in a dehydrated condition, and it is yet uncertain if enzyme catalysis takes place. It is hypothesized that the presence of enzymes would limit the range of substrates and products generated. Assessing volatile organic compounds (VOCs) is a convenient way to analyze chemical processes occurring in dry seeds, since these chemicals are a significant consequence of catabolic activities. An increasing amount of research has used GC or gas chromatography/mass spectrometry (GCMS) to non-invasively monitor metabolism by measuring VOCs. Major consequences of lipid peroxidation include volatile alkanes and aldehydes (Aldini et al., 2010; Grotto et al., 2009). Pentane and ethane are frequently produced through the oxidation of polyunsaturated fatty acids (PUFA), specifically linoleic and linolenic acid, which lead to an increase in ROS and denaturation. These compounds have been utilized as indicators of aging in various contexts, including plant tissue culture, human breath, and human or animal tissues. Due to the peroxidation of PUFA, reactive oxygen species (ROS) concentration increases and finally leads to protein oxidation, lipid peroxidation, and DNA damage. Saturated aldehydes, such as hexagonal, may serve as a biomarker for lipid peroxidation in seeds during storage (Colville et al., 2012; Mira et al., 2010), animal tissues (Orhan et al., 2006; Pignoli et al., 2009), and human milk (Elisia and Kitts, 2011). Aldini et al. (2010) and Grotto et al. (2009) found that lipid peroxidation processes also produce unsaturated aldehydes, such as hexenal and 4-hydroxynonenal. These aldehydes are typical indicators of oxidative stress and lipid peroxidation.

Several VOCs are highly reactive and potentially poisonous. These compounds can initiate reactions that cause deterioration and speed up the process of seed viability loss because of the production of ethanol and acetaldehyde. This has been demonstrated in studies with large quantities of ethanol and acetaldehyde (Akimoto et al., 2004). VOCs have the potential to be both a cause and a consequence of deterioration. Therefore, they could be used as a biomarker to indicate the loss of seed quality (Fielding and Goldsworthy, 1982; Hailstones and Smith, 1989). This study aims to further investigate the potential of VOC monitoring in dry seeds for detecting chemical activity.

3.2 VARIOUS VOC DETECTION METHODS

VOCs can be detected by thoroughly identifying all of their physiologically active components using suitable procedures. Most of the necessary processes for successfully carrying out work on these conditions rely on experimental analytical techniques, which provide numerous benefits in VOC characterization. Below, a description of several of these strategies is provided.

3.2.1 GAS CHROMATOGRAPHY (GC)

GC is a scientific method used to separate and analyze samples that include volatile compounds. Component partitioning can occur by introducing the sampled VOC mixture at the start of a chromatographic column and passing it through with the help of an inert gas flow. Subsequently, the components exhibit distinct interactions with the material of the inner column at the molecular level, based on their particular affinity for the compound-column combination under constant circumstances, thereby facilitating their separation. Therefore, the individual components are separated and released one by one from the column, enabling immediate identification and analysis of signals particular to each chemical. The process is enhanced through temperature regulation, either by maintaining constant levels or by implementing programmed increments. GC is the preferred method for analyzing VOCs due to its ability to separate and analyze these components simultaneously, with high sensitivity, reliability, consistency, and flexibility. This method is appropriate for volatile chemicals that are thermally stable, as it involves the utilization of high temperatures for both sample vaporization and chromatographic separation (Ernst et al., 2014; Xiao et al., 2021). GC analysis necessitates detectors that can recognize, identify, and quantify the separated components, offering a diverse range of options. The optimal selection of a detector depends on various factors such as the physical and chemical properties of the substance, its origin and concentration, and the scope of the study. In the study of VOCs that have an impact on seeds, the most commonly used detectors are the mass spectrometry detector (MSD) for identifying specific compounds at a molecular level, and the flame ionization detector (FID) for measuring the quantity of these compounds.

Furthermore, the GC not only serves the purpose of separating substances, but it also allows for the determination of the chemical composition of volatile semi chemicals based on retention data. Retention indices are widely acknowledged as an important metric for identifying VOCs in a standardized system (Héberger, 2007). The typical procedure for this identification involves determining the retention indices by measuring the retention times (tR) of each individual chemical in the GC profile. The calculation relies on the equations developed by Kováts for isothermal GC analysis, which are known as the retention index (I) or Kováts index (KI). For programmed GC analysis, the equations proposed by van den Dool and Kratz are used, which are referred to as the retention index (I), linear retention index (LRI), or programmed-temperature retention index (PTRI or IT) (Zellner et al., 2008). The terms Kovats Index (KI) and LRI have distinct meanings, although they are sometimes overused or interchanged in literature. Therefore, it is crucial to acknowledge this difference while conducting GC examination. The equations establish a correlation between the retention trend of VOCs and a consistent scale given by a homologous series. This scale exhibits a linear relationship between the logarithm of the retention time (tR) and the number of carbon atoms. Examples of such homologous series include n-alkanes and linear fatty acid methyl esters, among others (Castello, 1999). In order to ensure accurate identification, the RI calculations must be carried out on retention data obtained from separate runs using at least two GC capillary columns, involving a polar stationary phase (such as carbowax® (polyethylene glycols)) and a non-polar stationary phase (such as HP-5® (5%-phenyl)-methyl polysiloxane). Zellner et al. (2008) produced a comprehensive compilation that provides a thorough overview of the extent and features of LRI utilization.

Recording the chromatogram using multiple GC columns allows for the detection of a greater number of peaks. A significant breakthrough has been achieved with the implementation of a technique known as complete two-dimensional (2D) GC (GCxGC). The mixture is initially separated using a high-resolution capillary GC column, and the resultant fractions are then concentrated and injected into a second GC column. An interfacing device called a modulator is used to connect both columns and separate the eluate from the first column into smaller fractions. These fractions are limited to a width that is no more than one quarter of the peak width in order to ensure proper separation (Adahchour et al., 2006). Both columns can be accommodated in a single oven under identical temperature programming conditions. However, alternative designs may require two separate ovens, with each column having its own oven for distinct and adjustable temperature programming. In this type of analysis, a GC column with low or moderate polarity is commonly utilized. The column normally has dimensions of 15–30 m length, 0.25–0.32 mm inner diameter, and a film thickness of 0.1–1 µm. Temperature programming is employed, with a heating rate of 1–5 oC/min, and extended analysis times of 45–120 min are used for the first dimension. Next, a smaller and more compact GC column is employed as the second column. This column is typically 1–2 meters long with a diameter of 0.1 millimeters and a film thickness of 0.1 micrometers. It is designed to have a highly polar phase and a shape-selective nature. This column allows for exceptionally quick analysis, taking just 1–10 seconds (Adahchour et al., 2006; Dallüge et al., 2003). The resulting chromatogram is a two-dimensional figure that displays the retention times (in minutes) after the separations in the first and second columns. Therefore, this technique can also accomplish the reduction of overlapping peaks and the detection of small or trace compounds. These findings are crucial advancements, as small or trace chemicals would go unnoticed in one-dimensional GC, especially if larger peaks overlap with these constituents. The results obtained from utilizing GCxGC are highly encouraging, making it the preferred technique for GC analysis of complex mixtures in the identification of semi chemicals. This preference is due to the exceptional separation efficiency offered by this technique (Baeshen et al., 2023; Kalinová et al., 2006).

3.2.2 GAS CHROMATOGRAPHY COUPLED TO FLAME IONIZATION DETECTORS (GC-FID)

The FID is widely acknowledged as the most prominent detector for GC due to its exceptional sensitivity to volatile chemicals with hydrocarbon-based structures. Additionally, the FID exhibits a response that is dependent on the mass flow, making it highly suitable for the investigation of Volatile Organic chemicals (VOCs) (Poole, 2015). This detector is widely used for quantitative analysis, particularly after constructing calibration curves that plot peak areas against concentrations for each compound. However, it is not adequate for precise compound identification on its own. This limitation can be overcome by employing respective standards (co-chromatography), comparing linear retention indices with databases, or utilizing well-characterized samples. Thus, while conducting investigations of compounds that have biological activity on insects using GC-FID, it is necessary to use an extra detector, such as the MSD, to enhance the VOC profile of the sample (Rubiolo et al., 2010). If the analytical technique, including sampling and chromatographic analysis, is completely standardized and includes adequate quantitative standards, it is possible to determine the absolute concentrations of semiochemicals using GC-FID. This allows for trustworthy comparisons and confirmed results (Heuskin et al., 2010). Furthermore, while using SPME for collection, it is important to take certain analytical measures to ensure accurate quantitative analysis. This is because the recovery rate of each individual component might vary, and volatile compounds can also compete with each other (Bartelt, 1997). To that end, for both internal and external calibration, it is necessary to generate a mixture of the relevant standards in relative proportions that are comparable to those found in the target samples. The only approach to determine the absolute amounts of semiochemicals from SPME samples using GC peak regions is as described by Musteata and Pawliszyn (2007).

However, it is possible to compare the relative quantities of a specific volatile compound across samples, treatments, or species via GC relative abundance data obtained through SPME, as long as the collection and analysis are done under the same conditions. On the other hand, comparing unique volatiles throughout a chromatographic profile is not feasible due to the limitations of SPME mentioned earlier. Recent investigations in the field of GC-FID studies have focused on various topics. These include the detection of sex pheromones of insect pests (De Alfonso et al., 2012), the analysis of volatile compounds released by plants when a pollinator is present (Graham et al., 2017), the evaluation of changes in pest behavior when exposed to crops infested with phytopathogens (Buda et al., 2016), the evaluation of essential oils (Stashenko et al., 2010) with fumigant and repellent properties (Wang et al., 2014), and the profiling of plant VOCs (Özgenç et al., 2017; Szmigielski et al., 2012).

3.2.3 GAS CHROMATOGRAPHY COUPLED TO MASS SPECTROMETRY (GC-MS)

As a hyphenated instrumental technique, GC-MS combines the separation capabilities of GC with the sensitivity of MS. This makes it an excellent choice for studying molecules with low molecular mass and mid- to low polarity (Pocsfalvi et al., 2016). Because of its fast compound identification capabilities—in most cases, it can recognize the identity of a molecule with high efficiency—and its ability to compare analyte mass spectra to those recorded in databases, it is extensively used for VOCs (Tsizin et al., 2017). However, due to the complexity of the samples and the need for individual compound calibration curves caused by analyte response factor variation in the MSD, this method is rarely employed for VOC measurement (Ernst et al., 2014).

The sensitivity of gas chromatography-mass spectrometry (GC-MS) primarily relies on an analyzer called a mass spectrometric detector (MSD). The MSD separates and categorizes ions produced from the components into groups based on their mass-to-charge ratio (m/z) (De Villiers et al., 2016). Among the commercially available gas chromatography-mass spectrometry (GC-MS) analyzers, the quadrupole analyzer is the most frequently utilized due to its affordability and adequate accuracy in identifying compounds. Mossi et al. (2011) and Stashenko et al. (2010) characterized essential oils using GC-MS, while Besada et al. (2017), Casas et al. (2017), and Cui et al. (2017) reported metabolic profile analysis of plant VOCs released under diverse conditions. Furthermore, GC-MS has been employed to analyze the VOCs present in the seeds of *Lactuca sativa, Eruca vesicaria, Caram carvi* (Mira et al., 2016), as well as carrot, lettuce, peas, soybean, and rice seeds (Zhang et al., 1995). Additionally, GC-MS was used when analyzing the variations in taste profiles of corn seeds subjected to diverse treatments (native, washed, blanched, precooled, frozen, steamed, boiled, fried, and freeze-dried corn). The principal taste compounds in steamed and boiling corn were found to be identical. However, the abundance of flavor compounds in steamed corn was greater than that in boiled maize (Zhang et al., 2022).

Mohamed Farag and his colleagues (2019) examined the sensory and nutritional characteristics of grilled corn using a metabolomics method. They used SPME and GC-MS analyses to identify the chemical compounds involved and successfully enhanced the taste and flavor of grilled corn based on their extensive chemical studies. The researchers used gas chromatography-mass spectrometry (GC-MS) to detect and classify a total of 78 volatile chemicals present in both cooked brown rice and germinated brown rice, as reported by Xia et al. (2017). Błaszczak et al. (2007) examined the effect of high hydrostatic-pressure processing on the volatile composition of cooked Japonica rice and Jasmine rice using GC-MS.

This method has the potential to serve as a viable alternative to conventional pretreatment methods for enhancing the taste of cooked rice. GC-MS is a popular and efficient technique for assessing volatile chemicals in food samples. It combines gas-phase separation with mass spectrometry and solid-phase microextraction. GC-MS qualitative analysis determines volatile compound identification by comparing retention and indices and mass spectra to authentic standards under comparable chromatographic circumstances (Ghader et al., 2018; Rodríguez-Maecker et al., 2017).

3.2.4 MASS SPECTROMETRY NOT COUPLED WITH GC

Mass spectrometry (MS) is not limited to GC because it was developed independently. As a result, various innovations unrelated to GC have been made in the evolution of MS. It is noteworthy that despite this, MS has made significant contributions to the study of important VOCs in this field of research (Ho et al., 2017). The mass-to-charge ratio (m/z), which is calculated when molecules or their fragments are ionized using a device known as a mass spectrometer, is the basis of the mass spectroscopy (MS) technique. The mechanism involves three phases: ionizer (converts molecules to electrostatic particles), analyzer (filters, separates, and groups charged particles), and detector (quantifies particle m/z) (Svatoš, 2010).

3.2.5 GAS CHROMATOGRAPHY-ION MOBILITY SPECTROMETRY (GC-IMS)

Fast, sensitive, and accurate GC-IMS is an analytical method. The drift area is used for secondary separation of VOCs according to Liedtke et al. (2018). When compared to GC-MS, GC-IMS was discovered to separate components one-third as quickly (Taylor et al., 2017). Gerhardt et al. (2019), Li et al. (2021), and Liu et al. (2020) used this method to examine volatile compounds in dry-cured ham from six different areas, as well as in 1, 2, and 3-year-aged Jinhua ham from China. Wang et al. used GC-IMS to differentiate honey authenticity and observed that GC-IMS required less time in comparison to GC-MS (Wang et al., 2019). The GC-IMS technique works by first separating the volatile components of the sample using a chromatographic column. Then, for ion migration analysis, the components are eluted directly into the IMS ionization chamber (Nölting 2010, Wang et al., 2017).

The hyphenated technique adeptly merges the superior separation capabilities of GC with the heightened sensitivity of IMS, resulting in the analytical outcome of synergistically combining their respective advantages. The HS-GC-IMS three-dimensional spectrum includes the retention time, drift time, and signal intensity, which enhance the precision of the qualitative analysis. GC-IMS offers enhanced separation capabilities and a more cost-effective instrument, while also circumventing the drawbacks of complex operation and analytical challenges. Through the innovative and organic combination of gas-phase analysis technology and ion mobility spectroscopy, GC-IMS is able to analyze the test results in a quick, simple, intuitive, and accurate manner. Furthermore, it does not require any sample pretreatment or heating process. This allows the sample to retain its original flavor to the greatest extent possible, thereby enhancing the authenticity and convincingness of the analysis results (Zhang et al., 2016).

An outstanding case in point is a study conducted by Gao et al. (2023) that characterized volatile compounds derived from pomegranate seeds using a GC system equipped with an MXT-5 capillary column (15 m × 0.53 mm ID, 1 μm), with the GC column temperature set at 60°C and the IMS temperature at 45°C. The researchers used GC-IMS to identify the volatile chemicals present in white and yellowed rice. Additionally, they created a system for monitoring the yellowing process of rice during storage, which was based on GC-IMS (Zhang et al., 2020). Through the comparison and analysis of the shared taste compounds in maize subjected to various treatments, it was determined that the number of common flavor compounds discovered by both GC-MS and GC-IMS typically ranged from three to seven. The study conducted by Zhang et al. (2022) indicated that the GC-IMS method enables rapid and precise identification of maize seed fragrance. GC-IMS has been shown to be particularly successful in analyzing and characterizing volatile chemicals when combined with other methods (Fan et al., 2020). Guo et al. (2018) examined the impact of hot air treatment on the volatile components of *Tricholoma* matsutake using GC-IMS and GC-MS techniques. Employing a variety of methodologies reveals a more thorough, dependable, and rigorous understanding of food fragrance.

3.3 ULTRAFAST-GC-E NOSE

The traditional electronic nose (E-nose) is created to imitate the human sense of smell. It identifies substances based on a signal pattern (fingerprint) rather than the concentration of a specific molecule (Ghosh et al., 2016; Kodogiannis, 2017). An electronic nose, also known as an E-nose, is a very effective tool for quickly and precisely detecting the emission of VOCs that occur during the degradation of food. The objective of the research undertaken by Singh and Gaur (2023) is to create an E-nose using the Arduino-Uno R3 microcontroller. This E-nose will be equipped with MQ4, MQ5, MQ9, and MQ135 sensors to assess the shelf life of various edible seeds over a storage duration of 150 days. The olfactory distinctions among six silver carp samples, which were imperceptible to humans, were successfully discerned using signal fingerprints acquired from an electronic nose (Zhou et al., 2016). The use of this method has been extensive in evaluating the quality and verifying the authenticity of various food samples, including cherry tomato juice, wines, seeds, and meats (Hong and Wang, 2015; Nurjuliana et al., 2011). In contrast to conventional E-nose, which relies on sensors, ultrafast gas chromatography electronic-nose (ufGC E-nose) technology integrates GC and volatile fingerprints, enabling a prompt study of odor fluctuations (Rottiers et al., 2018). The uf-GC E-nose, a recently developed detection technique, has been used for the analysis of amino acids in rice bran (Yu et al., 2024).

3.4 NUCLEAR MAGNETIC RESONANCE (NMR)

The present status of VOC inquiry is well recorded, making it sufficient to employ technologies related to chemical databases for profiling compounds. This may be supplemented with further chemotaxonomic information. Nevertheless, there may be unidentified compounds that have not been previously described. It is crucial to conduct a comprehensive analysis of the compound's structure, which may be achieved via the use of NMR. This technique offers very accurate details about the structural characteristics of tiny molecules. Nuclear Magnetic Resonance (NMR) is a technique that involves subjecting a purified sample to a strong magnetic field. This magnetic field interacts with atomic nuclei that possess a non-zero spin, such as ^{1}H, ^{13}C, ^{15}N, and so forth, causing a reorientation of their orientation. A magnetic pulse is applied for a certain duration and then removed, causing the nuclear spins to revert to their original orientation in relation to the external magnetic field. This process results in radiofrequency emission that is affected by both the magnetic field and the chemical environment of the nuclei, specifically the spatial distribution and type of neighboring atoms.

The NMR equipment detects the emitted radiofrequencies from the sample and compares them to the values of the induced magnetic field. It records a spectrum that displays a pattern of signals, known as "chemical shifts," which typically fall within specific ranges based on the functional group and experimental conditions. The data is also analyzed to establish a relationship between the nuclei, allowing for the deduction of the whole chemical structure (Nozière et al., 2015). When studying naturally occurring volatile chemicals, NMR is often used as a distinct approach from other instruments like GC. However, it is worth noting that there are now GC systems connected to NMR (GC-NMR) that may be used together (Smelcerovic et al., 2013). The reason for this is the stringent sample requirements for NMR, which need high levels of purity, in contrast to other detection methods such as GC-MS, which require larger sample quantities in the range of micrograms. VOCs are often present in smaller amounts in natural sources and may be difficult to purify, recover, and analyze due to interference from other molecules in the matrix (Nojima et al., 2011). According to Reitz et al. (2015), NMR is a useful method for determining the structure of molecules, although it has the drawback of having limited sensitivity.

Consequently, the use of NMR in studies concerning physiologically active seed VOCs has been extensively cited for the purpose of determining their molecular structure after the extraction of VOCs. In the realm of seeds, there is extensive research focused on determining the structure and identifying purified substances found in both germinated and non-germinated rice seeds (Ishibashi et al., 2005). NMR analysis revealed the presence of many organic acids, including succinic acid, acetic acid, and syringic acid, in the Tongkat ali medicinal plant (Serang et al., 2023). Additionally, there are studies investigating the seed vigor of rice seeds (Song et al., 2018). Farang et al. (2021) subjected NMR-based metabolite fingerprinting on four main sprouts—*Cicer, Lens, Trigonella*, and *Vicia*. The results showed that *Trigonella* sprouts had greater levels of isoflavonoids, whereas *Cicer* sprouts had higher levels of trigonelline and 4-hydroxy-isoleucine. L-Dopa was the only substance present in *Vicia*, while acetate was more common in *Lens*.

3.5 CONCLUSION

There is a wide range of methods available for assessing VOCs such as ethylene, acetaldehyde, benzene, toluene that specifically target their effects on seeds, particularly those that are crucial in chemical interactions.

There are legitimate methods for the procedures involved in this kind of study, beginning with the collection of VOCs itself. Hence, the use of experimental, computational, and biological techniques for data collection, along with the application of statistical methodologies for interpretation, may result in dependable and conclusive statements. Regarding VOC analysis, identification, and characterization, there is a wide range of diverse approaches available, each with its own benefits and limits. As a result, these methods are mutually supportive and complementary to one another. The instrumental chemical analysis techniques enable the molecular identification and quantification of various VOCs with a high level of sensitivity, even when they are present at low quantities. These kinds of tests serve as the foundation for identifying the role that VOCs play in chemical interactions.

KEYWORDS

- **gas chromatography**
- **gas chromatography-ion mobility spectrometry**
- **mass spectrometry**
- **nuclear magnetic resonance**
- **seed metabolomics**
- **volatile organic compounds**

REFERENCES

Adahchour, M., Beens, J., Vreuls, R. J. J., & Brinkman, U. A. T. (2006). Recent developments in comprehensive two-dimensional gas chromatography (GC × GC): I. Introduction and instrumental set-up. *Trends in Analytical Chemistry, 25,* 438–454.

Aldini, G., Yeum, K. J., Niki, E., & Russell, R. M. (2010). *Biomarkers for Antioxidant Defense and Oxidative Damage: Principles and Practical Applications.* Ames: Wiley-Blackwell.

Baeshen, N. A., Almulaiky, Y. Q., Afifi, M., Al-Farga, A., Ali, H. A., Baeshen, N. N., Abomughaid, M. M., Abdelazim, A. M., & Baeshen, M. N. (2023). GC-MS analysis of bioactive compounds extracted from plant *Rhazya stricta* using various solvents. *Plants, 12,* 960.

Bailly, C. (2010). Active oxygen species and antioxidants in seed biology. *Seed Science Research, 14,* 93–107.

Bartelt, R. J. (1997). Calibration of a commercial solid-phase microextraction device for measuring headspace concentrations of organic volatiles. *Analytical Chemistry, 69,* 364–372.

Besada, C., Sanchez, G., Gil, R., Granell, A., & Salvador, A. (2017). Volatile metabolite profiling reveals the changes in the volatile compounds of new spontaneously generated loquat cultivars. *Food Research International, 100,* 234–243.

Błaszczak, W., Misharina, T. A., Yuryev, V. P., & Fornal, J. (2007). Effect of high pressure on binding aroma compounds by maize starches with different amylose content. *LWT–Food Science and Technology, 40,* 1841–1848.

Borrajo, P., Karwowska, M., & Lorenzo, J. M. (2022). The effect of *Salvia hispanica* and *Nigella sativa* seed on the volatile profile and sensory parameters related to volatile compounds of dry fermented sausage. *Molecules, 27*(3), 652.

Buda, V., Apšegaite˙, V., Blažyte˙-Čereškienė, L., Butkienė, R., Nedveckytė, I., & Pečiulytė, D. (2016). Response of moth *Plodia interpunctella* to volatiles of fungus-infected and uninfected wheat grain. *Journal of Stored Products Research, 69,* 152–158.

Casas, M. I., Vaughan, M. J., Bonello, P., McSpadden Gardener, B., Grotewold, E., & Alonso, A. P. (2017). Identification of biochemical features of defective *Coffea arabica* L. beans. *Food Research International, 95,* 59–67.

Castello, G. (1999). Retention index systems: Alternatives to the n-alkanes as calibration standards. *Journal of Chromatography A, 842,* 51–64.

Chan, H. W. S. (1987). *Autoxidation of Unsaturated Lipids.* London: Academic Press.

Colville, L., Bradley, E. L., Lloyd, A. S., Pritchard, H. W., Castle, L., & Kranner, I. (2012). Volatile fingerprints of seeds of four species indicate the involvement of alcoholic fermentation, lipid peroxidation, and Maillard reactions in seed deterioration during ageing and desiccation stress. *Journal of Experimental Botany, 63,* 6519–6530.

Cui, S., Wu, J., Wang, J., & Wang, X. (2017). Discrimination of American ginseng and Asian ginseng using electronic nose and gas chromatography–mass spectrometry coupled with chemometrics. *Journal of Ginseng Research, 41,* 85–95.

Dallüge, J., Beens, J., & Brinkman, U. A. T. (2003). Comprehensive two-dimensional gas chromatography: A powerful and versatile analytical tool. *Journal of Chromatography A, 1000,* 69–108.

De Alfonso, I., Hernandez, E., Velazquez, Y., Navarro, I., & Primo, J. (2012). Identification of the sex pheromone of the mealybug *Dysmicoccus grassii* Leonardi. *Journal of Agricultural and Food Chemistry, 60,* 11959–11964.

De Villiers, A., Venter, P., & Pasch, H. (2016). Recent advances and trends in the liquid chromatography–mass spectrometry analysis of flavonoids. *Journal of Chromatography A, 1430,* 16–78.

Elisia, I., & Kitts, D. D. (2011). Quantification of hexanal as an index of lipid oxidation in human milk and association with antioxidant components. *Journal of Clinical Biochemistry and Nutrition, 49,* 147–152.

Ernst, M., Silva, D. B., Silva, R. R., Vêncio, R. Z. N., & Lopes, N. P. (2014). Mass spectrometry in plant metabolomics strategies: From analytical platforms to data acquisition and processing. *Natural Product Reports, 31,* 784.

Fan, X., Jiao, X., Liu, J., Jia, M., Blanchard, C., & Zhou, Z. (2020). Characterizing the volatile compounds of different sorghum cultivars by both GC-MS and HS-GC-IMS. *Food Research International, 140*(4), 109975.

Farag, M. A., Ammar, N. M., El Gendy, A. N., & Mohsen, E. (2019). Effect of grilling as a processing method on *Zea mays* (corn) metabolite composition as analyzed via SPME GC-MS and chemometrics. *Journal of Food Processing and Preservation, 43*(11), e14165.

Farang, M. A., Sharaf El-Din, M. G., Selim, M. A., Owis, A. I., Abouzid, S. F., Porzel, A., Wessjohann, L. A., & Otify, A. (2021). Nuclear magnetic resonance metabolomics approach for the analysis of major legume sprouts coupled to chemometrics. *Molecules, 26*(3), 761.

Frankel, E. N. (1983). Volatile lipid oxidation products. *Progress in Lipid Research, 22,* 1–33.

Gao, L., Zhang, L., Liu, J., Zhang, X., & Lu, Y. (2023). Analysis of the volatile flavor compounds of pomegranate seeds at different processing temperatures by GC-IMS. *Molecules, 28*(6), 2717.

Gerhardt, N., Schwolow, S., Rohn, S., Pérez-Cacho, P. R., Galán-Soldevilla, H., Arce, L., & Weller, P. (2019). Quality assessment of olive oils based on temperature-ramped HS-GC-IMS and sensory evaluation: Comparison of different processing approaches by LDA, KNN, and SVM. *Food Chemistry, 278,* 720–728.

Ghader, M., Shokoufi, N., Es-Haghi, A., & Kargosha, K. (2018). Headspace solid-phase microextraction (HS-SPME) combined with GC-MS as a process analytical technology (PAT) tool for monitoring the cultivation of *C. tetani. Journal of Chromatography B: Analytical Technologies in the Biomedical and Life Sciences, 1083,* 222–232.

Ghosh, P. K., Chatterjee, S., Bhattacharjee, P., & Bhattacharyya, N. (2016). Removal of rancid-acid odor of expeller-pressed virgin coconut oil by gamma irradiation: Evaluation by sensory and electronic nose technology. *Food and Bioprocess Technology, 9*(10), 1724–1734.

Graham, K. K., Brown, S., Clarke, S., Röse, U. S. R., & Starks, P. T. (2017). The European wool-carder bee (*Anthidium manicatum*) eavesdrops on plant volatile organic compounds (VOCs) during trichome collection. *Behavioral Processes, 144,* 5–12.

Grosch, W. (1987). Reactions of hydroperoxides–products of low molecular weight. In H. W. S. Chan (Ed.), *Autoxidation of Unsaturated Lipids* (pp. 95–139). London: Academic Press.

Grotto, D., Santa Maria, L., Valentini, J., Paniz, C., Schmitt, G., Garcia, S. C., Juarez Pomblum, V., Rocha, J. B. T., & Farina, M. (2009). Importance of the lipid peroxidation biomarkers and methodological aspects for malondialdehyde quantification. *Química Nova, 32,* 169–174.

Guo, Y., Chen, D., Dong, Y., Ju, H., Wu, C., & Lin, S. (2018). Characteristic volatile fingerprints and changes of volatile compounds in fresh and dried *Tricholoma matsutake* Singer by HS-GC-IMS and HS-SPME-GC-MS. *Journal of Chromatography B, 1099,* 46–55.

Halliwell, B., & Chirico, S. (1993). Lipid peroxidation: Its mechanism, measurement, and significance. *American Journal of Clinical Nutrition, 57,* 715–724.

Halliwell, B., & Gutteridge, J. M. C. (1999). *Free Radicals in Biology and Medicine.* Oxford: Oxford University Press.

Héberger, K. (2007). Quantitative structure–(chromatographic) retention relationships. *Journal of Chromatography A, 1158,* 273–305.

Heuskin, S., Rozet, E., Lorge, S., Farmakidis, J., Hubert, P., Verheggen, F., Haubruge, E., Wathelet, J. P., & Lognay, G. (2010). Validation of a fast gas chromatographic method for the study of semiochemical slow-release formulations. *Journal of Pharmaceutical and Biomedical Analysis, 53,* 962–972.

Ho, Y. N., Shu, L. J., & Yang, Y. L. (2017). Imaging mass spectrometry for metabolites: Technical progress, multimodal imaging, and biological interactions. *Wiley Interdisciplinary Reviews: Systems Biology and Medicine, 9,* e1387.

Hong, X., & Wang, J. (2015). Use of electronic nose and tongue to track freshness of cherry tomatoes squeezed for juice consumption: Comparison of different sensor fusion approaches. *Food and Bioprocess Technology, 8*(1), 158–170.

Ishibashi, Y., Sueyoshi, R., Morita, T., Yoshimura, A., & Iwaya-Inoue, M. (2005). Detection of pre-harvest sprouting in rice seeds by using ^{1}H-NMR. *Environmental Control in Biology, 43*(2), 131–137.

Job, C., Rajjou, L., Lovigny, Y., Belghazi, M., & Job, D. (2005). Patterns of protein oxidation in *Arabidopsis* seeds during germination. *Plant Physiology, 138,* 790–802.

Kalinová, B., Jiroš, P., Žd'árek, J., Wen, X., & Hoskovec, M. (2006). GC × GC/TOF MS technique—A new tool in identification of insect pheromones: Analysis of the persimmon bark borer sex pheromone gland. *Talanta, 69,* 542–547.

Knutson, M. D., Handelman, G. J., & Viteri, F. E. (2000). Methods for measuring ethane and pentane in expired air from rats and humans. *Free Radical Biology and Medicine, 28,* 514–519.

Kodogiannis, V. S. (2017). Application of an electronic nose coupled with fuzzy-wavelet network for the detection of meat spoilage. *Food and Bioprocess Technology, 10*(4), 730–749.

Kranner, I., Minibayeva, F. V., Beckett, R. P., & Seal, C. E. (2010). What is stress? Concepts, definitions and applications in seed science. *New Phytologist, 188,* 655–673.

Labuza, T. P. (1980). The effect of water activity on reaction kinetics of food deterioration. *Food Technology, 34,* 36–41.

Li, W., Chen, Y. P., Blank, I., Li, F., Li, C., & Liu, Y. (2021). GC×GC-ToF-MS and GC-IMS based volatile profile characterization of the Chinese dry-cured hams from different regions. *Food Research International, 142,* 110222.

Liedtke, S., Seifert, L., Ahlmann, N., Hariharan, C., Franzke, J., & Vautz, W. (2018). Coupling laser desorption with gas chromatography and ion mobility spectrometry for improved olive oil characterization. *Food Chemistry, 255,* 323–331.

Liu, D., Bai, L., Feng, X., Chen, Y. P., Zhang, D., Yao, W., Zhang, H., Chen, G., & Liu, Y. (2020). Characterization of Jinhua ham aroma profiles specific to aging time by gas chromatography–ion mobility spectrometry (GC-IMS). *Meat Science, 168,* 108178.

Mendis, S., Sobotka, P. A., & Euler, D. E. (1994). Pentane and isoprene in expired air from humans: Gas-chromatographic analysis of single breath. *Clinical Chemistry, 40,* 1485–1488.

Mira, S., González-Benito, M. E., Hill, L. M., & Walters, C. (2010). Characterization of volatile production during storage of lettuce (*Lactuca sativa*) seeds. *Journal of Experimental Botany, 61,* 3915–3924.

Mira, S., Hill, L. M., González-Benito, M. E., Ibáñez, M. A., & Walters, C. (2016). Volatile emission in dry seeds as a way to probe chemical reactions during initial asymptomatic deterioration. *Journal of Experimental Botany, 67*(6), 1783–1793.

Mossi, A. J., Astolfi, V., Kubiak, G., Lerin, L., Zanella, C., Toniazzo, G., de Oliveira, D., Treichel, H., & Cansian, R. et al. (2011). Insecticidal and repellency activity of essential oil of *Eucalyptus* sp. against *Sitophilus zeamais* Motschulsky (Coleoptera, Curculionidae). *Journal of the Science of Food and Agriculture, 91,* 273–277.

Musteata, F. M., & Pawliszyn, J. (2007). *In vivo* sampling with solid phase microextraction. *Journal of Biochemical and Biophysical Methods, 70,* 181–193.

Nie, J. Y., Li, R., Jiang, Z. T., Wang, Y., Tan, J., Tang, S. H., & Zhang, Y. (2020). Antioxidant activity screening and chemical constituents of the essential oil from rosemary by ultra-fast GC electronic nose coupled with chemical methodology. *Journal of the Science of Food and Agriculture, 100*(8), 3481–3487.

Nojima, S., Kiemle, D. J., Webster, F. X., Apperson, C. S., & Schal, C. (2011). Nanogram-scale preparation and NMR analysis for mass-limited small volatile compounds. *PLoS ONE, 6,* e18178.

Nölting, B. (2010). *Ion Mobility Spectrometry* (pp. 177–197). Berlin/Heidelberg: Springer.

Nozière, B., Kalberer, M., Claeys, M., Allan, J., D'Anna, B., Decesari, S., Finessi, E., Glasius, M., Grgić, I., Hamilton, J. F., et al. (2015). The molecular identification of organic compounds in the atmosphere: State of the art and challenges. *Chemical Reviews, 115,* 3919–3983.

Nurjuliana, M., Che Man, Y. B., Mat Hashim, D., & Mohamed, A. K. S. (2011). Rapid identification of pork for halal authentication using the electronic nose and gas chromatography–mass spectrometer with headspace analyzer. *Meat Science, 88*(4), 638–644.

Orhan, H., Gurer-Orhan, H., Vriese, E., Vermeulen, N. P., & Meerman, J. H. (2006). Application of lipid peroxidation and protein oxidation biomarkers for oxidative damage in mammalian cells: A comparison with two fluorescent probes. *Toxicology In Vitro, 20,* 1005–1013.

Özgenç, Ö., Durmaz, S., Çelik, G., Korkmaz, B., & Yayli, N. (2017). Comparative phytochemical analysis of volatile organic compounds by SPME-GC-FID/MS from six coniferous and nine deciduous tree bark species grown in Turkey. *South African Journal of Botany, 113,* 23–28.

Pignoli, G., Bou, R., Rodriguez-Estrada, M. T., & Decker, E. A. (2009). Suitability of saturated aldehydes as lipid oxidation markers in washed turkey meat. *Meat Science, 83,* 412–416.

Pocsfalvi, G., Stanly, C., Fiume, I., & Vékey, K. (2016). Chromatography and its hyphenation to mass spectrometry for extracellular vesicle analysis. *Journal of Chromatography A, 1439,* 26–41.

Poole, C. F. (2015). Ionization-based detectors for gas chromatography. *Journal of Chromatography A, 1421,* 137–153.

Reitz, M., Gerhardt, H., Schmitt, C., Betz, O., Albert, K., & Lämmerhofer, M. (2015). Analysis of chemical profiles of insect adhesion secretions by gas chromatography–mass spectrometry. *Analytica Chimica Acta, 854,* 47–60.

Rodríguez-Maecker, R., Vyhmeister, E., Meisen, S., Rosales Martinez, A., Kuklya, A., & Telgheder, U. (2017). Identification of terpenes and essential oils by means of the static headspace gas chromatography–ion mobility spectrophotometer. *Analytical and Bioanalytical Chemistry, 409*(28), 6595–6603.

Rottiers, H., Tzompa Sosa, D. A., Van de Vyver, L., Hinneh, M., Everaert, H., De Wever, J., Messens, K., & Dewettinck, K. (2018). Discrimination of cocoa liquors based on their odor fingerprint: A fast GC electronic nose suitability study. *Food Analytical Methods, 12*(2), 475–488.

Rubiolo, P., Sgorbini, B., Liberto, E., Cordero, C., & Bicchi, C. (2010). Essential oils and volatiles: Sample preparation and analysis—A review. *Flavor and Fragrance Journal, 25,* 282–290.

Serang, A., Zayed, A., Medani, A., & Farag, M. A. (2023). Integrated comparative metabolic profiling via NMR and GC-MS analyses for *Tongkat ali* (*Eurycoma longifolia*) fingerprinting and quality control analysis. *Scientific Reports, 13,* 2533.

Singh, S., & Gaur, S. (2023). Development of rapid and non-destructive electric nose (E-nose) system for shelf-life evaluation of different edible seeds. *Food Chemistry, 426,* 136562.

Smelcerovic, A., Djordjevic, A., Lazarevic, J., & Stojanovic, G. (2013). Recent advances in analysis of essential oils. *Current Analytical Chemistry, 9,* 61–70.

Song, S., Song, P., Yang, H., Yang, T., Xu, J., & Wang, K. (2018). Detection of rice seed vigor by low-field nuclear magnetic resonance. *International Journal of Agricultural and Biological Engineering, 11*(6), 195–200.

Stashenko, E. E., Martínez, J. R., Ruíz, C. A., Arias, G., Durán, C., Salgar, W., & Cala, M. (2010). *Lippia origanoides* chemotype differentiation based on essential oil GC-MS and principal component analysis. *Journal of Separation Science, 33,* 93–103.

Svatoš, A. (2010). Mass spectrometric imaging of small molecules. *Trends in Biotechnology, 28,* 425–434.

Szmigielski, R., Cieslak, M., Rudziński, K. J., & Maciejewska, B. (2012). Identification of volatiles from *Pinus sylvestris* attractive for *Monochamus galloprovincialis* using an SPME-GC/MS platform. *Environmental Science and Pollution Research, 19,* 2860–2869.

Taylor, C., Lough, F., Stanforth, S. P., Schwalbe, E. C., Fowlis, I. A., & Dean, J. R. (2017). Analysis of *Listeria* using exogenous volatile organic compound metabolites and their detection by static headspace–multi-capillary column–gas chromatography–ion mobility spectrometry (SHS–MCC–GC–IMS). *Analytical and Bioanalytical Chemistry, 409*(17), 4247–4256.

Tsizin, S., Bokka, R., Keshet, U., Alon, T., Fialkov, A. B., Tal, N., & Amirav, A. (2017). Comparison of electrospray LC–MS, LC–MS with cold EI and GC–MS with cold EI for sample identification. *International Journal of Mass Spectrometry, 422,* 119–125.

Walters, C., Ballesteros, D., & Vertucci, V. A. (2010). Structural mechanics of seed deterioration: Standing the test of time. *Plant Science, 179,* 565–573.

Walters, C., Hill, L. M., & Wheeler, L. J. (2005). Dying while dry: Kinetics and mechanisms of deterioration in desiccated organisms. *Integrative and Comparative Biology, 45,* 751–758.

Wang, Q., Li, L., Wang, J., Zhang, J., Wang, Y., & Ding, W. (2017). Rapid detection of Ningxia small-tailed Han sheep meat adulterated with duck by electronic nose combined with GC-MS. *Food Science, 38*(20), 222–228.

Wang, X., Li, Q., Shen, L., Yang, J., Cheng, H., Jiang, S., Jiang, C., & Wang, H. (2014). Fumigant, contact, and repellent activities of essential oils against the darkling beetle, *Alphitobius diaperinus. Journal of Insect Science, 14,* 1–11.

Wang, X., Rogers, K. M., Li, Y., Yang, S., Chen, L., & Zhou, J. (2019). Untargeted and targeted discrimination of honey collected by *Apis cerana* and *Apis mellifera* based on volatiles using HS-GC-IMS and HS-SPME-GC–MS. *Journal of Agricultural and Food Chemistry, 67*(43), 12144–12152.

Wolfender, J. L., Marti, G., Thomas, A., & Bertrand, S. (2015). Current approaches and challenges for the metabolite profiling of complex natural extracts. *Journal of Chromatography A, 1382,* 136–164.

Xia, Q., Mei, J., Yu, W., & Li, Y. (2017). High hydrostatic pressure treatments enhance volatile components of pre-germinated brown rice revealed by aromatic fingerprinting based on HS-SPME/GC-MS and chemometric methods. *Food Research International, 91,* 103–114.

Xiao, S., Li, H., Xu, M., Huang, K., Luo, Z., & Xiao, L. T. (2021). A high-throughput method for profiling fatty acids in plant seeds based on one-step catalyzed methylation followed by gas chromatography–mass spectrophotometry. *Biotechnology and Biotechnological Equipment, 35*(1), 1076–1085.

Yu, X., Zhang, W., Xin, L., Xu, S., & Cheng, J. (2024). Evaluation of flavor substances of rice bran kvass based on electronic nose and gas chromatography–mass spectrometry. *Food Chemistry: X, 21,* 101161.

Zellner, B. A., Bicchi, C., Dugo, P., Rubiolo, P., Dugo, G., & Mondello, L. (2008). Linear retention indices in gas chromatographic analysis: A review. *Flavor and Fragrance Journal, 23,* 297–314.

Zhang, K., Gao, L., Zhang, C., Feng, T., & Zhuang, H. (2022). Analysis of volatile flavor compounds of corn under different treatments by GC-MS and GC-IMS. *Frontiers in Chemistry, 10,* 725208.

Zhang, L., Shuai, Q., Li, P., Zhang, Q., Ma, F., Zhang, W., et al. (2016). Ion mobility spectrometry fingerprints: A rapid detection technology for adulteration of sesame oil. *Food Chemistry, 192,* 60–66.

Zhang, M., Maeda, Y., Furihata, Y., Nakamaru, Y., & Esashi, Y. (1994). A mechanism of seed deterioration in relation to the volatile compounds evolved by dry seeds themselves. *Seed Science Research, 4,* 49–56.

Zhang, M., Yajima, H., Umezawa, Y., Nakagawa, Y., & Esashi, Y. (1995). GC-MS identification of volatile compounds evolved by dry seeds in relation to storage conditions. *Seed Science and Technology, 23,* 59–68.

Zhang, Z. D., Fan, X. J., Liu, M., Ma, J. F., Shang, W. T., Wenting, S., et al. (2020). A study on volatile metabolites screening by HS-SPME-GC-MS and HS-GC-IMS for discrimination and characterization of white and yellowed rice. *Cereal Chemistry, 97*(2), 496–504.

CHAPTER 4

Metabolic Changes of Carbohydrates, Proteins, Lipids, and Phytohormones During Seed Germination

HEMANTH KUMAR MANNE,[1,*] KODIDHALA VAISHNAVI,[2]
KUMARI PUNAM SINHA,[3] PARTHA MONDAL,[4]
YENKOKPAM SUPRIYA,[4] NISHA KUMARI,[1] and SUNAYANA RATHI[5]

[1]*Department of Biochemistry, Chaudhary Charan Singh Haryana Agricultural University, Hisar, Haryana, India*

[2]*Department of Plant Pathology, Uttar Banga Krishi Viswavidyalaya, Cooch Behar, West Bengal, India*

[3]*Department of Seed Science and Technology, Uttar Banga Krishi Viswavidyalaya, Cooch Behar, West Bengal, India*

[4]*Department of Agricultural Biochemistry, Bidhan Chandra Krishi Viswavidyalaya, Mohanpur, West Bengal, India*

[5]*Department of Biochemistry and Agricultural Chemistry, Assam Agricultural University, Jorhat, Assam, India*

*Corresponding author

ABSTRACT

Seed germination is a crucial stage in the development of plants and may be considered a decisive component in determining plant production. The physiological and biochemical alterations that take place during germination are intricately connected to the viability of seedlings and their subsequent vegetative development, eventually influencing the yield and caliber of the plants.

Seed Technology for Medicinal and Aromatic Plants: Production, Testing, Storage, and Marketing.
Priyanka Sharma, Kishore Kumar Sharma, and Gopal Shukla (Eds.)

DOI: 10.1201/9781779641199-4

The objective of this chapter is to understand the primary metabolic mechanisms involved in reserve mobilization, phytohormonal modulation, glyoxylate cycle, and respiration during germination. The process of seed imbibition triggers the activation of many metabolic processes, including the synthesis of hydrolytic enzymes. Consequently, this results in the breakdown of accumulated food reserves by hydrolysis, making it easier for the embryo to ingest. Abiotic stresses may affect the germination of seeds and the establishment of seedlings in a number of ways, such as decreased water availability, changed mobilization of reserves, imbalanced hormones, and changes to the structural makeup of proteins.

4.1 INTRODUCTION

One step of plant growth that is crucial and may influence plant production is seed germination. Water imbibition, food reserve mobilization, protein synthesis, and radicle protrusion follow (Hasanuzzaman et al., 2013). Lipids, carbohydrates, and proteins make up the majority of the food reserves that seeds store to support healthy seedling growth (Borgheretti et al., 2000). The main reserves in oilseed are protein and oil bodies, which serve as a source of energy, carbon, and nitrogen during the establishment of seedlings (Zienkiewicz et al., 2014).

Seeds store essential minerals in the form of sucrose or amino acids, which are later converted into starch or proteins during growth. These stored nutrients are then utilized for the early emergence of seedlings. Plants absorb phosphorus in the form of phosphate, which is then transported to mature seeds and stored primarily as phytic acid (about 75%). Starch can be broken down in two ways: hydrolytically and phosphorolytically. First, α-amylase hydrolyzes the amylose and amylopectin in the native starch grain by randomly breaking the α (1→4) glycosidic connections between the glucose residues in the chains. Glucose and maltose are created when α-amylase (or with the help of α-glucosidase) hydrolyzes the released oligosaccharides. Upon germination, seeds from a wide variety of species contain various versions of this enzyme. For instance, at specific pH levels, more than 20 α-amylase isoenzymes found in germinated wheat are divided into two categories by isoelectric focusing. The α (1→6) branch points of amylopectin cannot be hydrolyzed by α-amylase, leading to the production of limit dextrin's, which are highly branched cores of glucose units (Jones and Jacobsen, 1991).

Enzymes that are unique to the α(1→6) link, also known as debranching enzymes, are needed to release the tiny branches before they may be hydrolyzed to the monomer. The R-enzyme and limit dextranase are two names for the same enzyme; alternately, iso-amylase and the R-enzyme have been proposed. This rigid differentiation may not be warranted, since the former was thought to be specific for the tiny limit-dextrin's and incapable of releasing branches from amylopectin, while the latter could. Another type of amylase, known as β-amylase, is incapable of hydrolyzing native starch grains. Instead, it is responsible for cleaving successive maltose units off the nonreducing end of massive oligomers that have been liberated by α-amylolytic attack first. Once again, a debranching enzyme is required for the full hydrolysis of amylopectin.

α-Glucosidase catalyzes the hydrolysis of the disaccharide maltose, generated by amylase, into two glucose molecules. Starch phosphorylase catalyzes the release of glucose-I-phosphate by adding a phosphate group instead of water to the α(1→4) bond between the second-to-last and last glucose molecules at the nonreducing end of the polysaccharide chain. Through the process of phosphorolysis, it is theoretically feasible to entirely degrade amylose. Additionally, amylopectin can be degraded to around two or three residues of glucose of an α(1→6) branched connection. The enzyme is incapable of hydrolyzing starch granules, as they must first undergo partial degradation by other enzymes (Jones and Jacobsen, 1991).

α-glucosidase catalyzes the conversion of the disaccharide maltose, which is produced by amylase. Chrispeels and Varner (1967) discovered that the isolated aleurone did not produce α-amylase in the same amount as the distal half seeds. However, this deficiency was rectified by supplementing the medium with calcium. Calcium is likely to play a role in maintaining the stability of amylase and has a more intricate role in controlling enzyme activity (Jones and Jacobsen, 1991). *De novo* amylase production occurs during seed germination to increase the mobilization of stored starch, providing energy for the young plant until photosynthesis begins. Amylase has been found to have significant activity in this process (Filner and Varner, 1967). Parys et al. (1983) showed that the amylase activity in both cotyledons and axis is controlled by the concentration of reducing sugars *in vivo*. During the germination process, the amylase activity in the cotyledons gradually grew and reached its highest level on the 5th day.

Several studies have investigated the importance of α-amylase activity in seed germination under drought stress. These studies have found that high α-amylase activity directly promotes the germination of seeds under drought stress. However, this increased activity may also be an adaptive response to water deficit, as it is necessary for the accumulation of solutes and the reduction of osmotic potential (Kępczyński, 1986; Wilson, 1971). Furthermore, it is possible that the prevention of α-amylase production is not the mechanism by which dehydration inhibits the germination of *Agropyron desertorum* seeds (Kępczyński, 1986). Gibberellic acids (GAs) can mitigate the inhibitory effects of drought stress on seed germination by regulating the activity of α-amylase (Velasco et al., 1994).

4.2 STARCH METABOLISM

Within numerous species of the *Trifolieae*, there exists a fully developed endosperm that primarily consists of the storage carbohydrate galactomannan. This endosperm is located between the seed coat and the cotyledons. During seed development in fenugreek (*Trigonella foenum-graecum*), there is a significant accumulation of this polymer on the inner surface of the main walls. This leads to a gradual blockage of the live contents within the cells, eventually causing cell death in mature seeds. The aleurone layer, located on the outermost part of the endosperm, consists of live cells that do not have galactomannan reserves, unlike the rest of the endosperm. In many legumes with endosperm, such as carob, the cell walls of the endosperm do not fully enclose the cytoplasm, resulting in all cells containing live contents. Under optimal conditions of illumination and moisture, isolated fenugreek embryos exhibit germination and growth comparable to that of intact seed embryos. The growth is supported by the hydrolytic products derived from proteins and lipids that are stored in the cotyledons. Therefore, it is uncertain whether the endosperm is a necessary source of nourishment for the embryo, as it is challenging to draw conclusions from experiments conducted on isolated embryos that have been germinated and cultivated under optimal laboratory conditions, and apply them to the real-world environment. However, it is evident that the endosperm serves a purpose beyond being a storage organ. Galactomannans are highly viscous in nature and have a strong attraction to water, causing them to become mucilaginous when they absorb it. This property allows the endosperm to control the water levels of the embryo during germination. This is particularly significant for plants in their natural environment, as many species in the *Trifolieae* tribe are believed to have originated in the arid areas of the eastern Mediterranean. It can be inferred that, in natural circumstances, the endosperm has two main functions: controlling water levels during germination and acting as a stored energy source for the growing seedling after germination (Reid, 1971; Reid and Bewley, 1979).

In the fenugreek seed, the raffinose-series oligosaccharides, particularly raffinose and stachyose, found in the embryonic tissues are the initial carbohydrates to be used. This occurs through the breakdown of the α-galactosidic link, resulting in the production of sucrose and galactose. This begins shortly after absorption. Once the radicle emerges, the galactomannan stored in the endosperm starts to be released and utilized. A hydrolysis wave initiates near the aleurone layer and progresses towards the cotyledons until the reserves are exhausted. At least three crucial enzymes, namely α-galactosidase and β-mannosidase, are synthesized and released from this layer. α-Galactosidase is an enzyme that breaks the α(1→6) bond between the galactose side chains and the mannose backbone in polysaccharides (Reid, 1971).

The cotyledons absorb the galactose and mannose that are released. Galactose is absorbed through passive diffusion, while mannose requires active uptake with the help of a carrier-specific component. The cotyledons do not accumulate sugar, but instead undergo further metabolism, possibly through initial phosphorylation into Gal-1-P and Mn-6-P. If not directly utilized for energy metabolism, they undergo a transformation into sucrose and subsequently into starch. This starch is then mobilized when the sucrose levels in the cotyledons decrease after being transported to the axis. In the absence of the endosperm, a small quantity of starch accumulates in the cotyledons. However, the significant buildup of starch only happens when the endosperm is present, indicating that the breakdown products of galactomannan are utilized for the main synthesis of starch (Bewley, 1993).

The mechanism by which the embryo controls the mobilization of reserves inside the fenugreek endosperm is not fully understood. However, it is hypothesized that there is some form of temporal control involved, as the activity of hydrolytic enzymes does not increase and they are not released from layers of aleurone until after germination has finished. Abscisic acid (ABA) is found in the endosperm. When ABA is removed from individual endosperms, the activity of endo-mannosidases increases. There is no necessity for a hormone that stimulates growth, such as gibberellic acid (GA). The addition of ABA to the leached endosperms inhibits the enzymes from becoming more active. The endosperm of fenugreek contains compounds that are similar to diffuse saponins; these substances effectively prevent a rise in α-galactosidase activity. It is unclear how the germinated embryo counteracts these inhibitory effects or if these factors, together with ABA, have a role in preventing galactomannan hydrolysis until germination is finished.

As previously mentioned, raffinose family oligosaccharides function as store reserves in the seeds of many species. Their degradation requires the presence of α-galactosidase and invertases. During the process of seed development, raffinoses build up in the cytoplasm, as stated by Bewley et al. (2013), whereas the enzymes responsible for breaking down raffinoses are stored in the protein storage vacuole (PSV), according to Blochl et al. (2008). They unite during the process of germination. Due to their diminutive size and solubility, they serve as a more easily accessible storage reserve compared to starch. Blochl et al. (2008) found that pea seeds decompose 50% of raffinoses by radicle emergence. Germination in other species is not dependent on reduced raffinose breakdown, but in peas it significantly reduces the rate (Deinking and Bilyeu, 2009). The genes responsible for the production of enzymes that break down raffinose are activated during the beginning of the germination. These genes are also expressed with other genes involved in the production of trehalose, a similar compound (Sreenivasulu et al., 2008). Raffinose, acting as reserve food, contributes to the formation of the glassy state, known as vitrification, in seeds. This phenomenon has been proposed as a means to prevent total dehydration and safeguard cellular membranes (Bernal-Lugo and Leopold, 1995). During the process of imbibition, compounds such as raffinose may have a role in the absorption of water. According to Wattanakulpakin et al. (2012), the decreased storability of seeds obtained from osmopriming may be attributed to decreased amounts of raffinose. Similarly, due to the upregulation of alpha amylases, invertase, hexokinase, phosphoglucomutase, and fructokinase enzymes, starch depletion was accelerated during *Cinnamomum migao* medicinal plant seeds germination (Huang et al., 2021).

4.3 LIPID METABOLISM

Oilseeds typically consist of two components: the primary section known as the kernel, and the outer covering of the seed that encloses the kernel, referred to as the husk or tegument. The kernel consists of two components, namely the embryo and the endosperm. Oleosomes store triacylglycerols, which make up around 20 to 50% of the dry weight. During germination, triacylglycerols are broken down through hydrolysis to generate the energy necessary for the production of sugars, amino acids (particularly asparagine, aspartate, glutamine, and glutamate), and carbon chains that are essential for the growth of the embryo (Quettier and Eastmond, 2009). The lipid levels and lipase activity were examined in different germinating seeds. Research demonstrated that β-oxidation occurs four days following the germination of Castor bean seeds (Hutton and Stumpf, 1969), *Nelumbo nucifera* (Wang et al., 2016) and *C. migao* (Huang et al., 2021). The primary hydrolytic enzymes involved in lipid metabolism during germination are lipases. These enzymes catalyze the breakdown of ester carboxylate bonds, resulting in the release of fatty acids and organic alcohols. They can also facilitate the reverse reaction of esterification or various transesterification reactions. Lipases possess remarkable efficacy, stability, and adaptability, rendering them exceedingly appealing from a business perspective.

While mobilization causes the oil bodies of peanut (*Arachis hypogaea*) and castor bean seeds, which contain a lot of oil, to gradually shrink in size when the reserves are drained, they do not change overall appearance (Hills and Beevers, 1987; Muto and Beevers, 1974). Enzymes that hydrolyze storage TAGs should naturally be very close to the oil bodies where they are found, as these structures contain the TAGs. On the other hand, the oil bodies of peanuts and soybeans, for instance, have a negligible amount of lipase activity, with the majority of the activity being connected with the glyoximes. Huang et al. (2021) reported that seeds of *C. migao* contain TAG, and TAGs were broken down by lipases into free fatty acids (FFA) and glycerol, and the liberated glycerol is converted into glyoxysome and finally converted into acetyl Co-A. It is not yet clear how the TAGs are originally mobilized in these seeds, nor how the FFA, also known as monoacylglycerols, are transported to the glyoxysome, which is the necessary location for further conversion. Within the cotyledons of cotton, it is highly probable that a lipase is first produced in the cytoplasm, and then it is imported into the oil body thereafter. There is a correlation between lipases and oil bodies in castor bean, maize, and rapeseed as well; however, the enzyme is only found in the dry seed of castor bean. In other species, the activity of the enzyme increases after germination. However, even in castor bean, the acid lipase that exists in the dry seed is likely not engaged in lipolysis. This is because the enzyme's activity decreases upon imbibition, because the pH level of the cell is too high (more than 5.5) for the enzyme to operate efficiently. When TAG mobilization begins, two days after the beginning of imbibition, a second lipase that has optimal activity at neutral pHs begins to rise within the oil bodies. This occurs at the same time at which TAG mobilization begins. Approximately four days are required for the TAGs to be fully mobilized. The endosperm has dissolved and assimilates have been taken up by the growing root-shoot axis's expanding cotyledons. One possible function of a lipase linked to the glyoxysome membrane in *Arabidopsis* (Eastmond, 2006), medicinal plant of *C. migao* (Huang et al., 2021) and oil-storing seeds like castor bean is to convert the monoacylglycerols that diffuse out of the oil bodies into FFA and glycerol. Lipases have been researched very little in comparison to other mechanisms that are involved in the mobilization of TAGs, and they have only been found in the seeds of a select few species (Muto and Beevers, 1974).

Glycerol and FFA, the byproducts of lipolysis, are often broken down very quickly. There are a few possible uses for the released glycerol in persistent storage organs: respiration, reutilization for TAG synthesis, and conversion to sucrose for transport to the growth axis. Glycerol is mainly exported as sucrose and lost in respiratory processes in nonpersistent organs, thus it is not surprising that relatively little of it is reincorporated into TAGs. Glycerol kinase in the cytosol phosphorylates glycerol to produce α-glycerol phosphate, which is the initial step in the conversion to sucrose. In the mitochondria, dihydroxyacetone phosphate is produced through the oxidation of this by cytochrome-linked glycerol phosphate oxidoreductase. The last step is release into the cytosol for further hexose conversion. While the storage tissues of a small number of species do acquire FFA during lipolysis, the seeds of the vast majority of dicots and cereals do not. One example is the haustorium, a specialized structure in West African oil palm (*Elaeis guineensis*) seeds that invades the endosperm and transports the products of TAG catabolism to the growing root and shoot through its circulatory system. The endosperm accumulates FFA when oil degrades, and the haustorium absorbs the ones that do not get respired without first converting them to sucrose. The haustorium is another place where FFA build up; until the growing axis needs them, they may be converted to TAG and stored temporarily (Chapman and Trelease, 1991).

Villeneuve and other researchers have categorized lipases into three main groups based on their specificities. The first group is substrate specificity, where glycerol esters are the natural substrates. The second group is called regioselective and includes non-specific lipases that randomly hydrolyze triacylglycerols into fatty acids and glycerol, producing mono- and diacylglycerols as intermediate products. These intermediate products catalyze the hydrolysis at the glycerol bonds (C1 and C3) in triacylglycerols, resulting in the release of fatty acids and unstable intermediates. 2-monoacylglycerols and 1,2- or 2,3-diacylglycerols are types of molecules that can break the ester link of a specific fatty acid or set of fatty acids at any location of a triacylglycerol. This process is carried out by specific or selective fatty acids. The third category, which is enantioselective, has the ability to distinguish between enantiomers in a racemic mixture. The enantioselectivities of lipases are determined by the unique characteristics of the substrate (Castro and Anderson, 1995).

The activation of lipase activity during germination may rely on stimuli originating from the embryo (Tavener and Laidman, 1972). The initial investigation conducted by Shoshi and Reevers (1974) revealed the existence of two lipases in the endosperm of castor bean seeds. Specifically, an acid lipase was found in dried seeds, whereas an alkaline lipase was detected during the germination process. However, the storage tissues of all oilseeds, except for castor bean, only exhibited lipase activity that increased after germination (Anthony et al., 1978).

Due to sucrose serving as the substrate for lipid biosynthesis in growing seeds and being the end product of lipid breakdown, it is primarily regarded as a regulatory factor in the research of lipid metabolism mechanisms (Baud et al., 2008; Quettier and Eastmond, 2009). Furthermore, asparagine and nitrate are recognized as regulating elements in the lipid metabolism of lupine, as stated in reference (Ratajczak, 1986). During the germination of lupin seeds, the concentration of asparagine can reach up to 30% of the total dry weight. Asparagine serves as the primary means of transporting nitrogen from the source tissues to the sink tissues (Lehmann and Ratajczak, 2008). In their study, Borek et al. (2013) found that asparagine regulates carbohydrate metabolism by causing a notable reduction in soluble sugars and an increase in starch in the organs of sprouting lupin seeds. On the other hand, nitrate is not a preferred source of nitrogen in protein metabolism in lupin seeds (Ratajczak et al., 1996) and has no effect on carbohydrate metabolism (Borek et al., 2013). Nitrate, like sucrose, is considered a regulatory element that can influence plant metabolism through alterations in gene expression (Vidal and Gutiérrez, 2008).

Glycerol and FFAs are produced as a result of the hydrolysis of triacylglycerols in oleosomes by lipases. This is the first step in the process of storage lipid mobilization in germinating seeds. Fatty acids are subjected to β-oxidation within peroxisomes of the cell. Following this, the glyoxylate cycle will proceed in a manner that is partially carried out in the peroxisome and partially carried out in the cytoplasm. Citrate synthase, isocitrate lyase, and malate synthase are the three enzymes that are found in peroxisomes, whereas aconitase and malate dehydrogenase are the two enzymes that are found in the cytoplasm (Pracharoenwattana and Smith, 2008). The glyoxylate cycle is a metabolic process that involves the production of a number of different enzymes. While in mitochondria, succinate is carried from the peroxisome to the mitochondria, where it undergoes the Krebs cycle and is transformed into malate. Subsequently, malate undergoes a transformation into oxaloacetate once it has been transported to the cytoplasm. In conclusion, the processes that are a kind of carbon transport, particularly in the process of germinating seeds, include gluconeogenesis and the synthesis of sugars (Borek and Ratajczak, 2010).

The glyoxylate cycle is an important process in the breakdown of lipids in oilseeds and medicinal plants (*C. migao* and *H. annus*). It converts stored lipids into glucose, which is the primary source of energy for respiration during germination and the establishment of seedlings (Eastmond, 2000). Seed imbibition leads to a significant increase in oxygen consumption, indicating an acceleration in the oxidation of carbohydrates derived by the glyoxylate cycle (Muscolo et al., 2007). In addition to the glyoxylate cycle, the oxidative pentose phosphate pathway (OPPP) functions, including many enzymes and intermediates that are involved in both processes (ap Rees, 1980). It supplies NADPH to the cell for biosynthetic processes and seems to have a role in controlling germination (Perino and Come, 1991).

The glyoxylate cycle enzymes, malate synthase and isocitrate lyase, are crucial for oilseed germination because they bypass the TCA cycle's decarboxylation stages. Additionally, two moles of acetyl-CoA are added with each turn of the cycle, which results in the formation of succinate. This succinate is then carried from the glyoxysome into the mitochondrion, where it is transformed into malate via the TCA cycle. After being exported to the cytosol, this malate is changed into oxaloacetate by succinate. PEP-CK enzyme facilitates the transformation of oxaloacetate into phosphoenolpyruvate, which in turn provides the energy required for the production of soluble carbohydrates essential for germination (Muscolo et al., 2007).

Fatty acids are converted to sucrose via beta-oxidation and glyoxylate pathways, which is necessary for the TCA cycle to use them as an energy source (Pritchard et al., 2002). Imbibition is the first step in some species' preparations for an activity. Early imbibition increases barley and rice seed beta–oxidation enzyme transcripts and proteins (rice glyoxylate, barley lipases). Due to the fact that all glyoxylate pathway enzymes and calocin, an oil-body-guiding enzyme, are present in dry sugar beet seeds, it can be inferred that: (i) the glyoxylate pathway is crucial for early germination and (ii) *de novo* synthesis may not be necessary when metabolism begins (Catusse et al., 2008). Nonetheless, mutant sugar beet seeds deficient in calocin may germinate properly (Poxleitner et al., 2006). In a similar fashion, the seeds of wrinkled1, a mutant with minimal oil content, have a lower germination rate but are still capable of germinating when exposed to carbon sources from the outside. It is still unknown whether the observed germination phenotype resulted from structural modifications to the seed's morphology and development or from the decreased seed oil content itself.

4.4 PROTEIN METABOLISM DURING GERMINATION

Germination necessitates the process of *de novo* protein synthesis. The free amino acids found in dry seeds, on the other hand, are insufficient for the synthesis of proteins during the germination process (Bewley et al., 2013). *De novo* production of amino acids, excluding methionine, is weak in other species during the initial hours after imbibition. The methionine (Met) metabolic pathway in dry sugar beet seeds contains all necessary enzymes (Catusse et al., 2008). Remarkably, osmopriming leads to significantly more breakdown of storage proteins into their subunits compared to seeds imbibed in water in *Arabidopsis* and sugar beet (Wang et al., 2012). Proteins undergo initial degradation in the embryo and are subsequently used in the immediate vicinity (Bewley et al., 2013). Dicot embryonic axis and cotyledon cells, and cereal endosperm and aleurone layer include PSVs, which are a concentration point for store proteins. Based on the findings of Bewley et al. (2013), the process of germination involves the participation of a variety of enzymes, each of which has a distinct function in the breakdown of PSV. This process is initiated within the first few hours, and the peptidases that are responsible for it are stored in the seeds that are dry. Additionally, additional proteases accumulate throughout the imbibition phase (Yomo and Varner, 1973). Similarly, Wang et al. (2016) showed that most of the seed proteins (296) were changed during the seed germination of lotus.

4.5 PHYTIC ACID MOBILIZATION

The inositol hexophosphate (IP6) and phytic acid ($C_6H_{18}O_{24}P_6$) found in cereal and legume seeds are the most abundant forms of total phosphorus storage, accounting for 50–80% of the total (Jacela et al., 2010). Because of its propensity to interact with cations (particularly Fe, Ca, K, Mn, Mg, and Zn) via ionic association and produce a mixed salt known as phytin or phytate, which reduces their digestive availability, phytonutrients are considered antinutrients (Lott et al., 1995). Phytate, on the other hand, has the potential to protect against colon cancer and act as an antioxidant by generating an iron complex that reduces free radical production and membrane peroxidation (Thompson and Zuang, 1991). The phytate concentration in grains like beans must be measured and controlled since it is believed to be an antioxidant, anticarcinogen, or vitamin-like compound (Coelho et al., 2002; Okazaki et al., 2005). Creating crop cultivars with less phytin content in the seeds is a primary goal of breeding efforts. Seeds from maize mutants with a reduced phytic acid phenotype had more myo-inositol and fewer myoinositol phosphate intermediates, but these changes had minimal impact on plant development and growth (Shi et al., 2005). These results may imply that the viability, germination, and growth of seedlings do not require a high concentration of stored phytate. The majority of grains' phytin is housed in globoids, which are protein bodies found in the aleurone layer and scutellum cells of seeds. As with protein residue, phytic acid has a considerable chelating ability for multivalent metal ions, particularly zinc, calcium, and iron. Plant phosphorus fertilization is the primary environmental factor that determines the phytate content of seeds (Buerkert et al., 1998). There seem to be a number of genes that regulate the phytate level of beans, and research has demonstrated that this feature exhibits substantial genetic variability. The protein content of grains is another component that influences the phytate level. A correlation was also discovered between the two (Raboy et al., 1991) phytate concentration.

The process of hydrolyzing phytin in germinating seeds is carried out by an enzyme called phytase, which is an acid phosphatase (Hubela and Beck, 1996). This hydrolysis results in the release of phosphate, cations, and inositol, which are then utilized by the seedlings. Minimal alterations in extractible Pi levels were observed in hazel seeds during freezing, which coincided with the mobilization of IP6. This suggests a fast conversion of Pi into an organic form (Mukherji et al., 1971). The results were analyzed and interpreted as proof of a dynamic metabolic process occurring in growing seeds (Silva and Trugo, 1996). Phytase is significantly and competitively inhibited by Pi, and the reduction in phytase activity corresponds to the highest IP6 turnover (Andriotis and Ross, 2003). Ogawa et al. (1979) proposed that the initial breakdown of IP6 is crucial for the metabolic function of inactive tissue by providing phosphorus (Pi) and minerals for physiological and metabolic needs, such as enzymes involved in starch metabolism. Furthermore, IP6 and its related compounds, such as pyrophosphate containing inositol phosphates (PP-IP), have the potential to contribute to the provision of Pi for ATP synthesis in the initial phases of germination, before complete reliance on aerobic mitochondrial respiration, which is the primary source of ATP production.

Stressed seeds experience negative impacts on important processes like germination, growth, respiration, and other related activities. These effects can also influence metabolic activities, specifically the enzymes involved in phosphate metabolism, which are crucial for germination and seed development (Fincher, 1989). Phosphate metabolism is impaired under many stressful circumstances (Mihoub et al., 2005). During periods of stress, development is limited, and there is a decrease in the availability of phosphorus. This leads to an increase in the activity of phosphatases, which hydrolyze the insoluble phosphate form to create Pi. This process helps regulate the mechanism of free phosphate uptake. Olmos and Hellin (1997) found that acid phosphatase activity increased to maintain the level of Pi, allowing it to be co-transported with H^+ along a proton motive force gradient.

4.6 THE ROLE OF PHYTOHORMONES IN GERMINATION

The production of diverse growth regulators with unique chemical structures and physiological actions is a characteristic of plants. The compounds included in this group are auxins, ethylene (ET), gibberellins (GA), strigolactones, ABA, jasmonates (JA), brassinosteroids (BR), and salicylic acid (SA).

ABA, SA, JA, and ET are all identified as crucial factors in facilitating the plant's defensive response to various stimuli. *Arabidopsis* seeds exhibited a decline in JA and SA levels and a rise in auxin levels during the first stage of seed germination (Preston et al., 2009). Both JA and SA have been shown to function as inhibitory factors in seed germination (Dave et al., 2011). In conjunction with GAs, ABA, and ET, auxins are thought to regulate the germination process in seeds (Carrera et al., 2008). The brassinosteroids signal may promote germination by reducing the responsiveness to ABA (Steber and McCourt, 2001).

During an abiotic stress response, phytohormones play a crucial role in coordinating many signal transduction pathways and regulate a range of cellular functions in plants (Vob et al., 2014). The process of seed imbibition led to the activation of gibberellic acid (GA) biosynthesis and response pathways, resulting in the generation of bioactive GAs. Subsequently, gibberellic acids (GAs) activated the genes responsible for producing enzymes like endo-β-1,3 glucanase (Leubner-Metzger et al., 1995) and β-1,4 mannan endohydrolase (Nonogaki et al., 2014). These enzymes break down the endosperm and counteract the inhibitory effects of ABA on the potential development of the embryo (Koornneef and Karssen, 1994). The data demonstrate an antagonistic relationship between ABA and GA, indicating that seeds have high amounts of GA and low levels of ABA under favorable environmental circumstances, and the opposite ratio in unfavorable settings. GA-ABA ratio balances seed dormancy-germination crosstalk, a critical factor in early abiotic-stress adaptation.

ABA hinders the absorption of water by impeding the relaxation of the cell wall in the embryo, resulting in a decrease in the capacity for embryo development (Schopfer and Plachy, 1985). GAs directly increase embryo development during the late period. GAs suppress the ABA effect in both the early and late stages of germination by inducing the production of genes that relax the cell wall, which in turn produces remodeling enzymes such α-expansins in the early stages of germination. The combination of light and cold works in tandem to end the dormancy of soaked seeds and facilitate the process of seed germination by elevating the amounts of gibberellic acids (GAs). During Phase II, the effective completion of germination is influenced by many variables, one of which is a fast drop in the endogenous level of ABA (Weitbrecht et al., 2011). The significant release of cellular solutes during the first absorption suggests that the rehydration process has caused damage to the cellular membranes.

Furthermore, the process of drying and fast seed dehydration has an impact on the integrity of DNA (Matilla et al., 2005). Seeds have evolved many strategies to repair damage during the process of seed germination, such as repairing membranes, proteins, and DNA (Oge et al., 2008).

Phytohormones have a vital function in responding to stress circumstances via the mediation of stress-responsive proteins. It was discovered that the C1-domain, which is a member of the cysteine rich protein family, contains proteins that are involved in the hormone-mediated stress responses of plants (Bhaskar et al., 2015). Furthermore, a total of 72 proteins that respond to stress have been discovered in *Arabidopsis*. These proteins possess all three distinct signature domains. In barley and wheat, GA_3 affects a variety of hydrolytic enzymes' production and activities. Proteomic profiles of *Nelumbo nucifera* medicinal plant reveals that 296 proteins were changed during the various seed germination stages, and 103 (34.8%) out of the total 296 were related to the proteins of photosynthesis, glycolysis, hormones, nucleotides, and secondary metabolites (Wang et al., 2016).

4.7 RESPIRATION ENZYMES MOBILIZATION DURING GERMINATION

The primary release of stored food in seeds during the early stages of germination occurs mostly via anaerobic respiration. Anaerobic respiration is facilitated by enzymes, such as dehydrogenases, that do not rely on aerobic conditions (Turner and Turner, 1975). Dehydrogenase is an enzyme that helps move electrons from substrates to oxygen via the electron transport chain. It uses cofactors such as nicotinamide adenine dinucleotide (NAD^+) and riboflavin (Robert et al., 2009). The activities of dehydrogenases, including alcohol dehydrogenase, lactate dehydrogenase, and succinate dehydrogenase, have been shown to be involved in the conversion of stored lipids and carbohydrates via anaerobic respiration (Oaikhena et al., 2013). The complex enzyme known as succinate dehydrogenase, which is closely linked to the inner mitochondrial membrane, is responsible for the oxidation of succinate to fumarate (Devlin, 2011). Lactate dehydrogenase uses NAD+ to reversibly oxidize lactate to pyruvate. During periods of seed rest and the early phases of seed germination, anaerobic respiration was seen (Jones et al., 1999).

The respiration rate of growing seeds increases with glycolytic activity. The OPPP route feeds glycolysis intermediates back into glycolysis; hence, its activity also affects glycolysis flow (Podestá and Plaxton, 1994). Both α- and β-amylases degrade endosperm starch. The enzymes debranching enzyme, α- and β-amylases, and α-glucosidases (maltase) catalyze the hydrolysis of starch into glucose (Andriotis et al., 2016). Since amylases offer respiratory substrates for energy and carbon production before the seedling can photosynthesize, they are important. In addition, the transition from the quiescent stage to the active phase of embryo development is mostly dependent on the use of stored ATP and products of lipid breakdown that are stored (Nandi et al., 1995).

The phase of seed germination is an opportune time to research the formation of mitochondria. Prior transcriptome investigations have shown a significant rise in mitochondrial transcripts that encode proteins and protein content, together with alterations in their activities, during the first 3 hours following seed imbibition (Howell et al., 2009). From dry seed to seed that was soaked for 48 hours, 56 proteins showed differential expression, including members of the TIM17/22/23 and TOM40 families, which are involved in the inner and outer membranes, respectively.

Because the general import route in mitochondria has a much lesser capacity under anaerobic circumstances compared to aerobic ones, the idea that the capacity of the import pathway is completely reliant on the availability of O_2 is connected to this finding. A decrease in proteins involving import apparatus was detected in mature mitochondria, which may suggest that the accumulation of these import proteins in the dry seed could operate functions after two hours of imbibition; they act as substrates for TCA and ETC cycles (Howell et al., 2006). This suggestion is supported by the fact that three proteins belonging to the TIM17/22/23 family were found to be up-regulated by 6–14 folds after being exposed to anaerobic conditions (Howell et al., 2007).

4.8 CONCLUSION

One important step of plant growth that may influence plant production is seed germination. The morphological and physiological changes that occur during germination are closely linked to the rate of survival of seedlings and their vegetative development. Many changes occur in the metabolic processes, including respiration, glyoxylate cycle, reserve mobilization, and phytohormonal control as well. Initially, there is a surge in respiration to provide energy for growth. Enzymes activate to break down stored reserves like starch and proteins into simpler molecules such as glucose and amino acids; further, lipid reserves mobilize to provide energy. Additionally, there is an increase in the synthesis of plant hormones like gibberellins, which stimulate elongation and cell division. These metabolic changes are crucial for the successful germination and establishment of MAP.

KEYWORDS

- **germination**
- **lipid metabolism**
- **phytohormones**
- **protein metabolism**
- **seed germination**
- **starch metabolism**

REFERENCES

Andriotis, V. M. E., & Ross, J. D. (2003). Isolation and characterization of phytase from dormant *Corylus avellana* seeds. *Phytochemistry, 64,* 689–699.

Andriotis, V. M. E., Saalbach, G., Waugh, R., Field, R. A., & Smith, A. M. (2016). Maltase involved in starch metabolism in barley endosperm is encoded by a single gene. *PLoS ONE, 11*(3), e0151642.

Anthony, H. C., Huang, A. H. C., & Moreau, R. A. (1978). Lipases in the storage tissues of peanut and other oil seeds during germination. *Planta, 141,* 111–116.

Baud, S., Dubreucq, B., Miquel, M., Rochat, C., & Lepiniec, L. (2008). Storage reserve accumulation in *Arabidopsis*: Metabolic and developmental control of seed filling. *The Arabidopsis Book,* 6. American Society of Plant Biologists.

Bernal-Lugo, I., & Leopold, A. C. (1995). Seed stability during storage: Raffinose content and seed glassy state. *Seed Science Research, 5,* 75–80.

Bewley, J. D., Bradford, K. J., Hilhorst, H. W. M., & Nonogaki, H. (2013). *Seeds: Physiology of Development, Germination and Dormancy* (3rd ed.). New York: Springer.

Bewley, J. D., Leung, D. W. M., MacIsaac, S., Reid, J. S. G., & Xu, N. (1993). Transient starch accumulation in fenugreek cotyledons. *Plant Physiology and Biochemistry, 31,* 483–490.

Bhaskar, R. V., Mohanty, B., Verma, V., Wijaya, E., & Kumar, P. P. A. (2015). Hormone-responsive C1-domain-containing protein At5g17960 mediates stress response in *Arabidopsis thaliana. PLoS ONE, 10*(1), e0115118.

Blöchl, A., Peterbauer, T., Hofmann, J., & Richter, A. (2008). Enzymatic breakdown of raffinose oligosaccharides in pea seeds. *Planta, 228,* 99–110.

Borek, S., & Ratajczak, L. (2010). Storage lipids as a source of carbon skeletons for asparagine synthesis in germinating seeds of yellow lupine (*Lupinus luteus* L.). *Journal of Plant Physiology, 167,* 717–724.

Borek, S., Galor, A., & Paluch, E. (2013). Asparagine enhances starch accumulation in developing and germinating lupin seeds. *Journal of Plant Growth Regulation, 32,* 471–482.

Borgheretti, F., & Ferreira, A. G. (2000). *Germinação: Do básico ao aplicado* (p. 222). Porto Alegre: Artmed.

Buerkert, A., Haake, C., Ruckwied, M., & Marschner, H. (1998). Phosphorus application affects the nutritional quality of millet grain in the Sahel. *Field Crops Research, 57,* 223–235.

Carrera, E., Holman, T., Medhurst, A., Dietrich, D., Footitt, S., Theodoulou, F. L., et al. (2008). Seed after-ripening is a discrete developmental pathway associated with specific gene networks in *Arabidopsis*. *The Plant Journal, 53,* 214–224.

Castro, H. F., & Anderson, W. A. (1995). Fine chemicals by biotransformation using lipases. *Química Nova, 18*(6), 544–554.

Catusse, J., Strub, J. M., Job, C., Van Dorsselaer, A., & Job, D. (2008). Proteome-wide characterization of sugarbeet seed vigor and its tissue-specific expression. *Proceedings of the National Academy of Sciences of the United States of America, 105,* 10262–10267.

Chapman, K. D., & Trelease, R. N. (1991). Glyoxysome enlargement in cotyledons. *Journal of Cell Biology, 115,* 995–1007.

Chrispeels, M. J., & Varner, J. E. (1967). Gibberellic acid enhanced synthesis and release of α-amylase and ribonuclease by isolated barley aleurone layers. *Plant Physiology, 42,* 398–406.

Coelho, S. M., Taylor, A. R., Ryan, K. P., Sousa-Pinto, I., Brown, M. T., & Brownlee, C. (2002). Spatiotemporal patterning of reactive oxygen production and Ca^{2+} wave propagation in *Fucus* rhizoid cells. *The Plant Cell, 14,* 2369–2381.

Dave, A., M. L., He, Z., Andriotis, V. M., Vaistij, F. E., Larson, T. R., et al. (2011). 12-Oxophytodienoic acid accumulation during seed development represses seed germination in *Arabidopsis*. *The Plant Cell, 23,* 583–599.

Devlin, T. (2011). *Textbook of Biochemistry with Clinical Correlations* (Vol. 14, pp. 554–555).

Dierking, E. C., & Bilyeu, K. D. (2009). Raffinose and stachyose metabolism are not required for efficient soybean seed germination. *Journal of Plant Physiology, 166,* 1329–1335.

Eastmond, P. (2006). Sugar-dependent1 encodes a patatin domain triacylglycerol lipase that initiates storage oil breakdown in germinating *Arabidopsis* seeds. *The Plant Cell, 18,* 665–675.

Eastmond, P. J., Germain, V., Lange, P. R., Bryce, J. H., Smith, S. M., & Graham, I. A. (2000). Post-germinative growth and lipid catabolism in oilseeds lacking the glyoxylate

cycle. *Proceedings of the National Academy of Sciences of the United States of America, 97,* 5669–5674.
Filner, P., & Varner, J. E. (1967). A simple and unequivocal test for de novo synthesis of enzymes: Density labeling of barley α-amylase with H_2O^{18}. *Proceedings of the National Academy of Sciences of the United States of America, 58,* 1520–1526.
Fincher, G. B. (1989). Molecular and cellular biology associated with endosperm mobilization in germinating cereal grains. *Annual Review of Plant Physiology and Plant Molecular Biology, 40,* 305–346.
Hasanuzzaman, M., Nahar, K., Alam, M. M., Roychowdhury, R., & Fujita, M. (2013). Physiological, biochemical, and molecular mechanisms of heat stress tolerance in plants. *International Journal of Molecular Sciences, 14,* 9643–9684.
Hills, M. J., & Beevers, H. (1987). Lipases and oil mobilization in castor bean endosperm. *Plant Physiology, 84,* 272–276.
Howell, K. A., Cheng, K., Murcha, M. W., Jenkin, L. E., Millar, A. H., & Whelan, J. (2007). Oxygen initiation of respiration and mitochondrial biogenesis in rice. *Journal of Biological Chemistry, 282,* 15619–15619.
Howell, K. A., Millar, A. H., & Whelan, J. (2006). Ordered assembly of mitochondria during rice germination begins with promitochondrial structures rich in components of the protein import apparatus. *Plant Molecular Biology, 60,* 201–223.
Howell, K. A., Narsair, R., Carroll, A., Ivanova, A., Lohse, M., Usadel, B., et al. (2009). Mapping metabolic and transcript temporal switches during germination in rice highlights specific transcription factors and the role of RNA stability in the germination process. *Plant Physiology, 149,* 961–980.
Huang, X., Tian, T., Chen, J., et al. (2021). Transcriptome analysis of *Cinnamomum migao* seed germination in medicinal plants of Southwest China. *BMC Plant Biology, 21,* 270.
Hubel, F., & Beck, E. (1996). Maize root phytase: Purification, characterization, and localization of enzyme activity and its putative substrate. *Plant Physiology, 112,* 1429–1436.
Hutton, D., & Stumpf, P. K. (1969). Fat metabolism in higher plants: Characterization of the β-oxidation system from maturing and germinating castor bean seeds. *Plant Physiology, 44,* 508–516.
Jacela, J. Y., DeRouehey Tokach, M. D., Goodband, R. D., Nelssen, J. L., Renter, D., & Dritz, S. S. (2010). Feed additives for swine: Fact sheets—Prebiotics, probiotics, and phytogenics. *Journal of Swine Health and Production, 18,* 87–91.
Jones, J. D., Burneth, P., & Zollman, P. (1999). The glyoxylate cycle: Does it function in the dormant or active seed? *Comparative Biochemistry and Physiology Part B: Biochemistry and Molecular Biology, 124*(2), 177–179.
Jones, R. L., & Jacobsen, J. V. (1991). Mobilizing enzymes in cereals. *International Review of Cytology, 126,* 49–87.
Kępczyński, J. (1986). Ethylene-dependent action of gibberellins in seed germination of *Amaranthus caudatus. Physiologic Plantarum, 67,* 584–587.
Koornneef, M., & Karssen, C. M. (1994). Seed dormancy and germination. In *Arabidopsis Book/Cold Spring Harbor Monograph Archive* (Vol. 27, pp. 313–334).
Lehmann, T., & Ratajczak, L. (2008). The pivotal role of glutamate dehydrogenase (GDH) in the mobilization of N and C from storage material to asparagine in germinating seeds of yellow lupine. *Journal of Plant Physiology, 165,* 149–158.

Leubner-Metzger, G., Fründt, C., Vögeli-Lange, R., & Meins, F. J. (1995). β-1,3-Glucanases in the endosperm of tobacco during germination. *Plant Physiology, 109,* 751–759.

Lott, J. N. A., Greenwood, J. S., & Batten, G. D. (1995). Mechanisms and regulation of mineral nutrient storage during seed development. In J. Kigel & G. Galili (Eds.), *Seed Development and Germination* (pp. 215–235). New York: Marcel Dekker.

Matilla, A., Gallardo, M., & Puga-Hermida, M. I. (2005). Structural, physiological, and molecular aspects of heterogeneity in seeds: A review. *Seed Science Research, 15,* 63–76.

Mihoub, A., Chaoui, A., & El-Ferjani, E. (2005). Biochemical changes associated with cadmium and copper stress in germinating pea seeds (*Pisum sativum* L.). *Comptes Rendus Biologies, 328,* 33–41.

Mukherji, S., Dey, B., Paul, A. K., & Sircar, S. M. (1971). Changes in phosphorus fractions and phytase activity of rice seeds during germination. *Physiologia Plantarum, 25,* 94–97.

Muscolo, A., Sidari, M., Mallamaci, C., & Attinà, E. (2007). Changes in germination and glyoxylate and respiratory enzymes of *Pinus pinea* seeds under various abiotic stresses. *Journal of Plant Interactions, 2*(4), 273–279.

Muto, S., & Beevers, H. (1974). Triacylglycerol hydrolysis in castor bean endosperm. *Plant Physiology, 54,* 23–28.

Nandi, S., Das, G., & Sen-Mendi, S. (1995). β-Amylase activity as an index for germination potential in rice. *Annals of Botany, 75,* 463–467.

Nonogaki, H., Gee, O. H., & Bradford, K. J. (2000). A germination-specific endo-β-mannanase gene is expressed in the micropylar endosperm cap of tomato seeds. *Plant Physiology, 123,* 1235–1245.

Oaikhena, E. E., Ajibade, G. A., Appah, J., & Bello, M. (2013). Dehydrogenase enzyme activities in germinating cowpea (*Vigna unguiculata* (L.) Walp). *Journal of Biology, Agriculture and Healthcare, 3,* 32–36.

Ogawa, M., Tanaka, K., & Kasai, Z. (1979). Accumulation of phosphorus, magnesium, and potassium in developing rice grains: Followed by electron microprobe X-ray analysis focusing on the aleurone layer. *Plant and Cell Physiology, 20,* 19–27.

Oge, L., Bourdais, G., Bove, J., Collet, B., Godin, B., Granier, F., et al. (2008). Protein repair-L isoaspartyl methyltransferase 1 is involved in both seed longevity and germination vigor in *Arabidopsis*. *The Plant Cell, 20,* 3022–3037.

Okazaki, Y., & Katayama, T. (2005). Reassessment of the nutritional function of phytic acid, with special reference to myo-inositol function. *Journal of the Japan Society of Nutrition and Food Sciences, 58,* 151–156.

Olmos, E., & Hellin, E. (1997). Cytochemical localization of ATPase plasma membrane and acid phosphatase by cerium-based methods in a salt-adapted cell line of *Pisum sativum*. *Journal of Experimental Botany, 48,* 1529–1535.

Parys, E., Romanowska, E., & Poskuta, J. (1983). Amylase activities in attached and excised cotyledons and in embryonic axes of *Pisum sativum* L. *Plant and Cell Physiology, 21,* 181–188.

Perino, C., & Come, D. (1991). Physiological and metabolic study of the germination phases in apple embryo. *Seed Science and Technology, 19,* 1–14.

Podestá, F. E., & Plaxton, W. C. (1994). Regulation of cytosolic carbon metabolism in germinating *Ricinus communis* cotyledons. II. Properties of phosphoenolpyruvate carboxylase and cytosolic pyruvate kinase associated with the regulation of glycolysis and nitrogen assimilation. *Planta, 194,* 381–387.

Poxleitner, M., Rogers, S. W., Lacey Samuels, A., Browse, J., & Rogers, J. C. (2006). A role for caleosin in degradation of oil-body storage lipid during seed germination. *Plant Journal, 47,* 917–933.

Pracharoenwattana, I., & Smith, S. M. (2008). When is a peroxisome not a peroxisome? *Trends in Plant Science, 13,* 522–525.

Preston, J., Tatematsu, K., Kanno, Y., Hobo, T., Kimura, M., Jikumaru, Y., et al. (2009). Temporal expression patterns of hormone metabolism genes during imbibition of *Arabidopsis thaliana* seeds: A comparative study on dormant and non-dormant accessions. *Plant and Cell Physiology, 50,* 1786–1800.

Pritchard, S. L., Charlton, W. L., Baker, A., & Graham, I. A. (2002). Germination and storage reserve mobilization are regulated independently in *Arabidopsis*. *Plant Journal, 31,* 639–647.

Quettier, A. L., & Eastmond, P. J. (2009). Storage oil hydrolysis during early seedling growth. *Plant Physiology and Biochemistry, 47,* 485.

Raboy, V., Noaman, M. M., Taylor, G. A., & Pickett, S. G. (1991). Grain phytic acid and protein are highly correlated in winter wheat. *Crop Science, 31,* 631–635.

Ratajczak, W. (1986). Asparagine metabolism in developing seeds of *Lupinus luteus* L. *Biochemie und Physiologie der Pflanzen, 181,* 17–22.

Ratajczak, W., Gwóźdz, E. A., & Miądowicz, M. (1996). Effects of nitrogen nutrition on storage protein composition in yellow lupin cotyledons cultured *in vitro*. *Acta Physiologiae Plantarum, 18,* 295–304.

Rees, T. (1980). Integration of pathways of synthesis and degradation of hexose phosphates. In J. Preiss (Ed.), *The Biochemistry of Plants* (pp. 1–42). London: Academic Press.

Reid, J. S. G. (1971). Hydrolysis of the fenugreek endosperm. *Planta, 100,* 131–142.

Reid, J. S. G., & Bewley, J. D. (1979). Dual role of the fenugreek endosperm. *Planta, 147,* 145–150.

Robert, K. M., David, A. B., Katheleen, M. B., Peter, J. K., Victor, W. R., & Weil, P. A. (2009). *Harper's Illustrated Biochemistry*. Biologic oxidation (12th ed., pp. 99–100).

Schopfer, P., & Plachy, C. (1985). Control of seed germination by abscisic acid. *Plant Physiology, 77*(3), 676–686.

Shi, H., Bressan, R., Hasegawa, P. M., & Zhu, J. K. (2005). Sodium in plant nutritional genomics. In M. Broadlay & P. White (Eds.), *Nutritional Genomics* (pp. 127–149). London: Blackwell Publishing.

Shoshii, M., & Reevers, H. (1974). Lipase activity in castor bean endosperm during germination. *Plant Physiology, 54,* 23–28.

Silva, L. G., & Trugo, L. C. (1996). Characterization of phytase activity in lupin seed. *Journal of Food Biochemistry, 20,* 329–340.

Sreenivasulu, N., Usadel, B., Winter, A., Radchuk, V., Scholz, U., et al. (2008). Barley grain maturation and germination: Metabolic pathway and regulatory network commonalities and differences highlighted by new MapMan/PageMan profiling tools. *Plant Physiology, 146,* 1738–1758.

Steber, C. M., & McCourt, P. (2001). A role for brassinosteroids in germination in *Arabidopsis*. *Plant Physiology, 125,* 763–769.

Tavener, R. J. A., & Laidman, D. L. (1972). The induction of triglyceride metabolism in the germinating wheat grain. *Phytochemistry, 11,* 981–987.

Thompson, L. U., & Zhang, L. (1991). Phytic acid and minerals: Effect on early markers of risk for mammary and colon carcinogenesis. *Carcinogenesis, 2,* 2041–2045.

Turner, J. F., & Turner, D. H. (1975). The regulation of carbohydrate metabolism. *Annual Review of Plant Physiology, 26,* 159–186.

Velasco, R., Salamini, F., & Bartels, D. (1994). Dehydration and ABA increase mRNA levels and enzyme activity of cytosolic GAPDH in the resurrection plant *Craterostigma plantagineum. Plant Molecular Biology, 26,* 541–546.

Vidal, E. A., & Gutiérrez, R. A. (2008). A systems view of nitrogen nutrient and metabolite responses in *Arabidopsis. Current Opinion in Plant Biology, 11,* 521–529.

Vob, U., Bishopp, A., Farcot, A., & Bennett, M. J. (2014). Modelling hormonal response and development. *Trends in Plant Science, 19,* 311–319.

Wang, L., Fu, J., Li, M., Fragner, L., Weckwerth, W., & Yang, P. (2016). Metabolomic and proteomic profiles reveal the dynamics of primary metabolism during seed development of lotus (*Nelumbo nucifera*). *Frontiers in Plant Science, 7,* 750.

Wang, W. Q., Møller, I. M., & Song, S. Q. (2012). Proteomic analysis of embryonic axis of *Pisum sativum* seeds during germination and identification of proteins associated with loss of desiccation tolerance. *Journal of Proteomics, 77,* 68–86.

Wattanakulpakin, P., Photchanachai, S., Miyagawa, S., & Ratanakhanokchai, K. (2012). Loss of maize seed vigor as affected by biochemical changes during hydropriming. *Crop Science, 52,* 2783–2793.

Weitbrecht, K., Müller, K., & Leubner-Metzger, G. (2011). First off, the mark: Early seed germination. *Journal of Experimental Botany, 62,* 3289–3309.

Wilson, A. M. (1971). Amylase synthesis and stability in crested wheatgrass seeds at low water potentials. *Plant Physiology, 48,* 541–546.

Yomo, H., & Varner, J. E. (1973). Control of the formation of amylases and proteases in the cotyledons of germinating peas. *Plant Physiology, 51,* 708–713.

Zienkiewicz, Z., Zienkiewicz, A. K., Rejon, J. D., Alche, J. D., Castro, A. J., & Rodriguez-Garcia, M. I. (2014). Olive seed protein bodies store degrading enzymes involved in mobilization of oil bodies. *Journal of Experimental Botany, 65,* 103–115.

CHAPTER 5

Safe Seed Storage Including Ecofriendly Techniques Used in Medicinal and Aromatic Plants

CHAKRADHAR PATRA,[1] MANAS RANJAN NAYAK,[1,*] GOPAL SHUKLA,[2] N. BHOLA,[1] T. TAMRAKAR,[1] and H. K. DAS[1]

[1]*Odisha University of Agriculture and Technology, Bhubaneswar, Odisha, India*

[2]*Department of Forestry, North-Eastern Hill University, Tura Campus, Tura, Meghalaya, India*

*Corresponding author

ABSTRACT

Efficient seed storage is essential to preserve viability and vigor until sowing, especially for high-value medicinal and aromatic plants (MAPs). India's favorable environment and low-cost labor create unique opportunities for producing high-quality seeds for domestic use and export. This chapter outlines comprehensive strategies for eco-friendly MAP seed storage aimed at maintaining genetic diversity, promoting rural livelihoods, and reducing reliance on synthetic inputs. The pre-storage processes, including timely harvesting, thorough cleaning, and moisture adjustment via drying, are critical in ensuring seed health. Seed processing techniques, such as scarification, enhance germination rates, while safe treatment protocols help protect against pests and pathogens. Environmentally sustainable storage methods are then detailed: airtight containers with natural seal-ants, use of desiccants like silica gel or activated charcoal, temperature control in cool, dark environments, and refrigerated or frozen storage for

Seed Technology for Medicinal and Aromatic Plants: Production, Testing, Storage, and Marketing. Priyanka Sharma, Kishore Kumar Sharma, and Gopal Shukla (Eds.)

DOI: 10.1201/9781779641199-5

select seeds. Traditional Indigenous Knowledge (ITK) practices using ash, turmeric, neem, garlic, camphor, and sand are presented as viable, low-cost alternatives to synthetic pesticides. The roles of national gene banks, cryopreservation, and germplasm repositories in long-term genetic conservation are also explored. Through robust labeling, monitoring, and community engagement, this chapter promotes a holistic, environmentally responsible framework for safeguarding MAP seed resources now and for future generations.

5.1 INTRODUCTION

"Preservation of seed with minimum loss in viability and vigor until needed for sowing" is the definition of seed storage. Any type of crop can be produced in India with cheap labor and an environment that is suitable for producing vigorous, bold, and high-quality seeds for the home and international markets. This will not only earn foreign exchange but also save it and provide the rural poor with skills that will enable them to create jobs and income. The greatest option is to produce and export high-quality seeds.

Vegetable seeds and diverse crops, such as aromatic and medicinal crops, will always have a great chance of success and be crucial to the economy. In the past, people would gather therapeutic herbs from natural sources. Following several educational sessions, farmers are expressing interest in cultivating these plants, now in high demand on the global market for medicinal herbs. Commercial cultivation is used to grow medicinal herbs. It needs a high-quality seed to grow therapeutic plants. Quality seed production of various medicinal plants and production technology is being worked on by the Medicinal and Aromatic Plants (MAP) Research Station in Rajendranagar, Hyderabad. Techniques to be followed for quality seed production in crops like Senna, Aswagandha, Kalmegh, Velvet bean, Basil, Palmarosa, and Isabgol are advised. Removing unwanted material from the seed lot and upgrading the quality of the seed by removing damaged and undersized seed with maximum efficiency in seed processing—which includes basic cleaning, grading, drying, and pre-conditioning—are the main steps in producing seed for medicinal and aromatic crops.

Maintaining the genetic diversity and viability of MAPs requires safe seed storage. These highly valuable MAP must be preserved for future generations since they are rich in aromatic chemicals and natural treatments. Fortunately, several environmentally benign methods may be used to protect MAP seeds

5.2 PRE-STORAGE PREPARATION

5.2.1 TIMELY HARVESTING

Gathering seeds at their fullest potential, after they have naturally desiccated on the plant, helps inhibit the formation of mold and fungi while being stored.

5.2.2 CLEANING AND DRYING

Seeds need to be cleaned to get rid of dirt, chaff, and any young or ill seeds. Effective ways include air drying or sun drying in well-ventilated areas. Aim for an ideal moisture level of 10–14% while drying seeds to preserve seed viability and stop fungus development.

5.2.3 SEED PROCESSING

To encourage quicker and more consistent germination, seeds with hard coats can be scarified using methods like sandpaper rubbing or acid treatment with H_2SO_4.

5.2.4 SEED TREATMENTS

To shield seeds from pests and illnesses during storage, fungicides and insecticides can be used. But for organic seeds, natural substitutes like diatomaceous earth or neem seed powder are frequently chosen.

5.3 ECO-FRIENDLY STORAGE TECHNIQUES

These environmentally friendly methods will help to preserve the valuable genetic legacy of aromatic and medicinal plants for future generations.

Recall that using sustainable storage techniques helps create a more environmentally friendly future by protecting priceless seeds and reducing their negative effects on the environment.

For MAP to remain viable and retain their genetic variety, safe seed storage practices are essential.

5.3.1 AIRTIGHT CONTAINERS

Tight-fitting lids on glass jars, metal canisters, or ceramic pots efficiently shield seeds from light, moisture, and pests. Using natural sealants like clay or beeswax can improve security even further. Make use of airtight, moisture-proof containers such as laminated aluminum foil pouches of approximately 700 gauge, glass jars, or hermetically sealed tins. This prevents insects from getting inside, and oxygen from getting out, all of which can harm seeds (Figure 5.1).

FIGURE 5.1 Air tight glass jars for seed storage; Laminated aluminium foil pouch.

5.3.2 NATURAL DESICCANTS

Activated charcoal or silica gel pouches inserted into containers absorb extra moisture, control humidity up to 80%, and produce a dry atmosphere for extended storage.

5.3.3 TEMPERATURE CONTROL

Seed degeneration is considerably slowed down when stored in cold, dark spaces such as underground pits, pantries, or basements. Earthenware pots naturally insulate, assisting in the maintenance of steady temperatures in hot climates. Seeds should ideally be kept in a cool, dry environment with little light exposure. The ideal temperature range is 0 to 15°C, with a relative humidity of less than 50% (Figure 5.2).

FIGURE 5.2 Seed storage cold room.

5.3.4 FREEZING OR REFRIGERATION (IF APPROPRIATE)

To extend their shelf life, some aromatic and medicinal seeds are best kept in a freezer or refrigerator. But not all seeds are suited to cold weather; therefore, before utilizing this technique, find out what exactly each variety of seed needs. If freezing is advised for any particular seeds, make sure they are kept at a temperature between 0°F (−18°C) and 10°F (−12°C) and are packed in moisture-tight containers. Seeds must be dried properly to avoid harm from ice crystal development.

5.3.5 CONVENTIONAL TECHNIQUES

In many areas, tried-and-true methods such as storing seeds in sand, ash, or straw provide efficient and accessible solutions. These substances prolong the life of seeds by regulating moisture and acting as insulation (Figure 5.3).

FIGURE 5.3 Seeds stored in earthenware pots.

5.3.6 NATURAL CONTAINERS

Use of paper envelopes, cardboard boxes, or cloth bags are considered natural and biodegradable containers. Steer clear of plastic containers and bags since they are not environmentally friendly and may release chemicals that are detrimental to seeds.

5.3.7 SEED SAVING ENVELOPES

The use of recycled paper envelopes is advisable to hold seeds. These envelopes allow for air circulation to minimize moisture build-up and can be branded with information about seeds.

5.4 ADDITIONAL REQUIREMENTS

5.4.1 SPECIES-SPECIFIC REQUIREMENTS

Depending on the particular plant species, different MAPs have different storage requirements. Certain seeds include innate dormant mechanisms, necessitating particular temperature or light stimuli in order to initiate germination. For effective long-term preservation, it is essential to investigate each species' ideal temperature, humidity, and light levels.

5.4.2 LABELING AND RECORD-KEEPING

Accurate identification and tracking throughout time are ensured by clearly marking containers with species names, collection dates, and storage conditions. Keeping track of seed viability through regular germination tests enables prompt action when necessary.

5.4.3 REGULAR MONITORING

It is advisable to evaluate seeds regularly for indications of moisture damage, insect infestation, or viability loss. Testing for germination can be carried out periodically to evaluate the health of the seeds and, if required, modify storage conditions.

5.4.4 PREVENT FLUCTUATIONS

It is critical to maintain consistency when selecting a temperature. Steer clear of temperature swings that might harm seed viability. Seed damage can be caused by abrupt temperature fluctuations.

5.4.5 SEED BANKS OR COMMUNITY SEED EXCHANGES

To promote seed banks and community seed banks, it is recommended to get involved in initiatives related to seed banks and community seed exchanges. By promoting seed exchange and preservation, these programs guarantee a varied seed supply and lessen the demand for high personal storage. Preserving the genetic variety and long-term viability of highly valuable aromatic and medicinal plants for future generations by adhering to these safe seed storage practices is of utmost importance. To preserve seeds with the least amount of negative environmental impact, eco-friendly seed storage techniques are essential. It is important to remember that different kinds of aromatic and medicinal plants might require quite varied temperatures for effective preservation. As a result, it is essential to investigate and determine the particular storage needs for every kind of seed. Certain seeds may have particular requirements, such as requiring a certain combination of humidity and temperature for long-term storage. Therefore, it is important to store seeds dry, sealed, and labelled with pertinent details including the species name, collection date, and any specific handling or germination instructions. To preserve the viability of stored seeds, rotate them by utilizing older seeds first. Additionally, symptoms of dampness, pests, or deterioration must be carefully observed. Many synthetic pesticides have been used over the past few decades to prevent insect infestation of stored goods and other agricultural products, but their widespread usage harms the environment and intoxicated non-targeting creatures. The ideal technique to counteract the negative effects of synthetic pesticides is to use Indigenous Traditional Knowledge (ITK) to defend food grains and seeds against insect pest assault. Due to this, ITK ought to be preserved in all cases when it applies to both the current and future generations. The farmers used the following significant customary storage techniques (Table 5.1).

TABLE 5.1 Traditional Seed Storage Techniques

Sun drying of grains	Stepping method or Stamping method	Use of lime powder
• Use of ash	• Use of salt and chili powder	• Use of matchbox
• Red soil coating method	• Use of Neem (Margosa) leaves	• Fumigation of the godown /store rooms
• Plastering of storage bins with clay and cow dung	• Use of Neem (Margosa) leaves and dry chilies	• Use of Neem (Margosa) seed powder
• Storage of pulses with common salt	• Use of Neem Oil /Margosa oil	• Use of ginger rhizome
• Turmeric application method	• Use of camphor	• Use of custard apple seed powder
• Use of garlic cloves	• Use of castor powder	• Use of Tulsi (basil) seeds
• Mixing of leaves	• Sand mixture method	

5.4.5.1 SUN DRYING OF GRAINS

Before storing grains and pulses, sun drying is an ideal traditional method used by 92.0% of farmers. For storage, it is sun-dried for a longer time if the moisture content is high. Existing insect pests and their various phases, such as larvae or eggs, are destroyed by sun drying. In addition, grain dormancy is prolonged and damage can be reduced with the aid of sun drying. For seeds to be stored properly, the ideal moisture content is between 10 and 12%.

5.4.5.2 USE OF ASH

In clay pots, seeds are packed to a volume of 3/4, with the remaining 1/4 being filled with wood or cow dung ash. Seeds should be kept in storage for an extended length of time, and the pots are sun-dried and refilled with fresh ash after six months. About 78.00% of farmers are using this approach since ash includes silica; it prevents fungal pathogens from multiplying and obstructs insect eating. Dust dries the seed surface and lowers the storage conditions' relative humidity. Ash dust covers the seeds, which may inhibit the storage pests' ability to lay eggs and develop into larvae.

5.4.5.3 RED SOIL COATING METHOD

In a container, red soil and water are combined to make a paste. To ensure that the soil completely covers the seeds, the seeds are moved into this container and thoroughly mixed. Under shade, seeds are dried. After placing the seeds in a gunny bag, it is securely sealed and kept in the dark. Moreover, 60% of farmers are using this approach. The red soil serves as a barrier, thereby preventing the insect from feeding on the grains and seeds and allowing it to lay eggs. Soil has the potential to reduce moisture content from grains and seeds, prevent damage, in addition to enhancing germination.

5.4.5.4 PLASTERING OF STORAGE BINS WITH CLAY AND COW DUNG

In order to keep insects out of bamboo storage bins, it is to be coated with a mixture of cow dung and clay. Nearly 76.67% of farmers employ this technique because the upper and middle layer of soil helps to prevent further damage by absorbing moisture that is retained in the seeds. In this technique, seeds can be prevented from insects and pests by the application of soil, including cow dung paste, which acts as a barrier.

5.4.5.5 STORAGE OF PULSES WITH COMMON SALT

Pulses seeds were stored for six to eight months by physically mixing 200 grams of common table salt with one kilogram of pulses. About 61.33% of farmers use this approach. Insect development is limited in storage containers, bins, and boxes due to the abrasive impact of salt on insect skin, which prevents the insects from moving within. Salt has both insecticidal and hygroscopic properties. By absorbing moisture, salt helps to keep grains dry, preventing spoiling and promoting safe storage.

5.4.5.6 TURMERIC APPLICATION METHOD

Another effective substitute for treating the grains to keep insects and other pests away is to use powdered turmeric. Turmeric powder is combined with grains and seeds and then stored in jute bags or containers. This therapy is similarly safe to consume and offers protection for up to 6–8 months. This is the approach that 50% of farmers use. The ingredients in turmeric that have insect-repelling properties are called turmericnes and arturmerones. Food grains are kept insect-free by its strong scent and insecticidal qualities.

5.4.5.7 USE OF GARLIC CLOVES

The storage containers with seeds are used to store garlic cloves in layers. Cloves of garlic have anti-pest properties. The main ingredients consist of diallyl sulphide, diallyl trisulfide, and diallyl di-sulfide. Garlic possesses several qualities such as anti-feedant, bactericidal, fungicidal, insecticidal, nematicidal, and repellant.

5.4.5.8 MIXING OF LEAVES

Neem/Margosa (*Azadirchta indica*), Nirgandi/Chinese chaste tree (*Vitex nigundo*), Madar/Calotropis, and other leaves with insecticidal properties are gathered and dried in the shade until they get papery. These leaves are placed into the bags or storage container together with seeds. Leaves may be used to control a variety of storage pests and are easily obtained at a reasonable cost. The leaves of plants such as Neem (*Azadirchta indica*), Nirgandi (*Vitex nigundo*), Madar (*Calotropis*), and others have excellent insecticidal qualities. They also serve as antecedents, repellents, and growth inhibitors for pests that are stored.

5.5 STEPPING METHOD OR STAMPING METHOD

The plastic bags are filled with seeds and securely fastened with thread. Stamping the bag repeatedly not only keeps pests away but also kills any existing eggs and maggots in the seeds. This simple process does not need any components. Stamping kills both adult and juvenile insects and aids in their disturbance.

5.5.1 USE OF SALT AND CHILI POWDER

Twenty kg of seeds are combined with 250 g each of common salt and dried chili powder. Insects are repelled by the strong scent of chilies, and salt has both hygroscopic and insecticidal properties.

5.5.2 USE OF NEEM/MARGOSA LEAVES

The leaves of neem or margosa trees are harvested, dried in the shade, combined with grains or seeds, and then placed in containers or gunny sacks for storage. Neem/Margosa works as an anti-feed ant against a variety of pests and includes bitter components known as melanics, such as azardiracin, nimbin, salannin, meliantriol, and so forth. Azadirachtin, the active component of neem/Margosa leaves, has sterilizing, antifungal, nontoxic, and insect-repelling properties in addition to inhibiting insect eating.

5.5.3 USE OF NEEM/MARGOSA LEAVES AND DRY CHILIES

The bins or bags holding the seeds are filled with dried red chilies and neem leaves. Grain and seed can be shielded from diseases and pests. Azardiracin, nimbin, salannin, meliantriol, and other bitter compounds known as meliacins are found in neem/margosa and function as anti-feedants against a variety of pests. The active component of neem leaves, azadirachtin, has nontoxic properties and functions as an insect deterrent, sterilant, feeding inhibitor, and antifungal. The acrid taste of dried red chilies deters insect pests.

5.5.4 USE OF NEEM/MARGOSA OIL

On pulses, neem/margosa oil is carefully sprayed to ensure that each grain is evenly coated. Neem/Margosa seeds and grains are protected from pests and diseases by their bitter flavor. Numerous insects, including beetles, moths, and weevils, are repelled by neem/margosa oil. Neem oil preserves legumes by destroying insects while they are still in the egg stage. Neem/Margosa oil has a number of uses, including feeding, ovipositional deterrent, growth inhabitation, and repellence. It also almost kills insects even in the egg stage, preventing infestations from beginning at that early stage.

5.5.5 USE OF CAMPHOR

For every 5 kg of grain, 2 g of camphor are added to a jute gunny bag, which can be kept for up to three months. The grains are sun-dried once again every three months, and new camphor is stored in the bag. When camphor is kept, it gradually evaporates. To stop this, peppercorns and camphor are added to the container. Due to its strong scent, the camphor within the storage bag keeps pests away.

5.5.6 USE OF CASTOR POWDER

Bean seeds are sun-dried for 10 to 15 minutes. A small amount of castor seeds is roasted for a while in a dish before being pulverized into a powder. About 1 kg of bean seeds and 1/4 kg of castor powder are combined and kept in a clay pot. To prevent aeration, the pot's lid is covered and sealed with cow dung. When castor powder is applied on grains, its oily texture creates a slick surface that prevents female insects from depositing their eggs on the seeds. Additionally, it serves as a repellent and feeding deterrent, protecting grains against insect pest infestation.

5.5.7 *SAND MIXTURE METHOD*

Sun-dried seeds are distributed over a thick layer of sand that has been put to the mud pot's base. The seeds are covered with sand once again. Until the pot is completely filled, the same procedure of adding layers of the sand-seed combination is repeated. The container has a cover that closes, and cow dung paste is used to seal the air within. The sand particles destroy insect pests by acting as an abrasive agent on their cuticles. Additionally, it serves as a barrier against insects, shielding the seeds from pest damage. The top layer of cow dung in the pot serves as a repellent.

5.5.8 *USE OF DRIED RED CHILIES*

The dried chilies are stored in a container with pouring bags packed with seeds. The strong flavor of dried red chilies deters bugs.

5.5.9 *USE OF LIME POWDER*

Grain and lime (calcium carbonate) are combined evenly and kept in gunny bags in a dry location. For every kg of grains, 10–15 g of lime is often used. Lime inhibits the growth of insects and has repellent and antifeedant qualities.

5.5.10 *USE OF MATCHBOX*

It is the most traditional technique that women often utilize in their homes to store food grains. Layers of matchboxes are stored. About 8–12 matchboxes are usually placed in the middle, bottom, and top of the container, and the lid should be snugly closed. Strong repelling qualities of the phosphorus in matchsticks aid in preventing infection.

5.6 FUMIGATION OF THE GODOWN (WAREHOUSE)

The godowns, or warehouses, are fumigated with leaves of Vitex, Neem, Pomerania, and so forth, prior to the seeds or grains being stored there.

Fumigation produces a storage space devoid of inoculums and aids in eliminating the hibernation stages of stored insect pests that are found in cracks and crevices.

5.6.1 USE OF NEEM/MARGOSA SEED POWDER

One percent of the seed's volume is added to a mixture of neem seed powder. Neem/Margosa seeds and grains are protected from pests and diseases by their bitter flavor. Neem has a variety of qualities, including growth inhabitation, feeding and ovipositional deterrent, repellence, and so on.

5.6.2 USE OF GINGER RHIZOME

One kilogram of the pulse is combined with 30 grams of powdered ginger rhizome. Their seeds and grains are protected from pests and diseases by the pungent flavor of ginger. Because ginger contains alkaloids, anthocyanins, flavonoids, and tannins, it has anti-infective and pest-repelling properties.

5.6.3 USE OF CUSTARD APPLE SEED POWDER

To protect the pulse beetle from further damage, 50 g of custard apple seed powder and 1 kg of any kind of pulse are combined. Custard apple seeds' ability to repel and prevent oviposition shields grains and seeds against bruchid assault. Custard apple seeds contain acetogenins, which have vermicidal and insecticidal properties.

5.6.4 USE OF TULSI/BASIL SEEDS

Dried seeds and basil (*Ocimum sanctum*) leaves are combined with maize seeds. Tulsi/basil has excellent repelling qualities. Tulsi seeds include stearic acid, linolenic acid, linoleic acid, and palmitic acid, which have larvicidal and antifeedant properties.

5.7 FIELD GENE BANK (FIELD REPOSITORY/CLONAL REPOSITORY)

5.7.1 GENE BANK

Seeds are stored at –20°C for base collection and +4°C to 0°C for active collection. In India, three national gene banks (NGB) have been set up to preserve aromatic and therapeutic plants out of their natural habitat.

- National Bureau of Plants Genetic Resources (NBPGR), New Delhi, under ICAR.
- Central Institute of Medicinal and Aromatic Plants (CIMAP), Lucknow, Uttar Pradesh, under the Council of Scientific and Industrial Research (CSIR), Ministry of Science and Technology, Government of India.
- Tropical Botanical Gardens Research Institute, (TBGRI), Palade, Thiruvananthapuram (Kerala).

5.7.2 SEED GENE BANK

It is a more cost-effective method to save germplasm. The NGB, located at the NBPGR in New Delhi, is largely in charge of conserving the germplasm of agri-horticultural crops and their wild relatives to preserve seeds for future generations. These are kept in modules at –20°C and are known as the "Base Collection." After being dried to a moisture level of 4–6%, the seeds are hermetically packed in moisture-resistant aluminum foil packages. These seeds may be kept for 50–100 years and still be viable. Samples of seed that have more than 85% seed viability in most crops are solely treated. The FAO/IPGRI-recommended gene bank criteria are ideally followed for storing seeds in gene banks.

5.7.3 CRYOPRESERVATION (IN LIQUID NITROGEN AT –165°C TO –196°C)

Freezing or cryopreservation using liquid nitrogen:

- **Seed Preservation:** Based on how the seeds react to drying out, they have been roughly divided into two groups. Most of them, referred to as "Orthodox," may be kept for extended periods of time since they are resistant to desiccation. The second category of plant species is referred to as "Recalcitrant" because the seeds of these species become damaged during the drying process and cannot be kept at temperatures below zero.

Grain pests that are stored do significant harm to both seeds and cereal grains. There were many synthetic pesticides in use, but they have been demonstrated to have negative impacts on the environment, to linger longer in the form of residues, and to enter the food chain once items are consumed by living things. Thus, cleaner, more environmentally friendly, and farmer-friendly techniques were developed to replace these poisons. The majority of them are native methods that improve the use of resources that are readily available in the area. These techniques safeguard the food grains and pose no health risks, in addition to using less expensive, locally sourced, and environmentally friendly materials. Traditional farming methods and expertise are crucial to preserving and enhancing the sustainability of the ecosystem. In order to produce a wide range of new concepts and methods for the advancement of humanity, it also has to be integrated with current scientific understanding. While the previously mentioned traditional agricultural knowledge and methods were still available in tribal settings, their availability has not decreased. Therefore, in order to preserve the agricultural output, it is necessary to encourage the tribal farmers to continue using these methods. These ancient agricultural methods and expertise must be applied universally to all farming communities across the nation at the same time. Therefore, it is evident that seeds of highly valuable aromatic and medicinal plants have a wealth of untapped fragrant and therapeutic qualities in addition to the promise of future harvests.

KEYWORDS

- **botanical insecticides**
- **cryopreservation**
- **grain preservation**
- **neem**
- **seed storage**
- **traditional pest management**

REFERENCES

Dhaliwal, R., & Gurdeep, S. (2010). Traditional food grain storage practices of Punjab. *Indian Journal of Traditional Knowledge, 9*(3), 526–530.

Karthikeyan, C., Veeraragavathatham, D., & Karpagam, D. (2009). Traditional storage practices. *Indian Journal of Traditional Knowledge, 8*(4), 564–568.

Lakshman, C. D. (2016). Bio-diversity and conservation of medicinal and aromatic plants. *Advances in Plants and Agricultural Research, 5*(4), 561–566.

Padma, M. (2019). Quality seed production in medicinal and aromatic crops. *International Journal of Current Microbiology and Applied Science, 8*(7), 604–611.

Prakash, B. G., Raghavendra, K. V., Gowthami, R., & Shashank, R. (2016). Indigenous practices for eco-friendly storage of food grains and seeds. *Advances in Plants and Agricultural Research, 3*(4), 101–107.

Vanangamudi, K. (2014). *Seed Science and Technology*. New Delhi: New India Publication Agency (NIPA).

CHAPTER 6

Synthetic Seeds: Pioneering Biotechnology for Conservation and Sustainable Multiplication of Medicinal and Aromatic Plants

MANISH KUMAR SUTHA* and MANISH KUMAR MITTAL

ICAR–Directorate of Medicinal and Aromatic Plants Research, Boriavi, Anand, Gujarat, India

*Corresponding author

ABSTRACT

Synthetic seeds, an innovative outcome of biotechnology, hold significant promise in conserving, multiplying, and propagating medicinal and aromatic plants (MAP). These fabricated structures mimic real seeds but are formed through tissue culture techniques. In the context of preservation, synthetic seeds emerge as vital tools, serving as a lifeline for endangered or fragile MAP species. Encapsulating plant embryos or shoot tips safeguards these plants' genetic heritage, allowing storage in controlled environments, shielding against extinction, and eventual reintegration into natural habitats. Multiplication and propagation are cornerstones of sustainable plant utilization, with many MAP containing valuable bioactive compounds in demand by industries. Synthetic seeds enable controlled mass production, assuring consistent yields and compound quality in controlled environments that mitigate contamination risks and optimize growth conditions. Ensuring uniformity is paramount in plant propagation, a role that synthetic seeds fulfill by maintaining genetic consistency across generations and preserving specific traits, catering to industrial needs. Additionally, their aseptic

Seed Technology for Medicinal and Aromatic Plants: Production, Testing, Storage, and Marketing.
Priyanka Sharma, Kishore Kumar Sharma, and Gopal Shukla (Eds.)

DOI: 10.1201/9781779641199-6

production techniques foster disease-free propagation, especially crucial for susceptible medicinal plants, ensuring healthier yields. Synthetic seeds also surmount practical challenges due to their compactness and durability, facilitating efficient cross-regional distribution, and year-round production and storage, ensuring uninterrupted propagation regardless of seasons. Nonetheless, challenges persist, and success relies on meticulous species selection, encapsulation material formulation, and optimized growth conditions. In conclusion, synthetic seeds offer a promising route for conserving, multiplying, and propagating MAP, comprehensively addressing concerns of genetic diversity, mass production, uniformity, and disease control, thus presenting a holistic solution for sustainable utilization and safeguarding biodiversity in these pivotal plant resources.

6.1 INTRODUCTION

In the realm of agriculture and horticulture, a continuous stream of innovations has significantly enhanced productivity, crop quality, and sustainability. Among these innovations, the concept of synthetic seeds stands out as a pioneering advancement. Synthetic seeds are engineered entities designed to encase a plant embryo within a protective shell, complete with essential nutrients and growth-promoting elements (Gupta and Sood, 2022). These artificial "seeds" are crafted through a fusion of biotechnology and techniques derived from plant tissue culture. Unlike traditional seeds, synthetic seeds do not arise from sexual reproduction; instead, they are a product of asexual propagation methods. One of the primary merits of synthetic seeds is clonal propagation. These seeds can yield genetically identical plants to the parent, erasing the genetic variability found in conventional seeds. This is especially valuable for maintaining desired traits in crops and ornamental plants. Synthetic seeds also come to the aid of rare and endangered plant species. By preserving genetic material within synthetic seeds, we can safeguard biodiversity and ensure the survival of species threatened by habitat loss, climate change, and other challenges (Zarei et al., 2022). Conventional seeds may carry diseases or pests, posing potential threats to crop yields. Synthetic seeds, however, can be generated from disease-free plant material, reducing the risk of introducing harmful pathogens into agricultural systems. Additionally, synthetic seeds permit year-round cultivation by enabling the storage and transport of plant embryos. This capability is invaluable for crops with short growing seasons or plants sensitive to specific environmental conditions. Furthermore, the growth of plants from synthetic seeds is often swifter than from traditional propagation methods, expediting the development and production of new plant varieties. Creating synthetic seeds involves a series of carefully orchestrated steps. It commences with the selection of a healthy and desirable parent plant as the source of genetic material. Small pieces of plant tissue, such as leaves or stems, are then sterilized and nurtured in a nutrient-rich medium, fostering the growth of multiple plantlets derived from the same parent plant. Once the plantlets attain an appropriate size, they are encouraged to develop embryos, involving the formation of shoot and root structures. Subsequently, these formed embryos are encapsulated within a protective coating, which can comprise a gel matrix or other suitable materials (Ahmad et al., 2022; Kader et al., 2023). This coating contains essential nutrients, hormones, and other growth-promoting substances. Ultimately, the encapsulated embryos are left to desiccate, preserving them for extended durations. These dried synthetic seeds can be stored in seed banks or deployed for propagation.

The applications of synthetic seeds are wide-ranging and transformative (Ara et al., 2000). In agriculture, synthetic seeds have the potential to revolutionize crop production by facilitating the rapid multiplication of high-yielding, disease-resistant, and genetically uniform crop varieties. In horticulture, synthetic seeds represent a breakthrough, allowing for the mass production of unique and aesthetically appealing cultivars. Conservation efforts also benefit significantly from synthetic seeds, as they enable the preservation and reintroduction of endangered plant species into their natural habitats, bolstering biodiversity conservation. Forestry programs utilize synthetic seeds to propagate valuable tree species, thereby promoting reforestation and afforestation initiatives. Moreover, some synthetic seeds play a vital role in the production of medicinal plants, ensuring a consistent supply of bioactive compounds for pharmaceutical applications.

6.2 SYNTHETIC SEEDS AND MEDICINAL AND AROMATIC PLANTS

The scope of synthetic seeds in medicinal plant conservation and propagation is broad and holds great potential for addressing various challenges in this field (Gantait et al., 2015; Gupta et al., 2014; Saiprasad, 2001). Here are some key aspects that highlight the scope of synthetic seeds in medicinal plants.

6.2.1 CONSERVATION OF RARE AND ENDANGERED SPECIES

Synthetic seeds can play a vital role in the conservation of rare and endangered medicinal plant species (Corrie and Tandon, 1993; Devi et al., 2014; Khan et al., 2018; Lata et al., 2012). By preserving the genetic material of these plants in synthetic seed form, we can protect them from extinction. *Ginkgo biloba* (Ginkgo) is a well-known medicinal plant used for its cognitive and circulatory benefits. Due to its slow growth and high demand, the wild populations are at risk. Researchers have successfully used synthetic seeds to propagate Ginkgo, allowing for controlled cultivation and reducing the pressure on natural populations. *Taxus baccata* (European Yew) is a source of *taxol*, a compound used in cancer treatments. Overharvesting of yew trees for *taxol* extraction has threatened their survival. Synthetic seeds have been employed to produce large numbers of yew plants for *taxol* production, while also conserving wild populations. *Rauvolfia serpentina* (Indian Snakeroot) is known for its antihypertensive and sedative properties. It is under threat due to habitat loss and overexploitation. Researchers have used synthetic seeds to facilitate mass propagation, contributing to its conservation and ensuring a sustainable supply of traditional medicine. *Panax ginseng* (Ginseng) is a highly prized medicinal plant used for its adaptogenic properties. Wild ginseng populations have been overharvested, endangering the species. Synthetic seeds can be employed to meet the demand for ginseng while conserving natural habitats (Attia et al., 2018).

6.2.2 EFFICIENT PROPAGATION

Synthetic seeds have emerged as a game-changing innovation in the realm of medicinal plant propagation (Haque and Ghosh, 2016; Kim and Park, 2002). These artificial structures, comprising plant tissue, a nutrient-rich matrix, and a protective coating, offer a solution to several challenges faced in the cultivation and conservation of medicinal plants. One key advantage is their role in genetic preservation due to the use of clonally propagated (genetically similar) somatic embryos, nodal explants, and so forth (Gantait et al., 2017). Synthetic seed production relies on the development of somatic embryos or nodal explants through clonal methods, distinct from embryonic seeds that involve fertilization and genetic recombination. Consequently, synthetic seeds maintain a consistent genetic makeup. Medicinal plants often possess unique and valuable genetic traits, making the preservation of their genetic diversity critical for future research and traditional medicinal practices. Synthetic seeds ensure the continuity of these characteristics, safeguarding the identity of medicinal plant varieties.

Another notable role of synthetic seeds is in mass propagation. Medicinal plants often exhibit slow growth rates, which can hinder their availability for research and commercial purposes. Synthetic seeds, however, expedite the growth and multiplication of these plants, enabling large-scale production from a relatively small amount of starting material. It not only addresses the growing demand for medicinal plants but also reduces the pressure on wild populations, promoting their conservation and sustainable utilization. In essence, synthetic seeds have revolutionized the way we approach the propagation of medicinal plants, offering an effective and sustainable means to meet the demands of modern healthcare and traditional healing practices while ensuring the protection of these invaluable botanical resources.

6.2.3 DISEASE-FREE PLANTLETS

Synthetic seeds have paved a way towards the propagation of disease-free plantlets in the realm of MAP. These innovative structures play a pivotal role in ensuring the health and well-being of propagated plants. In the context of MAP, where the quality and therapeutic properties are of utmost importance, the advantages of synthetic seeds become evident. They are cultivated under aseptic conditions, eliminating the risk of disease transmission from contaminated sources. This characteristic is vital for preserving the purity of MAP varieties, ensuring that the plantlets start their growth free from potentially damaging pathogens and infections, which can compromise the quality of essential oils, medicinal compounds, and aromatic substances.

Furthermore, synthetic seeds facilitate the efficient elimination of known diseases from the source material, safeguarding the unique bioactive compounds that many MAP possess. The controlled environment in which they are produced ensures a consistent and uniform growth pattern, critical for maintaining the potency and quality of these plants.

As synthetic seeds expedite plant multiplication, they accelerate the availability of disease-free plantlets, contributing to the sustainability of these valuable plant resources. Overall, synthetic seeds represent a breakthrough in the cultivation and conservation of disease-free plantlets in MAP, supporting their critical roles in healthcare, wellness, and various industries. Furthermore, synthetic seeds offer the flexibility to produce medicinal plants on demand. This is particularly useful for scaling up production when there is an increased demand for specific species or bioactive compounds.

6.2.4 COMMERCIAL PRODUCTION

Commercial production of medicinal plants stands to gain significantly from the utilization of synthetic seeds. These artificial structures, encapsulating the genetic material of desired plant varieties, offer numerous advantages in the context of large-scale cultivation for the herbal medicine and pharmaceutical industries (Borner, 2006). The consistent and controlled growth provided by synthetic seeds ensures that each plantlet is genetically identical, guaranteeing uniformity in terms of bioactive compounds, therapeutic properties, and overall quality. This standardization is highly valuable for industries that depend on the reliability and predictability of plant materials for the production of herbal remedies and pharmaceutical drugs. Moreover, synthetic seeds expedite the multiplication of disease-free and healthy plantlets, ensuring a steady and efficient supply chain for medicinal plants, reducing the pressure on wild populations, and contributing to the sustainability of these valuable resources (Magray et al., 2017). In the competitive world of commercial herbal medicine and pharmaceutical production, the use of synthetic seeds is a game-changing approach that enhances the consistency, quality, and reliability of medicinal plant cultivation, meeting the stringent requirements of these industries while promoting responsible and sustainable resource management.

6.3 RESEARCH AND DEVELOPMENT

Synthetic seed technology serves as a powerful catalyst for research and development in the field of medicinal plants. By providing a controlled environment for the propagation and study of these plants, it offers a unique opportunity for scientists and researchers to delve into the genetic and biochemical aspects of medicinal species. This controlled environment eliminates the interference of external factors, such as disease and environmental stress, which can often complicate the research process. Researchers can precisely study the genetic traits, growth patterns, and bioactive compound production of medicinal plants, leading to a deeper understanding of their therapeutic properties. This knowledge, in turn, contributes to the development of new drugs, treatments, and therapeutic applications, potentially unlocking innovative solutions for various medical conditions. Synthetic seeds, by enabling consistent and disease-free plantlets, foster a research environment that is conducive to breakthrough discoveries and advancements in the pharmaceutical and healthcare industries.

Moreover, synthetic seeds support a sustainable approach to research and development because it eliminates the need for large-scale cultivation, reducing land, water, and nutrient requirements and contributing to more efficient resource utilization. Also, the process of generating synthetic seeds can be faster and more predictable than conventional seed production. By offering an alternative to wild harvesting, it reduces the ecological impact on natural populations of medicinal plants, which are often threatened due to over-exploitation. This conservation-conscious approach is especially important in the context of research and development, where the pursuit of new medical treatments and drug discovery should not compromise the preservation of plant biodiversity. In essence, synthetic seed technology not only enhances research outcomes but also aligns with ethical and environmental principles, promoting responsible scientific exploration of the potential of medicinal plants for the betterment of human health and well-being.

6.4 CLIMATE-RESILIENT VARIETIES

In the face of escalating climate change impacts, the development of climate-resilient varieties of medicinal plants through the utilization of synthetic seeds may emerge as a strategic imperative. Climate change brings about shifts in temperature, rainfall patterns, and the prevalence of pests and diseases, which can severely affect the growth and survival of various plant species, including those with medicinal value. Synthetic seed technology provides an avenue for creating plant lines with enhanced resilience to changing climatic conditions. Researchers and breeders can selectively target traits that make these plants more robust, such as drought tolerance, resistance to heat stress, or adaptability to altered growing seasons. By doing so, it ensures that these climate-resilient varieties of medicinal plants stand a better chance of thriving in the face of environmental upheaval, safeguarding their long-term survival and adaptability in the evolving landscape. As an illustration, consider the *Ocimum* species; a hybrid created through the combination of *O. basilicum* and *O. kilimandscharicum* exhibited enhanced adaptability to cold conditions (Gupta et al., 2024). Employing synthetic seed technology could facilitate the establishment of a stable line within this cross-pollinated plant species.

Moreover, climate-resilient medicinal plant varieties generated through synthetic seeds have a broader impact beyond mere survival. It offers a sustainable source of essential botanical resources in the face of climate-induced challenges, thereby bolstering the reliability and availability of these plants for traditional medicine, pharmaceuticals, and various industries. Additionally, their cultivation can reduce the pressure on natural populations of medicinal plants, which are often at risk due to the changing climate and overharvesting. In essence, the creation of climate-resilient medicinal plant varieties using synthetic seeds is not just a means of adaptation but a proactive approach that contributes to the resilience of ecosystems, the conservation of biodiversity, and the continued availability of these invaluable botanical resources in the context of a rapidly changing world.

6.5 PHYTOCHEMICAL AND QUALITY CONTROL

The application of synthetic seeds is instrumental in the meticulous control of phytochemical profiles and the maintenance of consistent quality in medicinal plants, making them an invaluable asset in the realm of herbal medicine and pharmaceuticals. Synthetic seeds offer a platform to tailor the genetic characteristics of propagated plants to ensure the desired phytochemical composition. This ability to select for specific bioactive compounds, such as alkaloids, flavonoids, or essential oils, holds great promise for the production of standardized medicinal plants. In the pharmaceutical industry, where dosage and quality control are paramount, this precision in phytochemical composition is pivotal. By cultivating plant lines with predictable phytochemical profiles through synthetic seeds, manufacturers can meet stringent quality and bioactivity requirements, ultimately producing pharmaceuticals with consistent efficacy and safety.

Furthermore, synthetic seeds empower quality control and assurance in the herbal medicine and pharmaceutical sectors. The standardized and uniform growth of plants derived from synthetic seeds ensures that the final product remains consistent in terms of bioactive compounds, therapeutic properties, and overall quality. This is a crucial factor in both industries, where patients and consumers rely on the predictability and reliability of herbal remedies and pharmaceutical drugs. The use of synthetic seeds not only simplifies quality control procedures but also reduces the risk of batch-to-batch variation, leading to enhanced safety and efficacy in the use of medicinal plants for therapeutic purposes. In summary, synthetic seeds offer an innovative and efficient approach to securing consistent phyto-chemical profiles and maintaining stringent quality control in medicinal plant cultivation, facilitating the production of high-quality herbal medicines and pharmaceuticals with reliable therapeutic outcomes.

In summary, the scope of synthetic seeds in medicinal plant conservation and propagation is extensive and multifaceted. It addresses the challenges of conservation, commercial production, quality control, and research, while also fostering collaboration, education, and sustainable resource management. As technology and techniques continue to advance, the potential for synthetic seeds to contribute to the preservation of medicinal plants and the advancement of healthcare remains promising. MAPs have been esteemed throughout history for their diverse array of uses. These plants, treasured for their therapeutic properties and aromatic compounds, have played vital roles in traditional medicine, modern pharmaceuticals, perfumery, and culinary arts. However, their sustainable cultivation poses multifaceted challenges, from genetic variability to disease susceptibility. Enter synthetic seeds—a cutting-edge innovation poised to revolutionize the cultivation of MAPs. This chapter aims to explore the potential of synthetic seeds as a transformative solution to the challenges associated with the cultivation of MAP. We delve into its applications, advantages, production techniques, challenges, and future prospects. Our goal is to illuminate how synthetic seeds can unlock the potential of MAPs and usher in a new era of sustainable cultivation. The present chapter provides an in-depth examination of the use of synthetic seeds in the cultivation of MAP. It spans various dimensions, including the historical and contemporary significance of MAPs, the advantages and intricacies of synthetic seed production, practical applications, real-world case studies, challenges, and a glimpse into the future. Through this comprehensive exploration, we endeavor to showcase the potential of synthetic seeds in promoting sustainable MAP cultivation.

6.6 PLANT TISSUE CULTURE AND SYNTHETIC SEED PRODUCTION

In the realm of modern agriculture, ensuring efficient and sustainable methods of crop propagation is of paramount importance (Prasad et al., 2014). Traditional methods of seed production have limitations when it comes to maintaining the genetic purity of elite plant varieties, ensuring the availability of disease-free planting material, and rapidly multiplying desirable plants. A game-changing solution to these challenges lies in the synergy between tissue culture and synthetic seed production (Cheruvathur et al., 2013; Chithra et al., 2005). The journey of creating synthetic seeds commences with the meticulous selection of healthy and desirable parent plants. These parent plants should exhibit the desired traits and serve as reliable sources of high-quality MAP materials. The production of synthetic seeds is intricately linked to tissue culture techniques—an indispensable component of modern plant biotechnology. Key steps encompass:

- **Initiation:** Small segments of plant tissue, such as leaves, stems, or shoot tips, are carefully excised from the selected parent plant. These tissues undergo rigorous sterilization to eliminate contaminants.
- **Multiplication:** The sterilized plant tissues are placed on a nutrient-rich medium containing essential macronutrients, micro-nutrients, vitamins, and plant growth regulators (hormones). This stage fosters the multiplication of cells and the development of multiple plantlets.
- **Embryo Formation and Protective Encapsulation:** Once the plantlets attain an appropriate size, they are induced to form embryos. This stage involves the development of shoot and root structures, mimicking the natural process of seed development. Subsequently, the former embryos are encapsulated within a protective coating, which can take the form of a gel matrix or other suitable materials.

- **Ensuring Longevity:** Following encapsulation, synthetic seeds undergo desiccation. This drying process preserves the embryos, enabling long-term storage (Dhir and Shekhawat, 2013). These desiccated synthetic seeds can be safely stored in seed banks or deployed for propagation when needed.

6.7 A GLIMPSE INTO THE FUTURE

In this section, we provide specific case studies that offer a glimpse into the practical application of synthetic seeds in the cultivation of MAP. These case studies underscore the advantages, challenges, and outcomes of employing synthetic seeds for various species.

6.7.1 LAVENDER

Lavandula angustifolia, commonly known as lavender, is renowned for its aromatic and therapeutic properties. The essential oil extracted from lavender is used in aromatherapy and cosmetics due to its calming and soothing effects. Lavender also boasts antimicrobial properties, making it valuable for various medicinal preparations. Lavender is sensitive to soil-borne diseases, and conventional propagation methods can result in disease transmission. Synthetic seeds offer disease-free plant material, ensuring the quality of lavender essential oil. Lavender requires well-drained soil and is susceptible to root rot. Effective cultivation practices are necessary to optimize growth. The use of synthetic seeds has enabled efficient propagation of lavender, ensuring a disease-free and genetically uniform source of essential oil.

6.7.2 PEPPERMINT

Mentha piperita, commonly known as peppermint, is prized for its aromatic leaves and essential oil, which have digestive and soothing properties. Peppermint is extensively used in herbal teas and therapeutic formulations. Peppermint is susceptible to diseases such as rust and mildew. Synthetic seeds provide a means to propagate disease-free plants, ensuring the quality of peppermint oil and leaves. Peppermint requires regular pruning to promote leaf production and maintain oil quality. Proper cultivation practices are essential. Synthetic seeds have facilitated the efficient propagation of peppermint, ensuring a consistent supply of high-quality leaves and essential oils for various applications.

6.7.3 BASIL

Ocimum basilicum, commonly known as basil, is celebrated for its culinary and medicinal uses. Basil contains essential oils with antimicrobial properties and is used in traditional medicine systems to alleviate various ailments. Basil is susceptible to various diseases and pests. Synthetic seeds provide a disease-free starting material, reducing the need for pesticides in cultivation. Basil requires specific environmental conditions, including adequate sunlight and warmth. Meeting these conditions is essential for successful cultivation. The utilization of synthetic seeds has streamlined the production of basil, ensuring a consistent supply of high-quality, disease-free leaves for culinary and medicinal purposes.

6.7.4 DAMASK ROSE

Rosa damascena, known as the Damask rose, is highly regarded for its fragrant blossoms, which are the source of rose essential oil, widely used in perfumery and cosmetics. This species is known for its exquisite aroma and therapeutic properties. The cultivation of Damask roses can be challenging due to susceptibility to diseases like powdery mildew and black spot. Synthetic seeds offer a reliable source of disease-free rose plants, ensuring the quality of rose oil. Damask roses require specific care, including regular pruning and pest management, to maintain optimal oil production. The adoption of synthetic seeds has streamlined the propagation of Damask roses, ensuring a disease-free and genetically uniform source of fragrant blossoms for the perfume industry.

6.8 CHALLENGES AND LIMITATIONS IN SYNTHETIC SEED APPLICATIONS

6.8.1 THE COST CONUNDRUM

The adoption of synthetic seed technology can be associated with significant costs, which may include setting up infrastructure for tissue culture laboratories, hiring skilled personnel, and procuring specialized growth media and equipment. These expenses can be prohibitive for small-scale growers and traditional medicinal practitioners (Badr-Elden, 2018). Efforts should be made to make synthetic seed technology more accessible and affordable. Initiatives such as government subsidies, technology transfer programs, and capacity-building for small-scale growers can help overcome financial barriers.

6.8.2 REGULATORY ROADBLOCKS

The use of synthetic seeds, especially in the case of genetically modified MAP, raises regulatory and ethical considerations. Ensuring compliance with national and international regulations is essential to avoid legal and ethical challenges. Collaboration between regulatory authorities, biotechnologists, and plant scientists is crucial for the development of clear guidelines and ethical frameworks for the use of synthetic seeds in the cultivation of MAP.

6.8.3 SCALING UP THE SYNTHESIS

Scaling up the production of synthetic seeds to meet the global demand for MAP materials is a considerable challenge. Developing large-scale production methods that maintain the quality and genetic uniformity of synthetic seeds is essential. Research and development efforts should focus on optimizing synthetic seed production methods to achieve economies of scale. Collaboration between academia and industry can accelerate the scaling-up process.

6.8.4 NAVIGATING ETHICAL FRONTIERS

The genetic uniformity promoted by synthetic seeds raises concerns about the potential loss of genetic diversity in cultivated populations of MAPs. This can impact the adaptability and resilience of these plants in the face of changing environmental conditions. Efforts should be made to strike a balance between genetic uniformity and genetic diversity. Strategies such as maintaining *ex situ* gene banks of diverse germplasm and encouraging the cultivation of wild and traditional varieties can help address these concerns.

6.9 THE DAWN OF A NEW ERA: FUTURE PROSPECTS AND RESEARCH DIRECTIONS

Continuous research and development in synthetic seed technology hold the promise of further enhancing the efficiency and scalability of synthetic seed production (Rihan et al., 2017; Saha et al., 2015). Innovations in encapsulation materials, nutrient formulations, and cryopreservation techniques can expand the applicability of synthetic seeds.

6.9.1 TOWARD SUSTAINABLE PRODUCTION MODELS

Sustainable production models that integrate synthetic seed technology with agroforestry, organic farming, and sustainable harvesting practices can contribute to the long-term conservation of MAPs (Raju et al., 2016). These models should prioritize environmental stewardship and community engagement.

6.9.2 INTEGRATION WITH CONSERVATION EFFORTS

The integration of synthetic seed technology with conservation initiatives can play a crucial role in the preservation of endangered MAPs. Efforts to collect and store genetic material from threatened species and ecosystems should be intensified.

6.9.3 GENETIC MODIFICATION: THE SYNBIO FRONTIER

The field of synthetic biology (SynBio) offers opportunities for enhancing the therapeutic properties of MAPs through genetic modification. However, ethical and regulatory considerations must be carefully navigated in the pursuit of genetically engineered medicinal plants.

6.10 DEVELOPMENT OF SYNTHETIC SEEDS

The concept of synthetic seed production dates back to the 1970s when researchers began experimenting with encapsulation techniques. The initial focus was on creating artificial seeds for economically valuable crops. Over the years, the technology has undergone significant refinement, with advancements in somatic embryogenesis, encapsulation materials, and desiccation methods (Aquea et al., 2008; Attia et al., 2018;

Baskaran et al., 2015; Cartes et al., 2009; Grzegorczyk and Wysokińska, 2011). Synthetic seeds contain the genetic material necessary for plant development, including the embryonic axis, cotyledons, and shoot apical meristem. They possess a protective coating or encapsulation that shields the enclosed embryos from external stressors and desiccation. Synthetic seeds are capable of germination and producing whole plants under suitable conditions. They are produced under sterile conditions, minimizing the risk of disease transmission. Further, synthetic seeds can be customized to suit specific plant species and propagation needs.

Synthetic seed production is a remarkable technique that combines the principles of plant tissue culture and encapsulation technology to create artificial seeds. The process involves several critical steps to encapsulate plant embryos or somatic embryos in a protective coating, thus mimicking natural seeds (Figure 6.1). These encapsulated propagules can be stored, transported, and sown like traditional seeds. The essential steps in synthetic seed production can be summarized as in below sections.

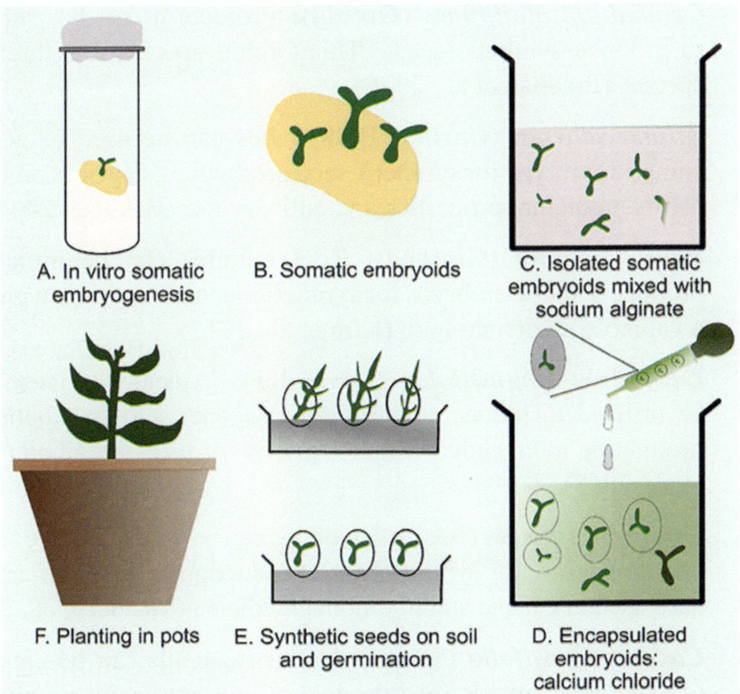

FIGURE 6.1 General outline of various steps involved (A-F) in synthetic seed development.

6.11 SELECTION OF PARENT PLANTS

Before embarking on synthetic seed production, it is essential to carefully select the parent plants. The choice of parent plants plays a pivotal role in determining the quality and characteristics of the synthetic seeds. Parent plants should exhibit genetic purity (in the case of economically important varieties) and stability to ensure uniformity in the resulting synthetic seeds, as genetic purity in a plant variety ensures the uniformity and stability of its essential traits across successive generations. Preference is given to parent plants with inherent resistance to diseases and pests. The selection of parent plants with high-yielding traits contributes to the productivity of the synthetic seeds.

In synthetic seed production for MAP, specific plant parts and tissues are used to generate somatic embryos, which can be encapsulated to create synthetic seeds. These plant parts are selected based on the species and the desirable traits they possess. Here are examples of MAPs along with the plant parts commonly used in synthetic seed production:

- ***Cymbidium aloifolium* (Orchid):** Protocorm can be employed to produce synthetic seeds. This Orchid species is a threatened species (Pradhan et al., 2016).
- ***Allium sativum* (Garlic):** Bulb scales can be used to generate somatic embryos for synthetic seed production. Garlic is renowned for its medicinal properties and culinary use (Bekheet, 2006).
- ***Panax ginseng* (Ginseng):** Root cultures can be initiated to produce somatic embryos for synthetic seed production in ginseng, a valuable medicinal herb (Kim et al., 2019).
- ***Lavandula angustifolia* (Lavender):** Apical meristems can be utilized to induce somatic embryogenesis for synthetic seed production in lavender, which is prized for its essential oil (Salehi et al., 2018).
- ***Ocimum basilicum* (Basil):** Shoot tips can be used to initiate somatic embryogenesis for synthetic seed production in basil, an aromatic herb used in cuisine and traditional medicine (Mandal et al., 2000).
- ***Cassia angustifolia* (Senna)**: Nodal segments can be employed to generate synthetic seed production in senna, a plant valued for sennosides production (Parveen and Shahzad, 2014).

- ***Withania somnifera*** **(Ashwagandha):** Nodal explants can be used to produce somatic embryos for synthetic seed production in ashwagandha, a medicinal herb in traditional Ayurvedic medicine (Fatima et al., 2013).
- ***Rauvolfia serpentina*** **(Indian Snakeroot)**: Nodal segments can be employed for somatic embryogenesis and synthetic seed production in this medicinal plant known for its alkaloid content (Ray and Bhattacharya, 2008).
- ***Curcuma longa*** **(Turmeric):** Leaf base explant can be used to generate somatic embryos for synthetic seed production in turmeric, a spice with medicinal properties (Soundar et al., 2015).

6.12 ENCAPSULATION

Encapsulation techniques are crucial components of synthetic seed production, facilitating the creation of artificial seeds that house somatic embryos or propagules within protective coatings. These techniques are fundamental for safeguarding the viability and successful delivery of somatic embryos, effectively mimicking the natural seed structure (Ahmad and Anis, 2010; Ahmad et al., 2012, 2015; Alatar and Faisal, 2012; Dhir et al., 2014; Gupta and Sood, 2022; Kader et al., 2023; Redenbaugh et al., 1984). In this comprehensive guide, we will delve into various encapsulation techniques, elucidating their underlying principles, advantages, and versatile applications in synthetic seed production. Alginate bead encapsulation stands as one of the most widely employed techniques in synthetic seed production. This method utilizes sodium alginate, a natural polysaccharide derived from brown algae, as the primary encapsulation material. In alginate bead encapsulation, somatic embryos or propagules are mixed with a sodium alginate solution, forming a suspension. Droplets of this suspension are then carefully extruded into a solution of calcium chloride or calcium nitrate. Upon contact with calcium ions, the sodium alginate undergoes gelation, resulting in the formation of solid beads that encapsulate the somatic embryos. This approach boasts high encapsulation efficiency, biocompatibility, and a rapid gelation process, making it ideal for a broad spectrum of plant species, including those with fragile or sensitive embryos. Another variation of alginate encapsulation involves sodium alginate-coated synthetic seeds. In this method, somatic embryos are pre-cultured to attain a suitable stage before being coated with a sodium alginate solution. These coated embryos are then immersed in a calcium solution, yielding a protective alginate layer around them. This approach offers precise control over the size and shape of synthetic seeds, enabling uniform coating of somatic embryos. It is particularly valuable for plant species with delicate embryos where the uniformity of synthetic seeds is paramount.

In conclusion, encapsulation techniques represent a cornerstone of synthetic seed production, enabling the protection, storage, and controlled release of somatic embryos or propagules (Sharma et al., 2013). These techniques offer versatility and customization options, making them applicable to a wide range of plant species, including MAPs. The choice of encapsulation method depends on the specific requirements of the plant species and the desired outcomes in terms of uniformity, protection, and environmental sustainability. As technology continues to advance, further innovations in encapsulation techniques are likely to enhance the efficiency and applicability of synthetic seed production across diverse plant species.

6.13 DEHYDRATION PROCESS

The dehydration process is a critical stage in the cryopreservation of synthetic seeds, where the water content of encapsulated somatic embryos or propagules is reduced to a level compatible with long-term storage without metabolic activity (Gonzalez-Arnaol and Engelman, 2006; Tolga and Yesmin, 2017). This process is crucial for maintaining the viability and shelf life of synthetic seeds, making it a pivotal step in the overall methodology of synthetic seed production. The dehydration process in synthetic seed production serves several vital purposes. Firstly, it minimizes the water activity within the synthetic seeds, which is essential for preventing microbial and fungal growth during storage. By reducing water content, the risk of microbial contamination is significantly diminished, ensuring the long-term preservation of somatic embryos or propagules. Secondly, dehydration protects the genetic material within the synthetic seeds from damage caused by ice crystal formation during freezing. This is particularly important for species that require cryopreservation, a method used to store synthetic seeds at extremely low temperatures

(LN, –196°C). Furthermore, dehydration helps to maintain the structural integrity of synthetic seeds, preventing rupture or damage during handling and storage. Several techniques are employed in the dehydration process of synthetic seed production, depending on the characteristics of the plant material and the desired storage conditions. Controlled drying is one of the most commonly used methods where synthetic seeds are subjected to controlled drying using various approaches. Air drying, for instance, involves exposing the synthetic seeds to ambient air under laminar flow or by using silica gel. The gradual reduction in moisture content occurs naturally as water evaporates from the seeds.

6.14 STORAGE CONDITIONS

Storage conditions play a paramount role in the success and longevity of synthetic seeds, which are artificial structures designed to encapsulate somatic embryos or propagules for extended periods. Proper storage conditions are crucial for preserving the viability and genetic integrity of the encapsulated material, allowing for long-term availability and utilization in various applications (Micheli et al., 2007; Mohanty et al., 2013). Generally, synthetic seeds are stored at cool temperatures, often ranging from –20°C to 4°C. Lower temperatures, such as –20°C or below, are typically used for long-term storage, while temperatures around 4°C are suitable for short to medium-term storage. These low temperatures slow down metabolic processes, effectively preserving the embryos or propagules within synthetic seeds. However, it's crucial to avoid freeze-thaw cycles, as they can damage the structure and viability of synthetic seeds.

6.15 FACTORS AFFECTING SYNTHETIC SEED PRODUCTION

The success of synthetic seed production is contingent on various environmental, nutritional, hormonal, and quality control factors affecting the somatic embryo or other propagules (da Silva and Malabadi, 2012; Etienne et al., 2018; Ghanbarali et al., 2016; Hatzilazarou et al., 2021; Phanomchai et al., 2022). Understanding these factors is crucial for optimizing the process.

6.15.1 ENVIRONMENTAL FACTORS

Synthetic seed production, a biotechnological process for encapsulating somatic embryos or propagules within protective coatings, is significantly influenced by various environmental factors that impact the growth and development of the encapsulated material (Rai et al., 2009). Understanding and carefully managing these environmental factors are crucial for the success of synthetic seed production. Temperature plays a pivotal role in synthetic seed production. The temperature at which somatic embryos or propagules are cultured during the encapsulation process can greatly affect their development. Optimal temperature conditions promote efficient somatic embryogenesis and the production of healthy embryos. Furthermore, temperature control is vital during the subsequent storage of synthetic seeds, as it helps preserve their viability. Synthetic seeds may be stored at specific temperatures to minimize metabolic activity and ensure their long-term shelf life. Light conditions are another environmental factor that can influence the success of synthetic seed production. Light exposure can affect the growth and development of somatic embryos. While some plant species may require light during certain stages of somatic embryogenesis, others may be sensitive to it and need to be cultured in darkness. The choice of light conditions is species-specific and must be carefully tailored to meet the needs of the encapsulated material. Relative humidity (RH) is a critical environmental factor that affects both the encapsulation process and the storage of synthetic seeds. During the encapsulation process, maintaining an appropriate RH level in the culture environment is essential to prevent desiccation of the embryos and ensure their successful encapsulation. In storage, RH control is crucial to prevent moisture-related issues that could compromise the quality and viability of synthetic seeds. Storage rooms or containers are often equipped with humidity control systems to maintain optimal RH levels. Air quality is another consideration in synthetic seed production. Contaminants in the air, such as fungi and bacteria, can lead to contamination of the cultures and adversely affect the success of the process. Proper ventilation and air filtration systems are employed to maintain a controlled and sterile environment during the encapsulation process. Carbon dioxide (CO_2) concentration is an environmental factor that can be manipulated to optimize somatic embryogenesis and synthetic seed production. In some cases, adjusting the CO_2 levels in the culture environment can enhance the efficiency of somatic embryo formation and growth. This controlled manipulation of CO_2 concentration is particularly important when working with species that are sensitive to changes in atmospheric CO_2.

6.15.2 NUTRIENT REQUIREMENTS

Nutrient requirements are a critical aspect of synthetic seed production, as they directly influence the growth and development of somatic embryos or propagules within the synthetic seeds (Naik and Chand, 2006). Macro and micronutrients such as nitrogen (N), phosphorus (P), potassium (K), calcium (Ca), magnesium (Mg), sulfur (S), iron (Fe), manganese (Mn), zinc (Zn), copper (Cu), and boron (B) play crucial roles in somatic embryo development and overall plant growth. Growth regulators, including auxins (IAA) and cytokinin's (BAP), are essential components of the culture medium used in synthetic seed production. Carbon sources, primarily in the form of carbohydrates, are vital for providing energy and carbon skeletons for the growth and development of somatic embryos. Vitamins and organic supplements, such as thiamine (vitamin B1), myo-inositol, and casein hydrolysate, are often included in culture media to support the growth and development of somatic embryos. The pH (5.8) of the culture medium is a critical aspect of nutrient management.

6.15.3 HORMONAL REGULATION

Hormonal regulation plays a pivotal role in the intricate process of synthetic seed production, where somatic embryos or propagules are encapsulated within protective coatings to mimic natural seeds (Kumar et al., 2005; Malabadi and Staden, 2005; Manjkhola et al., 2005). The precise control and manipulation of plant hormones are essential for inducing and regulating somatic embryogenesis, as well as guiding the subsequent development and conversion of somatic embryos into synthetic seeds. Auxins and cytokinin's are two key types of plant hormones that are central to hormonal regulation in synthetic seed production. Auxins, such as indole-3-acetic acid (IAA) and 2,4-dichlorophenoxyacetic acid (2,4-D), are instrumental in initiating somatic embryogenesis. They stimulate the formation of embryogenic callus from explants, which serves as the starting point for somatic embryo development. On the other hand, cytokinin's, including kinetin and benzyl adenine (BA), promote the proliferation and differentiation of somatic embryos from the callus tissue. The precise balance between auxins and cytokinin's, along with their concentrations, is carefully adjusted to optimize the somatic embryogenesis process. Abscisic acid (ABA) is another hormone that plays a critical role in synthetic seed production, particularly during the encapsulation phase. ABA is involved in regulating seed dormancy and desiccation tolerance. In synthetic seed production, ABA is often applied to the somatic embryos before encapsulation to induce desiccation tolerance, making the synthetic seeds more robust and suitable for storage. Gibberellins (GAs) are hormones that can influence various aspects of synthetic seed production. They are involved in regulating embryo elongation and germination. Depending on the target plant species and desired outcomes, the application of specific gibberellins or their inhibitors can be employed to control the size and development stage of somatic embryos within synthetic seeds. Ethylene, a gaseous plant hormone, can also impact synthetic seed production. Ethylene is associated with processes like fruit ripening and senescence, but it can influence the growth and development of somatic embryos. The presence or absence of ethylene, as well as its concentration, may need to be carefully managed to achieve optimal results in synthetic seed production.

Hormonal regulation is not a one-size-fits-all approach. It requires a deep understanding of the target plant species, the specific goals of synthetic seed production, and the developmental stages of somatic embryos. Hormonal treatments are meticulously applied during various stages of the process to induce somatic embryogenesis, guide the development of embryos, and prepare them for encapsulation. Achieving the right hormonal balance is critical to producing high-quality synthetic seeds that can germinate successfully and grow into mature plants when planted in the field or under controlled conditions.

6.15.4 DISEASE AND CONTAMINATION CONTROL

Maintaining a sterile and contamination-free environment is of paramount importance in synthetic seed production, where somatic embryos or propagules are encapsulated within protective coatings. Disease and contamination can jeopardize the viability and quality of synthetic seeds, rendering them ineffective for various applications, from crop propagation to conservation and biotechnology. Robust disease and contamination control measures are essential to ensure the success of the synthetic seed production process.

6.15.4.1 STERILIZATION OF EXPLANTS

The process begins with the sterilization of explants, which are plant tissues used as the starting material for somatic embryogenesis. Explants are typically surface-sterilized using chemical agents like bleach (0.5–1%) or ethanol (70–80%) to eliminate surface microorganisms. Ensuring the sterility of explants is critical because any contaminants introduced at this stage can proliferate and compromise the entire process.

6.15.4.2 ASEPTIC CULTURE TECHNIQUES

Throughout somatic embryogenesis, aseptic culture techniques are rigorously employed. These techniques involve working in a controlled environment, such as a laminar flow hood, to minimize the risk of contamination. All tools, equipment, and culture vessels are sterilized, and culture media are autoclaved (121°C and 15 psi) to ensure they are free from contaminants.

6.15.4.3 ANTIMICROBIAL AGENTS

In some cases, synthetic seed production may require the incorporation of antimicrobial agents, such as antibiotics (e.g., 0.5–1 μg mL^{-1} Streptomycin /tetracyclin) or fungicides (e.g., 1–20 mg L^{-1} Bavistin / Miconazole) into the culture medium. These agents help prevent the growth of bacteria, fungi, or other microorganisms that could jeopardize the development of somatic embryos.

6.15.4.4 ENCAPSULATION MATERIAL STERILIZATION

The materials used for encapsulating somatic embryos must also be sterile. Alginate, gelatin, or other encapsulation materials are sterilized through autoclaving or filtration methods to ensure they do not introduce contaminants during encapsulation.

6.15.4.5 QUALITY CONTROL AND MONITORING

Regular quality control and monitoring procedures are essential to detect and address any signs of contamination. Contaminated cultures or synthetic seeds must be isolated and discarded to prevent the spread of contaminants to healthy embryos or propagules.

6.15.4.6 CLEANROOM FACILITIES

Some synthetic seed production facilities use cleanroom facilities equipped with high-efficiency particulate air (HEPA) filters to maintain a sterile environment. Cleanrooms provide an additional layer of protection against airborne contaminants, ensuring the integrity of the process.

6.15.4.7 HYGIENE PRACTICES

The personnel involved in synthetic seed production must adhere to stringent hygiene practices. This includes wearing appropriate protective clothing, such as lab coats and gloves, and thoroughly washing hands and tools before entering the controlled environment.

6.15.4.8 QUARANTINE AND ISOLATION

When introducing new plant material for synthetic seed production, it is advisable to subject it to a quarantine period to monitor for potential diseases or contaminants. Isolation of cultures or propagules from different plant sources can prevent cross-contamination.

6.16 ADVANTAGES AND DISADVANTAGES

As with any technology, synthetic seed production comes with a set of advantages and disadvantages:

6.16.1 ADVANTAGES

- **Rapid Multiplication:** Synthetic seeds facilitate the rapid multiplication of desirable plant material, reducing the time required for mass propagation.
- **Germplasm Preservation:** They offer a reliable means of preserving germplasm, ensuring the conservation of plant genetic diversity.
- **Disease-Free Propagation:** Synthetic seeds are produced under sterile conditions, minimizing the risk of disease transmission during planting.
- **Year-Round Availability:** They provide a solution to seasonal constraints by enabling year-round availability of planting material.
- **Reduced Storage and Transportation Costs:** Synthetic seeds are compact and lightweight, reducing storage and transportation expenses.

6.16.2 DISADVANTAGES

- **Genetic Stability:** Maintaining genetic stability in synthetic seed production remains a challenge, with concerns about somaclonal variation and genetic drift.
- **High Initial Costs:** The technology demands significant initial investments in infrastructure, equipment, and research.
- **Ethical and Regulatory Issues**: Ethical concerns and regulatory complexities, including intellectual property rights and biosafety regulations, may hinder progress.
- **Acceptance by Farmers**: Convincing traditional farmers to adopt synthetic seeds can be challenging due to their unfamiliarity with the technology.

6.17 FUTURE PROSPECTS AND INNOVATIONS

The future of synthetic seed production holds promise and potential for further innovations that can address existing challenges and broaden its applications. Advancements in genetic modification techniques will enable the engineering of synthetic seeds for specific traits, including enhanced resistance to biotic and abiotic stresses. Synthetic seed technology can be seamlessly integrated into precision farming practices, allowing for customized nutrient and water delivery systems. In the realm of biopharmaceuticals, synthetic seed technology can serve as a platform for the production of transgenic medicinal plants engineered to enhance the yield of bioactive compounds or introduce novel therapeutic molecules. This opens up new avenues for the development of personalized medicine and the extraction of pharmaceuticals from plant sources. Synthetic seeds have the potential to contribute significantly to sustainable agriculture by reducing the reliance on chemical inputs and enhancing crop productivity.

6.18 CONCLUSION

The utilization of synthetic seeds in the cultivation of MAPs marks a groundbreaking innovation in the realm of agriculture and biotechnology. These artificial seeds have the potential to revolutionize the way we grow and propagate valuable plant species, offering solutions to challenges such as genetic variability, disease susceptibility, and environmental sensitivity. The advantages of synthetic seeds are manifold. They allow for rapid and efficient propagation, ensuring a consistent supply of genetically uniform plants. This is particularly critical in industries like pharmaceuticals, traditional medicine, perfumery, and culinary arts, where the quality and consistency of plant materials are paramount. Synthetic seeds also play a vital role in conservation efforts by enabling the controlled cultivation of endangered MAP species. By preserving genetic material and reducing the pressure on wild populations, they contribute to biodiversity conservation. While there are challenges to overcome, the future of synthetic seeds in the cultivation of MAP is promising. As research and development in this field continue to advance, the adoption of synthetic seeds is likely to become more widespread, ensuring a sustainable supply of these valuable plant resources for generations to come. As we venture into the future, the fusion of biotechnology and traditional horticultural knowledge will continue to unlock new possibilities in the conservation and sustainable multiplication of MAPs, offering a brighter, greener, and healthier world for all.

KEYWORDS

- **conservation**
- **genetic stability**
- **germplasm preservation**
- **medicinal plants**
- **synthetic seeds**
- **tissue culture**

REFERENCES

Ahmad, N., & Anis, M. (2010). Direct plant regeneration from encapsulated nodal segments of *Vitex negundo. Biologia Plantarum, 54*, 748–752.

Ahmad, N., Faisal, M., Fatima, N., & Anis, M. (2012). Encapsulation of microcuttings for propagation and short-term preservation in *Ruta graveolens* L., a plant with high medicinal value. *Acta Physiologiae Plantarum, 34*, 2303–2310.

Ahmad, Z., Yadav, V., Shahzad, A., Emamverdian, A., Ramakrishnan, M., & Ding, Y. (2022). Micropropagation, encapsulation, physiological, and genetic homogeneity assessment in *Casuarina equisetifolia. Frontiers in Plant Science, 13*, 905444.

Ahmed, M. R., Anis, M., & Al-Etta, H. A. (2015). Encapsulation technology for short-term storage and germplasm exchange of *Vitex trifolia* L. *Rendiconti Lincei, 26*, 133–139.

Alatar, A. A., & Faisal, M. (2012). Encapsulation of *Rauvolfia tetraphylla* microshoots as artificial seeds and evaluation of genetic fidelity using RAPD and ISSR markers. *Journal of Medicinal Plants Research, 6*, 1367–1374.

Aquea, F., Poupin, M. J., Matus, J. T., Gebauer, M., Medina, C., & Arce-Johnson, P. (2008). Synthetic seed production from somatic embryos of *Pinus radiata. Biotechnology Letters, 30*, 1847–1852.

Ara, H., Jaiswal, U., & Jaiswal, V. (2000). Synthetic seed: Prospects and limitations. *Current Science, 78*, 1438–1444.

Attia, A. O., Ismail, I. A., Dessoky, E. D. S., & Alotaibi, S. S. (2018). Synthetic seeds as an *in vitro* conservation method for Al-Taif rose plant *Rosa damascena trigintipetala* Dieck. *Bioscience Research, 15*, 1113–1119.

Badr-Elden, A. M. (2018). New approaches for reducing the cost of synthetic seed storage using sugarcane bagasse and different additives to the gel matrix for sugarcane plant *in vitro*. *Egyptian Journal of Botany, 58*, 1–10.

Baskaran, P., Kumari, A., & Van Staden, J. (2015). Embryogenesis and synthetic seed production in *Mondia whitei*. *Plant Cell, Tissue and Organ Culture, 121*, 205–214.

Bekheet, S. A. (2006). A synthetic seed method through encapsulation of *in vitro* proliferated bulblets of garlic *Allium sativum* L. *Arab Journal of Biotechnology, 9*, 415–426.

Börner, A. (2006). Preservation of plant genetic resources in the biotechnology era. *Biotechnology Journal, 1*, 1393–1404.

Cartes, P., Castellanos, H., Ríos, D., Sáez, K., Spierccolli, S., & Sánchez, M. (2009). Encapsulated somatic embryos and zygotic embryos for obtaining artificial seeds of rauli-beech *Nothofagus alpina* Poepp. & Endl. oerst. *Chilean Journal of Agricultural Research, 69*, 112–118.

Cheruvathur, M. K., Kumar, G. K., & Thomas, T. D. (2013). Somatic embryogenesis and synthetic seed production in *Rhinacanthus nasutus* L. Kurz. *Plant Cell, Tissue and Organ Culture, 113*, 63–71.

Chithra, M., Martin, K. P., Sunandakumari, C., & Madhusoodanan, P. V. (2005). Somatic embryogenesis, encapsulation, and plant regeneration of *Rotula aquatica* Lour., a rare rheophytic woody medicinal plant. *In Vitro Cellular & Developmental Biology–Plant, 41*, 28–31.

Corrie, S., & Tandon, P. (1993). Propagation of *Cymbidium giganteum* Wall. through high frequency conversion of encapsulated protocorms under *in vivo* and *in vitro* conditions. *Indian Journal of Experimental Biology, 31*, 61–64.

da Silva, J. A. T., & Malabadi, R. B. (2012). Factors affecting somatic embryogenesis in conifers. *Journal of Forestry Research, 23*, 503–515.

Devi, P. S., Satyanarayana, B., Arundhati, A., & Rao, T. R. (2014). Plant regeneration from encapsulated nodal segments of *Sterculia urens* Roxb., an endangered gum-yielding tree. *Journal of Crop Science and Biotechnology, 17*, 21–25.

Dhir, R., & Shekhawat, G. S. (2013). Production, storability, and morphogenic response of alginate-encapsulated axillary meristems and genetic fidelity evaluation of *in vitro* regenerated *Ceropegia bulbosa*, a pharmaceutically important threatened plant species. *Industrial Crops and Products, 47*, 139–144.

Dhir, R., Shekhawat, G. S., & Alam, A. (2014). Improved protocol for somatic embryogenesis and calcium alginate encapsulation in *Anethum graveolens* L., a medicinal herb. *Applied Biochemistry and Biotechnology, 173*, 2267–2278.

Etienne, H., Breton, D., Breitler, J. C., Bertrand, B., Déchamp, E., Awada, R., Marraccini, P., Léran, S., Alpizar, E., Campa, C., Courtel, P., Georget, F., & Ducos, J. P. (2018). Coffee somatic embryogenesis: How did research, experience gained, and innovations promote the commercial propagation of elite clones from the two cultivated species? *Frontiers in Plant Science, 9*, 1630.

Fatima, N., Ahmad, N., Anis, M., & Ahmad, I. (2013). An improved *in vitro* encapsulation protocol, biochemical analysis, and genetic integrity using DNA-based molecular markers in regenerated plants of *Withania somnifera* L. *Industrial Crops and Products, 50*, 468–477.

Gantait, S., Kundu, S., Ali, N., & Sahu, N. C. (2015). Synthetic seed production of medicinal plants: A review on influence of explants, encapsulation agent, and matrix. *Acta Physiologiae Plantarum, 37*, 98.

Gantait, S., Kundu, S., Yeasmin, L., & Ali, M. N. (2017). Impact of differential levels of sodium alginate, calcium chloride, and basal media on germination frequency of genetically true artificial seeds of *Rauvolfia serpentina* L. Benth. ex Kurz. *Journal of Applied Research on Medicinal and Aromatic Plants, 4*, 75–81.

Ghanbarali, S., Abdollahi, M. R., Zolnorian, H., Moosavi, S. S., & Seguí-Simarro, J. M. (2016). Optimization of the conditions for production of synthetic seeds by encapsulation of axillary buds derived from minituber sprouts in potato *Solanum tuberosum*. *Plant Cell, Tissue and Organ Culture, 126*, 449–458.

Gonzalez-Arnao, M. T., & Engelmann, F. (2006). Cryopreservation of plant germplasm using the encapsulation-dehydration technique: Review and case study on sugarcane. *CryoLetters, 27*(3), 155–168.

Grzegorczyk, I., & Wysokińska, H. (2011). A protocol for synthetic seeds from *Salvia officinalis* L. shoot tips. *Acta Biologica Cracoviensia Series Botanica, 53*, 80–85.

Gupta, A. K., Harish, R., Phulwaria, M., Agarwal, T., & Shekhawat, N. S. (2014). *In vitro* propagation, encapsulation, and genetic fidelity analysis of *Terminalia arjuna*, a cardioprotective medicinal tree. *Applied Biochemistry and Biotechnology, 173*, 1481–1494.

Gupta, P., Dhawan, S. S., Lal, R. K., Mishra, A., & Chanotiya, C. S. (2024). Low-temperature perception and modulations in *Ocimum basilicum* commercial cultivar CIM-Shishir: Biosynthetic potential with insight towards climate-smart resilience. *Gene, 896*, 148041.

Gupta, R., & Sood, H. (2022). Artificial seed production and cryopreservation by encapsulation dehydration for medicinal herb of Himalayan region, *Swertia chirayita*. *CryoLetters, 43*(5), 295–302.

Haque, S. M., & Ghosh, B. (2016). High-frequency somatic embryogenesis and artificial seeds for mass production of true-to-type plants in *Ledebouria revoluta*, an important cardioprotective plant. *Plant Cell, Tissue and Organ Culture, 127*, 71–83.

Hatzilazarou, S., Kostas, S., Nendou, T., & Economou, A. (2021). Conservation, regeneration, and genetic stability of regenerants from alginate-encapsulated shoot explants of *Gardenia jasminoides* Ellis. *Polymers, 13*, 1666.

Kader, A., Sinha, S. N., & Ghosh, P. (2023). A standardized protocol for genetically stable artificial seed production of *Ficus religiosa* L. using ISSR fingerprinting. *Journal of Applied Genetics, 64*, 275–287.

Khan, M. I., Ahmad, N., Anis, M., Alatar, A. A., & Faisal, M. (2018). *In vitro* conservation strategies for the Indian willow *Salix tetrasperma* Roxb., a vulnerable tree species via propagation through synthetic seeds. *Biocatalysis and Agricultural Biotechnology, 16*, 17–21.

Kim, J. Y., Adhikari, P. B., Ahn, C. H., Kim, D. H., Kim, Y. C., Han, J. Y., Kondeti, S., & Choi, Y. E. (2019). High frequency somatic embryogenesis and plant regeneration of interspecific ginseng hybrid between *Panax ginseng* and *Panax quinquefolius*. *Journal of Ginseng Research, 43*, 38–48.

Kim, M. A., & Park, J. K. (2002). High frequency plant regeneration of garlic *Allium sativum* L. calli immobilized in calcium alginate gel. *Biotechnology and Bioprocess Engineering, 7*, 206.

Kumar, M. B. A., Vakeswaran, V., & Krishnasamy, V. (2005). Enhancement of synthetic seed conversion to seedlings in hybrid rice. *Plant Cell, Tissue and Organ Culture, 81*, 97–100.

Lata, H., Chandra, S., Mehmedic, Z., Khan, I. A., & ElSohly, M. A. (2012). *In vitro* germplasm conservation of high Δ^9-tetrahydrocannabinol yielding elite clones of *Cannabis sativa* L. under slow growth conditions. *Acta Physiologiae Plantarum, 34*, 743–750.

Magray, M. M., Wani, K., Chatto, M., & Ummyiah, H. (2017). Synthetic seed technology. *International Journal of Current Microbiology and Applied Sciences, 6*, 662–674.

Malabadi, R. B., & Van Staden, J. (2005). Storability and germination of sodium alginate encapsulated somatic embryos derived from the vegetative shoot apices of mature *Pinus patula* trees. *Plant Cell, Tissue and Organ Culture, 82*, 259–265.

Mandal, J., Pattnaik, S., & Chand, P. K. (2000). Alginate encapsulation of axillary buds of *Ocimum americanum* L. (hoary basil), *O. basilicum* L. (sweet basil), *O. gratissimum* L. (shrubby basil), and *O. sanctum* L. (sacred basil). *In Vitro Cellular & Developmental Biology–Plant, 36*, 287–292.

Manjkhola, S., Dhar, U., & Joshi, M. (2005). Organogenesis, embryogenesis, and synthetic seed production in *Arnebia euchroma*—a critically endangered medicinal plant of the Himalaya. *In Vitro Cellular & Developmental Biology–Plant, 41*, 244–248.

Micheli, M., Hafiz, I. A., & Standardi, A. (2007). Encapsulation of *in vitro*-derived explants of olive *Olea europaea* L. cv. Moraiolo: II. Effects of storage on capsule and derived shoots performance. *Scientia Horticulture, 113*, 286–292.

Mohanty, P., Nongkling, P., Das, M. C., Kumaria, S., & Tandon, P. (2013). Short-term storage of alginate-encapsulated protocorm-like bodies of *Dendrobium nobile* Lindl., an endangered medicinal orchid from North-east India. *3 Biotech, 3*, 235–239.

Naik, S. K., & Chand, P. K. (2006). Nutrient-alginate encapsulation of *in vitro* nodal segments of pomegranate *Punica granatum* L. for germplasm distribution and exchange. *Scientia Horticulture, 108*, 247–252.

Parveen, S., & Shahzad, A. (2014). Encapsulation of nodal segments of *Cassia angustifolia* Vahl. for short-term storage and germplasm exchange. *Acta Physiologiae Plantarum, 36*, 635–640.

Phanomchai, S., Bodhipadma, K., Noichinda, S., & Leung, D. W. M. (2022). Short-term storability of alginate-encapsulated Persian violet microshoots for germplasm exchange. *Plants, 11*, 185.

Pradhan, S., Tiruwa, B. L., Subedee, B. R., & Pant, B. (2016). Efficient plant regeneration of *Cymbidium aloifolium* L. Sw., a threatened orchid of Nepal through artificial seed technology. *American Journal of Plant Sciences, 7*, 1964–1974.

Prasad, A., Singh, M., Yadav, N. P., Mathur, A. K., & Mathur, A. (2014). Molecular, chemical, and biological stability of plants derived from artificial seeds of *Centella asiatica* L. Urban—an industrially important medicinal herb. *Industrial Crops and Products, 60*, 205–211.

Rai, M. K., Asthana, P., Singh, S. K., Jaiswal, V. S., & Jaiswal, U. (2009). The encapsulation technology in fruit plants—A review. *Biotechnology Advances, 27*, 671–679.

Raju, C. S., Aslam, A., & Shajahan, A. (2016). Germination and storability of calcium-alginate coated somatic embryos of mango ginger *Curcuma amada* Roxb. *Horticulture, Environment, and Biotechnology, 57*, 88–96.

Ray, A., & Bhattacharya, S. (2008). Storage and plant regeneration from encapsulated shoot tips of *Rauvolfia serpentina*—An effective way of conservation and mass propagation. *South African Journal of Botany, 74*, 776–779.

Redenbaugh, K., Nichol, J., Kossler, M. E., & Paasch, B. D. (1984). Encapsulation of somatic embryos for artificial seed production. *In Vitro, 20*, 256–257.

Rihan, H., Kareem, F., El-Mahrouk, M., & Fuller, M. (2017). Artificial seeds: Principle, aspects, and applications. *Agronomy, 7*, 71.

Saha, S., Sengupta, C., & Ghosh, P. (2015). Encapsulation, short-term storage, conservation, and molecular analysis to assess genetic stability in alginate-encapsulated microshoots of *Ocimum kilimandscharicum* Guerke. *Plant Cell, Tissue and Organ Culture, 120*, 519–530.

Saiprasad, G. V. S. (2001). Artificial seeds and their applications. *Resonance, 6*, 39–47.

Salehi, B., Mnayer, D., Özçelik, B., Altin, G., Kasapoğlu, K. N., Daskaya-Dikmen, C., Sharifi-Rad, M., Selamoglu, Z., Acharya, K., Sen, S., Matthews, K. R., Fokou, P. V. T., Sharifi-Rad, F., Setzer, W. N., Martorell, M., & Sharifi-Rad, J. (2018). Plants of the genus *Lavandula*, from farm to pharmacy. *Natural Product Communications, 13*, 1934578X1801301.

Sharma, S., Shahzad, A., & Teixeira da Silva, J. A. (2013). Synseed technology—A complete synthesis. *Biotechnology Advances, 31*, 186–207.

Soundar, R. C., Aslam, A., & Shajahan, A. (2015). High-efficiency direct somatic embryogenesis and plant regeneration from leaf base explants of turmeric *Curcuma longa* L. *Plant Cell, Tissue and Organ Culture, 122*, 79–87.

Tolga, Y., & Yesim, Y. M. (2017). The usage of cryopreservation and synthetic seeds on preservation for plant genetic resources. *International Journal of Cell Science & Molecular Biology, 2*(2), 555583.

Zarei, A., Feyissa, B. A., Davis, B., & Tavakouli Dinani, E. (2022). Cannabis synthetic seeds: An alternative approach for commercial scale of clonal propagation and germplasm conservation. *Plants, 11*, 3186.

CHAPTER 7

Exploring Optimal Seed Storage and Germination Techniques for Medicinal Plants: A Review

PATRUSH LEPCHA,[1,2,*] BIKRAM SUBBA LIMBOO,[2] RAKSHA MUKHIA,[2] DEEPAK CHETTRI,[2] and NIMA TSHERING LEPCHA[3]

[1]*Animal Husbandry and Veterinary Services Department, Government of Sikkim, Krishi Bhawan, Tadong, Sikkim, India*

[2]*Department of Botany, Sikkim University, Tadong, Gangtok, Sikkim, India*

[3]*Department of Botany, Nar Bahadur Bhandari Degree College, Tadong, Gangtok, Sikkim, India*

*Corresponding author

ABSTRACT

Medicinal plants have been an integral part of the human healthcare system for centuries, providing bioactive compounds with therapeutic properties. The diverse range of bioactive compounds present in medicinal plants has led to their extensive utilization in traditional and modern medicine. However, the escalating demand for herbal medicine, overexploitation, habitat destruction, and human-induced biodiversity losses have risked the natural population of these plants, and some of them are on the verge of local extinction. Seeds, as the fundamental dispersal and propagative unit of plants, can serve as the essential source of regeneration for plants in their natural habitat. The storage of the seeds of medicinal plants for a prolonged period is crucial for their conservation and can serve as a source of several noble drugs of pharmaceutical importance in the future. Thus, long-term seed viability retention is necessary to maintain the genetic

Seed Technology for Medicinal and Aromatic Plants: Production, Testing, Storage, and Marketing.
Priyanka Sharma, Kishore Kumar Sharma, and Gopal Shukla (Eds.)

DOI: 10.1201/9781779641199-7

integrity of the stored sample. However, the seeds of the medicinal plants degrade quickly and lose their viability within a short span of time after harvest. Therefore, optimizing an effective seed storage environment and the germination strategies are the first steps in the successful propagation of these plant species in *ex situ* and *in situ* conditions. Specifically, this chapter aims to provide a comprehensive overview of the past, present, and future prospects of different seed storage techniques, the influence of storage conditions on seed viability, strategies for enhancing germination, including studies on some Himalayan medicinal plants, and the importance of maintaining genetic diversity in medicinal plants. Along with the value of ongoing study in this field, this chapter provides implications for conservation efforts and sustainable medicinal plant production.

7.1 INTRODUCTION

Medicinal plants have been an integral part of human civilization for millennia and serve as an important source of bioactive compounds with therapeutic properties. Each of the medicinal plants possesses unique chemical compounds with specific properties that have been utilized for their medicinal value. Currently, 30–50% of drugs used in pharmaceuticals are derived from medicinal plants (Anand et al., 2019). According to the International Union for Conservation of Nature (IUCN) and the World Wildlife Fund, there are about 50,000–80,000 flowering plants that are used for medicinal purposes (Chen et al., 2016). However, due to increasing population, plant consumption, overharvesting, and habitat destruction, nearly 15,000 medicinal plant species are threatened with extinction (Bentley, 2010) and 20% of their wild resources are on the verge of exhaustion (Ross, 2005). Habitat destruction of medicinal plants on a global scale has increased the risk of their extinction in countries like India (Hamilton, 2008; Heywood and Iriondo, 2003), China (Heywood and Iriondo, 2003; Nalawade et al., 2003), Nepal (Hamilton, 2008), Tanzania (Zerabruk and Yirga, 2012), Kenya (Hamilton, 2008), and Uganda (Zerabruk and Yirga, 2012).

Medicinal plant seeds are known to degrade rapidly, losing their viability in a relatively short period following harvest and storage. Consequently, high-quality seeds are essential for both the production and preservation of medicinal plant genetic resources (PGR) (Pradhan and Badola, 2012). During storage, seeds undergo an aging process that ultimately leads to a reduction in quality, rendering them unable to produce viable offspring under unfavorable storage conditions (Harrington, 1972). The negative effects of seed aging are primarily attributed to oxidative mechanisms (Walters et al., 2010), which can cause the deterioration of proteins, including enzymes (Goel et al., 2003), lipids, and subsequently cellular membranes (Harman and Mattick, 1976), RNA (Fleming et al., 2019), and DNA (El-Maarouf-Bouteau et al., 2011). These factors collectively compromise the cellular and metabolic integrity of both seeds and seedlings (Kranner, 2013). Therefore, the retention of long-term seed viability is necessary to uphold the genetic integrity of stored samples (Tiwari et al., 2022).

Seeds are required to be stored in an appropriate storage material (e.g., paper bags, polythene bags, cloth bags, sealed specimen tubes or jars, etc.) under optimal environmental conditions (such as seed moisture content, storage temperature, humidity, and gaseous environment) to avoid low germination and to maintain seed viability for a prolonged period. Storage of seeds in inappropriate conditions such as room temperature (Schmidt, 2000) and moisture (Afzal et al., 2020) naturally causes the loss of seed viability, low seed germination, and seed degradation. Seed moisture content, storage temperature, humidity, and duration of storage are the major factors that influence seed degradation, viability, and germination. For example, a slight increase in storage temperature and humidity may favor fungal (Roberts, 1972) and insect growth in seeds (Christensen, 1972). Overdying and long-term storage of seeds (6–30 months, depending on plant species) may reduce the germination or even cause the death of seeds (Tiwari et al., 2022). However, optimal storage conditions complement the seeds to retain significant viability and high germination over a long period of time (Chen et al., 2007; Pradhan and Badola, 2008b; 2012).

Seed germination is a crucial stage and process in the life cycle of plants that influences plant density and yield (Li et al., 2023; Mohammadi et al., 2023). Germination and seedling emergence are influenced by several factors such as genetics, vigor, seed size, seed skin, burial depth of seed, temperature, light, soil moisture, pH, water stress, and salt stress (Chachalis and Reddy, 2000; Chauhan et al., 2006; Koger et al., 2004; Reed et al., 2022). Therefore, maintaining the high viability (based on tetrazolium or 2,3,5-tryphenyl tetrazolium chloride, TTC test) of seeds during storage is essential for improved seedling establishment and sustainable production (Belmehdi et al., 2018; Finch-Savage, 1995).

Due to the increasing demand for medicinal plants in pharmaceuticals, studies on their seed germination are now gaining worldwide attention for their commercial production (Chandra et al., 2010; Elhindi et al., 2016; Khalaki et al., 2021). Therefore, in this chapter, we have discussed the influence of storage conditions on seed viability, different seed storage techniques, and strategies for enhancing germination that can be utilized in medicinal plants. Further, we have also included some of the studies on Himalayan medicinal plants such as costus or kuth (*Aucklandia costus* Falc.), spikenard or jatamansi (*Nardostachys jatamansi* (D. Don) DC.), chirayita (*Swertia chirayita* (Roxb.) H. Karst.), and Himalayan mayapple (*Podophyllum hexandrum* Royle), along with prioritizing the importance of maintaining genetic diversity in medicinal plants.

7.2 EXPLORING OPTIMAL SEED STORAGE CONDITIONS FOR MEDICINAL PLANTS

Seeds are the most commonly used storage material to maintain the genetic diversity of plants (Trusiak et al., 2023). Storage of seeds as *ex situ* germplasm, as in the National Bureau of PGR, India; the Indian Seed Vault, India; the Svalbard Global Seed Vault, Norwegian island of Spitsbergen; and so forth, (where the seeds of various cultivated plant species and their wild relatives are preserved) is an essential step for the long-term conservation of PGR. Preservation of quality seed stock in storage for the next season is as important as producing quality seeds every year; thus, retaining the potential germinability and vigor of seeds is of major concern under ambient storage conditions.

The seeds of the medicinal plants degrade fast, lose their viability within a short span of time, and show low germination rates and long dormancy. This is because of internal (genetic, physiological, and structural) and external (mainly storage conditions such as humidity and temperature) factors that cause the aging of the stored seeds (Kurek et al., 2019), which leads to the deterioration of the seed metabolic system (Vistis et al., 2020), specifically by the oxidation of crucial biomolecules like proteins, lipids, RNA, DNA, and so forth, eventually resulting in the death of the seeds. Therefore, it is pivotal to develop optimal seed storage conditions and germination strategies for the effective propagation and conservation of the economically important medicinal plants. However, the storability of the seeds is influenced by two major groups of factors: (i) internal or seed factors, such as genetic makeup, growing condition (biotic or abiotic stress or nutrient deficiency) especially during seed maturity, seed quality (maturity, size, and physical damage), moisture content of seeds (MC) during storage, and presence or absence of pathogens and internal infestation by pests; (ii) external or storage factors, such as relative humidity (RH), temperature, gaseous environment, packaging materials, and condition of seed stores (Dadlani et al., 2023; Doijode, 1997). Among all factors, relative humidity, temperature, and gaseous composition of the storage condition are the crucial environmental factors that affect the deterioration of seeds and their viability (Dadlani et al., 2023; Tiwari and Das, 2014; Vitis et al., 2020). However, seeds of different plant species respond to the environment differently before and during storage (Tiwari and Das, 2014). Therefore, we should first experimentally determine the seed behavior and storage factors of the species of interest and then formulate their optimal storage conditions based on their responses to different environmental factors.

Reducing moisture content to certain thresholds increases seed longevity in a specific way for about 90% of the plant species (Roberts, 1973). Thus, one of the crucial issues to focus on before storing seeds is determining their desiccation tolerance, which is defined as the reduction in moisture content of seed to a certain level that retains the seed viability (Trusiak et al., 2023). Based on their resistance to desiccation, seeds are grouped into three categories: orthodox seeds (tolerate desiccation up to 5% moisture level) (Roberts, 1972), recalcitrant seeds (do not tolerate drying below 20–40% of moisture content) (Baskin et al., 2006; Hong and Ellis, 1996), and intermediate seeds (tolerate desiccation to a moisture content of 10%) (Ellis et al., 1990). The viability of orthodox seed can be maintained in *ex situ* conditions for a long duration of time under an optimal environment (Tiwari and Das, 2014) and withstand freezing temperatures (e.g., rice – *Oryza sativa* L., maize – *Zea mays* L., etc.) (Parimala et al., 2013). Intermediate seeds can also be stored for a long duration, but they cannot tolerate low temperatures below 10°C (e.g., papaya – *Carica papaya* L., lemon – *Citrus limon* (L.) Osbeck, Legumes, etc.), and recalcitrant seeds are storable only for a short span of time as they do not enter dormancy but consecutively enter germination (e.g., mango – *Mangifera indica* L., jackfruit – *Artocarpus heterophyllus* Lam, etc.) (Parimala et al., 2013). Therefore, determining the desiccation tolerance of the seeds of the medicinal plants will assist us in choosing storage for long duration or short duration and normal temperature or freezing temperature.

Once the desiccation is known, the next step is to determine the appropriate temperature for storage. According to "the rules of thumb," for every 1% decrease in moisture content of seed (between 5% and 14%) and for every 5°C reduction in the storage temperature (between 0°C and 50°C), the life span of seeds doubles, and the optimum seed storage condition is achieved when the sum of the values of relative humidity (RH; in%) and temperature (≤ 50°F) in the storage environment should not exceed 100 (Harrington, 1960 and 1972). Thus, orthodox seed storage conditions require low moisture content (3–7%), an airtight container, and a low temperature below 5°C (Vitis et al., 2020). For short-term storage (<18 months) and long-term storage (more than 18 months), temperatures of 0°C to 5°C and −18°C to −20°C, respectively, are sufficient to maintain the viability of dry seeds (Hong and Ellis, 1996). To maintain the viability of recalcitrant seeds, the seed of the plant needs to be slightly dry (moisture content between 23 and 61.5%), the storage environment should have free availability of oxygen, and storage temperatures should be 7°C to 17°C for tropical origin, and between −3°C and 5°C for temperate species (Hong et al., 1996). Polyethene bags with 0.075–1.0 mm (3–4 mils) thickness can be beneficial storage materials as they allow gaseous exchange, such as water vapor transmission rates for seeds (Bonner and Karrfalt, 2008). The optimal seed moisture content depends on the storage temperature, and it decreases as the temperature rises. Therefore, RH of 20–25% is considered optimal for seed storage at an ambient temperature (Vertucci et al., 1994), and for cryogenic seed storage (-150°C), the RH should be about 70% (Vertucci and Ross, 1990). Afzal et al. (2020) reported that the moisture condition during storage of *Moringa oleifera* Lam. (drumstick tree) seeds is a major factor in their deterioration and suggested that seed storage in hermetically sealed superbugs with 8% seed moisture content is highly preferable. In *Desmodium gangeticum* (L.) DC. (commonly known as Salparni), seed storage with 8% moisture level, packed in polythene bags, and kept in a refrigerated environment ensured high germination (Gayathri and Kanakamany, 2018). Nevertheless, if desiccation tolerance (seed moisture content), optimal temperature, and RH of storage are not available, then it needs to be optimized experimentally. Various studies aiming to determine storage conditions and packaging material and their ability to retain a high germination rate in some medicinal plants are provided in Table 7.1.

TABLE 7.1 Effect of Seed Storage Condition and Packaging Material on Seed Germination of Some Medicinal Plants

Medicinal plant	Storage Condition	Storage Material	Germination	Duration	Source
Abrus precatorius L., (jequirity bean or rosary pea)	25 ± 3°C	Paper bags	83%	9 months	Tiwari and Das (2014)
Plumbago zelaynica L., (Ceylon leadwort, doctorbush or wild leadwort)	4°C	Cloth bags and paper bags	80%	9 months	Tiwari and Das (2014)
Baccharoides anthelmintica (L.) Moench, (ironweed)	25 ± 3°C	Paper bags	83%	9 months	Tiwari and Das (2014)
Swertia chirayita (Roxb. exFleming) (chirayita or chiretta)	4°C	Sealed specimen tube	50%	24 months	Pradhan and Badola (2012)
Aucklandia costus Falc., (costus or kuth)	25 ± 3°C	Airtight-sealed polyethylene jars	82%	30 months	Sharma et al. (2014)
Nardostachys jatamansi (D. Don) DC. (spikenard or jatamansi)	0–4°C	Airtight polythene bags	63%	24 months	Chauhan and Nautiyal (2007)

7.2.1 FUTURE PROSPECTIVE

Developing seed storage techniques by reducing moisture content and storage temperature has many constraints. Other new approaches have been investigated to develop optimal storage conditions. One such effort is the evaluation of the modulation of gases (oxygen) present in the storage environment, as the aging process of seeds is enhanced by the availability of oxygen in the storage condition (Groot et al., 2014). Thus, seed storage at low concentrations of oxygen has been investigated as a technique to enhance seed viability (Trusiak et al., 2023). Groot et al. (2012) demonstrated that the aging of the dry seeds of plants such as barley (*Hordeum vulgare* L.), cabbage (*Brassica oleracea* L.), lettuce (*Lactuca sativa* L.), and soybean (*Glycine max* (L.) Merr.) was accelerated when stored under high-pressure oxygen, and eventually the seeds deteriorated. The presence of an accelerated level of oxygen during seed storage causes the accumulation of DNA oxidation, resulting in chromosomal aberrations during cell division (Abdalla and Roberts, 1968), which impacts the viability and germination ability of the seeds. In the follow-up study, Groot et al. (2014) reported that anoxia and low moisture conditions improved the seed viability of celery and celeriac (different crop types of the species *Apium graveolens* L.) seeds. Such studies have suggested either a positive or neutral influence of limited oxygen in storage conditions on seed longevity (Groot et al., 2014; Roberts and Ellis, 1989), indicating that limited oxygen conditions improve seed quality only under certain combinations of temperature and moisture content (Trusiak et al., 2023). Recent studies such as the use of oxygen absorbers and/or silica gel to modulate the gaseous environment (Han et al., 2021) and the beneficial effects of nitrogen (Monchini et al., 2020) in storage conditions have the potential to improve seed longevity during storage. Nonetheless, new methods for improving seed storage, including the invigoration of seeds or the use of nanoparticles (NPs) such as zinc oxide (ZnO-NPs), titanium dioxide (TiO_2-NPs), calcium oxide (CaO-NPs), and so forth, to protect seeds from microbial infections and preventing destructive physiological processes by reactive oxygen species (ROS) scavenging activity (Adetunji et al., 2021; Hu et al., 2020; Xia et al., 2020) appeared to be very promising. These emerging seed storage conditions and techniques can be applicable to medicinal plants; however, more investigation is required to support this statement.

7.3 SEED PACKAGING

Seed storage materials greatly influence the seed germination of medicinal plants (Tiwari and Das, 2014; Tiwari et al., 2022). Therefore, after drying and achieving an optimum level of moisture in the seed, it has to be packed in containers or packaging materials (e.g., paper bags, polythene bags, cloth bags, sealed specimen tubes or jars, etc.) that prevent the moisture content of the seeds from exceeding the safe level during storage. A characteristic of packaging materials is their ability to resist the permeability of vapor and gases through them (Dadlani et al., 2023). Moisture-proof packs, that is, packs with a low moisture penetration rate, composed of polyvinyl, polyethene, cellophane, polyester, aluminum foil, and so forth, are generally used as seed packaging material. The ultra-hermetic UV-resistant cocoon composed of polyvinyl chloride (GrainPro®) has low permeability to air and moisture and prevents considerable insect activity (Dadlani et al., 2023).

Several studies have reported high germination of seed stored in moisture-proof storage materials (Khalequzzaman et al., 2012; Rahman and Rahman, 1997; Saisanthosh and Patil, 2018) because low moisture contributes to reducing the respiration rate of the seeds during storage (Doijode, 2007). However, in some species, porous storage materials like cloth bags also demonstrated high germination after storage, indicating a need to identify species-specific storage materials for higher germinability of stored seeds (Netra et al., 2015; Saisanthosh and Patil, 2018). The selection of suitable materials for seed storage has been advocated by several medicinal plants, and some of the examples are depicted in Table 7.1.

7.4 EXPLORING OPTIMAL SEED STORAGE TECHNIQUES FOR MEDICINAL PLANTS

Seed storage is a notable process that plays a vital role in maintaining the viability and vitality of seeds throughout a specified duration of storage. Therefore, studying the various seed preservation techniques is beneficial for conserving genetic resources of different species and contributes to rescuing a species from extinction (de Oliveira et al., 2021). Seed moisture, storage environment, and type of storage container are the important factors that determine seeds' longevity during storage. Interaction of these factors generally leads to physiological and biochemical changes in the stored seeds, resulting in deterioration of quality and quantity of seeds, especially in countries located in tropical and sub-tropical regions (Dadlani et al., 2023). Different storage techniques (indigenous or modern) can be used depending upon the duration of storage (Parimala et al., 2013). Indigenous or traditional storage techniques include gourd casing, kuthir, kodambae, thombarai, mara thombai, earthen pots, Kalangiyam, puri, gunny bags, and so forth, while modern techniques include pusa bins, storage in metal drums, storage in metal bins, pucca kothi, gharelu thekka, and reinforced cement concrete ring bins (Parimala et al., 2013). According to Dadlani et al. (2023), based on the purpose and seed material, different storage techniques or structures are used to maintain seed quality for commercial purposes.

7.4.1 STORAGE WITH OR WITHOUT VENTILATION

In temperate climates, seeds can be stored with regulated natural ventilation, whereas in warm and humid climatic conditions, natural ventilation could be detrimental (Copeland and McDonald, 2001). However, measures must be taken to protect seeds from insects, rodents, and birds. This type of unconditioned storage technique is preferred for short storage durations (6–8 months) from harvest until the next cropping season.

7.4.2 STORAGE WITH REGULATED MOISTURE

This type of storage is suitable in a temperate climate where the temperature remains low all the time. Seed moisture is regulated either by packing dried and low-moisture-containing seeds (6–8%) in containers or bags impermeable to vapor or by packing seeds in sacks or bags permeable to vapor and utilizing a dehumidifier. For warmer regions, in-the-bag desiccants and dehumidifiers are used during the wet season to maintain storage RH < 70%, which maintains seed quality at least for one season.

7.4.3 STORAGE WITH REGULATED RH AND TEMPERATURE

This type of storage technique is generally used to store bulk seeds for 2–3 years, where 50% RH and a temperature of 20°C are maintained for seed storage (Copeland and McDonald, 2001). This technique is commonly used for vegetables (hybrids or small seeds) and flower seeds. It requires a continuous power supply to maintain a consistent temperature and RH.

7.4.4 MOIST-COLD STORAGE WITH REGULATED TEMPERATURE

Recalcitrant seeds with high moisture content (>20–40%) can be stored at temperatures just above freezing point, especially during research. The storage condition should have near saturation RH, and periodic moistening of the seed is required.

7.4.5 STORAGE IN LOW-TEMPERATURE

In warmer regions, low-temperature storage techniques are often used in which seeds are dried to low MC, packed in hermetically sealed containers (impermeable to moisture or vapor), and stored in low temperature.

The storage containers should be airtight, but vacuum-sealed is not preferred in order to avoid seed damage (Robertson and Labate, 2006). Such conditions generally reduce the metabolic activity of cells, which enables the preservation of seeds (Kulus, 2019). About 5–10°C at reduced humidity is preferred for short-term storage of seeds, and freezing temperatures (–20 or –80°C) for mid-term storage (Kulus, 2019).

7.4.6 CRYOPRESERVATION

Cryopreservation, that is, storage in liquid nitrogen (–196°C) or cold nitrogen vapor (–150°C), is an effective long-term seed storage technique that can be utilized to conserve genetic material and requires minimal space (de Oliveira et al., 2021). Vegetative parts, such as buds, shoot tips, pollen grains, calli, and seeds, are preserved in cryopreservation techniques (Dinato et al., 2020; Jenderek and Reed, 2017; Souza et al., 2016; Wang et al., 2020). The ultra-low temperatures minimize the rate of cellular metabolism and pause tissue development; however, regeneration is still allowed when the optimal temperature is restored (Jenderek and Reed, 2017; Panis and Lambardi, 2006). In the case of seeds, the most crucial factor required for cryopreservation is the optimal moisture content, which contributes to maintaining the seed viability after freezing and thawing and thus affects the final rate of seed germination (Generoso et al., 2019; Silva et al., 2021; Vettorazzi et al., 2019). Therefore, it is necessary to determine the desiccation tolerance that could support the cryopreservation of the plant seeds. de Oliveira et al. (2021) implemented the cryopreservation technique to conserve the seeds of the 20 *Tillandsia* species by desiccating the seeds to a water content of about 7% (dry mass basis).

7.5 EXPLORING SEED GERMINATION TECHNIQUES FOR MEDICINAL PLANTS

The seed's ability to germinate remains the most important factor in maintaining the optimum population of the plants. The interaction between environmental factors (e.g., temperature, light, and water) and plant growth regulators or phytohormones (e.g., gibberellic acid (GA_3), abscisic acid, ethylene, etc.) plays a vital role in dormancy and germination (Nautiyal et al., 2023). However, the germination of seeds and the establishment of seedlings of medicinal plants in the field are difficult due to dormancy, low seed viability, limited emergence, and fast deterioration.

Therefore, the development of seed enhancement techniques (i.e., physical or chemical treatments used on seeds to promote germination) is crucial for conservation and commercial production. Various physical, chemical, or physiological treatments can be used to release the dormancy and enhance germination (Table 7.2).

Scarification is the most frequently used method to break seed dormancy in seeds with a hard coat (Baskin and Baskin, 2004). In this process, scars are created on the seed coat to permit the entry of water into the seed, which hydrates the embryo (Nautiyal et al., 2023), thereby allowing germination. Seeds are scarified either by physical or chemical treatments. In physical treatment, heating (Probert, 2000), boiling (Li et al., 1999), chilling (Baskin and Baskin, 2004), or mechanical abrasion (Morrison et al., 1998) are used to elevate the hardness of the seeds. In chemical treatment, they are treated with concentrated H_2SO_4, or a strong alkali, like NaOH (20%), for an appropriate timing (Nautiyal et al., 2023). Further, techniques such as seed priming, seed coating, and seed conditioning are used to enhance the germination and vigor of germinating seedlings, which augments the yield of the plants. In priming, seeds are hydrated to a level that allows for pre-germinative metabolic activity but inhibits the actual emergence of the radicle (TeKrony, 2016). Depending upon the solvent used for hydration of the seeds, the priming can be hydro-priming (use of water for seed hydration), osmo-priming (use of osmotic solution, e.g., PEG, NaCl, KCl, KNO_3, K_3PO_4, etc.), solid matrix-priming (use of moist sand), chemo- or hormonal-priming (use of chemicals, e.g., tocopherol, melatonin, hydrogen peroxide, mannose, fungicide, etc., or phytohormones, e.g., ABA, auxins, gibberellins (Figure 7.1), kinetin, ethylene, etc.).

TABLE 7.2 Effect of Different Treatments on Germination of Some Medicinal Plants

Medicinal Plant	Treatment	Germination/ Survival Rate	Sources
Aucklandia costus Falc. (costus or kuth)	Chilling, GA_3 and KNO_3	83%	Bano et al. (2018); Sharma et al. (2014, 2006)
Carum carvi L. (Caraway, meridian fennel or Persian cumin)	Chilling and GA_3	72%	Sharma et al. (2006)
Inula racemosa Hook. f. (Pushkarmool or pushkarmula)	Chilling and GA_3	80%	Sharma et al. (2006)

TABLE 7.2 *(Continued)*

Medicinal Plant	Treatment	Germination/ Survival Rate	Sources
Paris polyphylla Smith (paris or satuwa)	Combination of low temperature treatment and GA_3	95%	Chen et al. (2007); Zhou et al. (2003)
Podophyllum hexandrum Royle (Himalayan Mayapple)	GA_3, combination of H_2SO4/GA_3 and GA_3/BA (6-benzyladenine)	84%	Kumar et al. (2017); Nadeem et al. (2000); Peng et al. (2023); Sharma et al. (2006)
Rheum austral D. Don. (Himalayan Rhubarb or red-veined pie plant)	GA_3 and KNO_3	88%	Sharma et al. (2006)
Picrorhiza kurrooa Royle ex Benth. (Picrorhiza, katuka or kutki)	GA_3, BAP (6-benzylaminopurine) or a combination of both	83%	Chandra et al. (2006)
Aconitum heterophyllum Wall. (aconite, Devil's helmet, wolfsbane, Atis or Ativisha, etc.)	KNO_3, IAA, Water soaking, GA_3	43%	Pandey et al. (2000); Srivastava et al. (2011)

FIGURE 7.1 Germination of seeds of some medicinal plants of Sikkim Himalaya in different concentrations of GA_3 (A) *Tetradium fraxinifolium* (Hook.) T.G. Hartley; common name: Euodia, Evodia, or Bee bee tree (B) *Oroxylum indicum* (L.) Benth. ex Kurz; common name: Indian trumpet flower, midnight horror, broken bones plant.

Plant growth regulators, or plant hormones or phytohormones, are the chemicals synthesized by plants that are essential for the germination, growth, and development of the plants (Table 7.3). Priming of seeds with an optimal concentration of phytohormones such as gibberellins/gibberellic acids (GAs/ GA_3), auxins (IAAs), cytokinin's (CKs), ABA, ethylene (ET), and salicylic acid (SA) is known as hormonal priming, which is a common technique used to improve seed germination, seedling growth, and yield of the plants.

TABLE 7.3 Commonly Used Phytohormone in Seed Priming and Their Role in Seed Germination

Regulators/ Effectors	Enhancements
Gibberellin/ Gibberelic acids (GAs/ GA_3)	GA_3 has a positive influence on seed germination, flowering initiation, stem elongation, and flower and fruit development (Richards et al., 2001; Yamaguchi, 2008). It stimulates the synthesis of hydrolytic enzymes, especially α-amylase in seeds resulting in the seed germination (Bhardwaj et al., 2016; Chen et al., 2008; Wciślińska, 1977).
Cytokinins (CKs)	CKs alleviate the stresses like drought, heavy metals salinity, and oxidative stress, and thus, enhance the seed germination (Atici et al., 2005; Khan and Ungar, 1997; Nikolic et al., 2006, Peleg and Blumwald, 2011).
Ethylene (ET)	ET affects tissue growth and development, and seed germination, however, knowledge of its effect on germination is obscure (Miransari and Smith, 2014). The concentration of ET rises during the seed germination of many plants like rice, wheat, corn, and soybean, and influences the rate of seed germination (Pennazio and Roggero, 1991; Zapata et al., 2004). It also promotes seed germination by stimulating the synthesis of GAs in certain species (Kumar et al., 2022).
Auxin (IAA, NAA)	Although, auxin itself is not necessarily required for seed germination, however, it is present on the tip of the radical of a seed during and after germination (Miransari and Smith, 2014). Auxin is required for the growth of young seedlings (Bialek et al., 1992, Hentrich et al., 2013). Low concentration of auxins can promote seed germination in certain plant species (Jan et al., 2010).
Brasinosteriods (BRs)	BRs controls the inhibitory effects of the ABA on germination, (Finkelstein et al., 2008, Zhang et al., 2009a, b) and contribute to the growth of embryo (Leubner-Metzger, 2001), thus, enhances the seed germination. Promotes seed germination in certain species (Asci et al., 2019; Kaur et al., 2009)
Abscisic acid (ABA)	Priming of seeds with ABA enhances germination and tolerance to variety of stresses by different crops (Gurmani et al., 2013).
Salicylic acid (SA)	SA enhances the productivity of maize (*Zea mays* L.) under cold stress, as well as the growth germination of the garden cress (*Lepidium sativum* L.) under salinity stress (Habibi and Abdoli, 2013).

7.5.1 FUTURE PERSPECTIVE

Recently, new seed germination enhancement techniques and technologies have emerged with promising implications. Techniques such as pulse electric field processing (PEF) have been recently used for germination enhancement, where the electric field is employed to change the physiology of the seeds and improve water absorption through dipole-dipole interaction (Songnuan and Kirawanich, 2012). The use of UV rays (UV-A and UV-C), which can easily penetrate plant tissue, is also beneficial to enhance the germination of seeds and can also elevate the vigor of the seeds (Neelamegam and Sutha, 2015). Similarly, the utilization of non-thermal plasma (NTP) to break dormancy, enhance germination, and increase growth rate has also been reported (Puligundla et al., 2017). Treating seeds with magnetic fields also improves germination, as magnetic fields elevate the level of alpha-amylase in the seeds, which in turn enhances seed germination (Afzal et al., 2012). Ozone treatment of the seeds reduces the amount of ABA in the seeds, thereby helping to escape its inhibitory effect, which results in the breaking down of the seed dormancy (Sudhakar et al., 2008). Ultrasound treatment of the seeds of the soybeans increased activities of amylase, lipase, protease, and peroxidase, thus, improved the germinability of soybean (*Glycine max* (L.) Merr.) seeds (Chen et al., 2023). These are some of the crucial emerging germination enhancement technologies that need to be implemented in medicinal plants for research, conservation, and commercial production.

7.6 STUDIES ON HIMALAYAN MEDICINAL PLANTS

The Himalayan region is known for its rich biodiversity due to its diverse range of climate, elevation, and soil conditions; therefore, it harbors a huge diversity of economically important medicinal plant species. The Indian Himalayan Region (IHR) alone has approximately 8000 species of angiosperms, 600 species of pteridophytes, and 44 species of gymnosperms (Singh et al., 1996), of which 1748 species are known to possess medicinal properties (Samant et al., 1998). However, due to their commercial implications, these medicinal plants are under severe threat of illegal collection, overexploitation, habitat fragmentation, and climate change, which can endanger their natural population. Also, regeneration of medicinal plants in their natural habitat is slow because their seeds degrade fast, lose their viability within a short span of time, show a low germination rate, and have a long dormancy. Therefore, at present, medicinal plants are highly prioritized for conservation because future trade in pharmaceutical industries solely depends on the natural abundance and diversity of plants in the wild (Mehta et al., 2020). Several studies have been conducted to develop optimal seed germination and seed techniques for *ex situ* conservation and cultivation of many threatened, endangered, and critically endangered medicinal plant species. We have summarized the studies on four commercially valued but IUCN red-listed medicinal plant species of the Himalayan region in the following section.

7.6.1 NARDOSTACHYS JATAMANSI (D. DON) DC.

N. jatamansi (Family - Caprifoliaceae) is a commercially important rhizomatous medicinal herb distributed in the alpine region of the Himalayas in India (Sikkim, Himachal Pradesh, Uttarakhand, and Arunachal Pradesh), Nepal, Bhutan, China, and Myanmar, at an elevation ranging from 2200 to 5000 m asl (Ghimire et al., 2005; Ved et al., 2015). The roots and rhizomes of *N. jatamansi* contain essential oils (coumarins and sesquiterpenes), which exert different medicinal properties (Liu and Liu, 2014; Mishra et al., 1995; Nautiyal, 2013; Rekha et al., 2013; Singh et al., 2015). Due to its wide utilization in traditional medicine, overexploitation, small reproductive phase, and low germination rate (10–20%) (Nautiyal et al., 2013), it is listed as critically endangered on the IUCN Red List of Threatened Species (IUCN, 2020). Chauhan and Nautiyal (2007) suggested that germination of *N. jatamansi* seeds can be enhanced with optimum temperature, photoperiod, hormonal treatment, soil composition, sowing months, and depth of seed sowing in soil. According to their study, *N. jatamansi* seeds stored in airtight polythene bags in the refrigerator (0–4°C) retain its viability twice as much as those stored at room temperature (10–35°C). About 15°C temperature with continuous light and alternate temperature (25°C in light and 10°C in dark for 12 h each) favors the high germination of *N. jatamansi* seeds in laboratory conditions, and treatment with GA_3 produces higher germination than treatment with other hormones (Chauhan and Nautiyal, 2007). Seed-sowing depth of 0.5 cm with a 1:1:1 and 1:2:1 proportion of soil, sand,

and FYM mixture is favorable for germination at middle altitude (1800 m) during October and February and during May at high altitudes (3600 m) (Chauhan and Nautiyal, 2007). In the plant biotechnological context, only a few protocols have been developed for *in vitro* regeneration in *N. jatamansi* with different explants (roots, leaves, petioles, and shoot tips) (Bose et al., 2016; Mathur, 1992, 1993; Rawat et al., 2018). However, much work needs to be done to develop efficient somatic embryogenesis methodology and synthetic seed technology that could be beneficial for the commercialization and conservation of critically endangered high-value medicinal plants like *N. jatamansi.*

7.6.2 SWERTIA CHIRAYITA (ROXB.) H. KARST

S. chirayita (Family – Gentianaceae) is a highly marketed medicinal herb distributed in temperate regions of the Himalaya at elevations between 1200 m and 3000 m asl, from Kashmir (India) to Bhutan (Baskey et al., 2016; Pradhan and Badola, 2012). It is effective in treating several aliments such as fever, skin diseases, intestinal worms, cold, cough, diarrhea, stomachache, gastric ulcers, liver diseases, and so forth, and also has anti-inflammatory, antibacterial, antifungal, hypoglycemic, antimalarial, and antioxidant properties (Joshi and Dhawan, 2005; Pradhan and Badola, 2008). This species is used to produce herbal drugs like *Himoliv, Diabecon*, and D-400 (Pradhan and Badola, 2012). *S. chirayita* is critically endangered (Joshi and Dhawan, 2005) and has a low seed viability and germination rate (Badola and Pal, 2002; Chandra et al., 2012). Therefore, appropriate storage techniques that maintain seed viability for a long period of time and methods that could produce a high germination rate are critical for its mass propagation (*in situ* or *ex situ*). According to Pradhan and Badola (2012), storage of *S. chirayita* seeds with 10–40% moisture content at 4°C in a sealed specimen tube is an effective storage condition to retain seed viability for at least two years. Their study also suggested that seed storage rooms (25 ± 5°C) and −15°C temperatures can also be useful for short-term storage up to one year. Among the various plant hormones used in seed germination studies (GA_3, IAA, and KNO_3), GA_3 treatment showed the best result in *S. chirayita* seeds, reducing the mean germination time (MGT) by up to 8 days (Baskey et al., 2016). Synthetic seed technology, an applied modern plant biotechnology, can also provide an opportunity for micropropagation and germplasm conservation through cryopreservation (Ara et al., 2000; Gantait et al., 2015; Perveen and Anis, 2014; Sharma et al., 2013a). The micropropagation of *S. chirayita* through tissue culture has the potential to reduce pressure on the natural population and augment the conservation of *S. chirayita*. Synthetic seed production and plant regeneration techniques from somatic embryos have also been developed in *S. chirayita* (Kumar et al., 2014). However, more detail and intense studies are needed to improve such technologies that can be potentially used for commercial purposes.

7.6.3 *PODOPHYLLUM HEXANDRUM* ROYLE

P. hexandrum (Synonym – *Sinopodophyllum hexandrum* (Royle) T. S. Ying, Family – Berberidaceae) is distributed in the Himalayan areas (alpine region) of India, Nepal, Bhutan, Afghanistan, Pakistan, and south China (Airi et al., 1997; Choudhary et al., 1998) at elevations between 3000–4200 m asl (Nag et al., 2020). It is a recognized medicinal plant in traditional Indian, Ayurveda, Chinese, Unani, Siddha, and Tibetan medicine, used in the treatment of cancer, condylomas, and warts (Gowdey et al., 1995; Wong et al., 2000) and used to treat various ailments such as cancer, condylomas, and warts (Anand et al., 2022). It can also be utilized to treat radiation-induced cell and tissue damage (Verma and Gupta, 2015). A resin called podophyllin is an important chemical constituent of *P. hexandrum* which contains several lignans that possess pharmacological properties (Anand et al., 2022). This plant has been overexploited due to commercialization (Nag et al., 2020), which has resulted in habitat fragmentation (Young et al., 1996). Due to overexploitation, overgrazing, long dormancy, and a low rate of natural regeneration, *P. hexandrum* is listed as an endangered species (Kala, 2005) and at risk of extinction (Chaurasia et al., 2012). Therefore, it is urgently necessary to develop and implement conservation (*in situ* and *ex situ*) and regeneration (natural and *in vitro*) strategies for this commercially valuable medicinal plant for its sustainable utilization. Seeds of *P. hexandrum* remain viable for 30 months (based on TTC reduction), but show a low germination rate of 20–32% and a long MGT of 46–60 days (Shama et al., 2010). Sodium hypochloride treatment (Nadeem et al., 2000), GA_3 treatment (Kharkwal et al., 2008; Kumar et al., 2017; Nadeem et al., 2000; Peng et al., 2023; Sharma et al., 2006, 2010), and combination of GA_3 with BA (6-benzyladenine) (Nadeem et al., 2000) and sulfuric acids (Sharma et al., 2006, 2010) are some of the effective methods to break dormancy and enhance germination in *P. hexandrum* seeds. Continuous light and temperatures (20, 25, and 25/15°C) and cold stratification (3 and 5 months) can also improve its germination (Peng et al., 2023). However, all these germination treatments become ineffective and the germination rate declines with increasing storage periods (Sharma et al., 2010). Therefore, obtaining an optimal long-term storage strategy for *P. hexandrum* is crucial for *ex situ* production and *in situ* conservation. To minimize the pressure on the natural population, several biotechnological techniques have been utilized in *P. hexandrum* such as *in vitro* podophyllotoxin production (Chattopadhyay et al., 2001, 2002; Giri and Narasu, 2000; Lin et al., 2003), production of lignans (Ahmad et al., 2007), micropropagation (Chakraborty et al., 2010; Gou et al., 2012; Dhiman et al., 2016; Sultan et al., 2006), and embryogenesis (Rajesh et al., 2014a, b; Sultan et al., 2009). Synthetic seed development techniques can also be further explored, which will be beneficial in terms of sustainable production and utilization of *P. hexandrum*.

7.6.4 AUCKLANDIA COSTUS FALC.

A. costus (Synonym – *Saussurea costus* (Falc.) Lipsch., Family – Asteraceae) is an economically important and critically endangered medicinal plant species native to the Himalayan region. It is found in India (Kashmir, Ladakh, Himachal Pradesh, and Uttarakhand) (Khan et al., 2021; Shah, 2019; Singh et al., 1957) and China (Xinzhu Village, Ludian Town, Yulong County, and Yunnan Province) (Kumari et al., 2023) in elevations ranging between 2800 and 4000 m asl. Traditionally, it is used to treat various ailments such as ulcers, stomach problems, asthma, and inflammatory diseases (Bano et al., 2018). It has musk-scented roots, which have medicinal properties with antimicrobial, antifungal, antiulcer, anthelmintic, antidiabetic, antihepatotoxic, immunostimulant, and anti-inflammatory characteristics (Shah, 2019). The essential oil (costus oil) derived from the roots of *A. costus* is an important and costly drug in the international market (Vishvamitera et al., 2023) that is used to prepare high-grade perfume, hair oil, insect repellent, and incense (Rathore et al., 2021). *A. costus* is a critically endangered species (IUCN, 2020) and among the 37 Himalayan endangered medicinal plants that are prioritized for both *in situ* and *ex situ* conservation in India (Sharma et al., 2014). Therefore, understanding the storage-dependent physiological modifications of *A. costus* seed is crucial for its cultivation and *ex situ* conservation. According to Sharma et al. (2014), seed stored in airtight-sealed polyethylene jars at room temperature (average: 20 ± 3°C) can retain 95% viability for at least 18 months and 82% viability for 30 months. Seed stored in hermetic containers at –35°C with 6 ± 1% of seed moisture content also depicted a similar result (Butola and Samant, 2006). The fresh seeds of *A. costus* exhibit 65% germination with a MGT of 7 days (Sharma et al., 2014). Chilling treatment (4°C for 30 days) enhances the germination by 25% and reduces MGT to 4 days, and GA_3 (0.1 and 1 mM) treatment for 24 hours increases germination to 80 and 90%, respectively, compared to 65% germination in the controlled set (Sharma et al., 2014). Bano et al. (2018), Mohi-ud-din et al. (2005), and Sharma et al. (2006), and also suggested that chilled and GA_3 treatments are the best methods to achieve high seed germination in *A. costus.* Temperature regimes of 20 and 22°C can also enhance germination up to 83 and 75%, respectively (Shankar, 2015). Micropropagation techniques have also been developed in *A. costus,* such as direct shoot generation from shoot apices (Johnson et al., 1997) and *in vitro* callus induction (Khan et al., 2021; Warghat et al., 2016). However, somatic embryo production and synthetic seed development can also strengthen the *ex situ* production and conservation of *A. costus*.

7.7 CONSERVATION OF GENETIC DIVERSITY IN MEDICINAL PLANTS

The demand for medicinal plants has increased drastically throughout the world due to the resurgence of interest in herbal medicines, which are known to have very few or no side effects (Sewell and Rafieian-Kopaei, 2014). They make a substantial contribution to the advancement of life-saving drugs, emphasizing their potential for future findings and providing viable and economically efficient healthcare remedies, particularly in areas with limited resources (Fabricant and Farnsworth, 2001; UNEP, 2015). The trade of medicinal plants serves as a crucial means of generating revenue and facilitating employment, especially within rural regions (Bodeker et al., 2014). However, this high demand for medicinal plants has led to overexploitation, illegal collection, and habitat destruction. About 15,000 medicinal plant species are threatened with extinction (Bentley, 2010), and 20% of their wild resources are on the verge of exhaustion (Ross, 2005). Therefore, the conservation and sustainable utilization of medicinal plants becomes an essential obligation for us to maintain a natural source of drugs for the future of the pharmaceutical industry. Storage or preservation of seeds as *ex situ* germplasm is an essential step for the long-term conservation of PGR, especially of medicinal plants. However, presently, unlike cultivated vegetables and crop species, seed banks or germplasm conservation institutes for medicinal plants are not available. As mentioned above, exploring different storage conditions, storage materials, germination enhancement techniques, and so forth, becomes a crucial part of conserving medicinal plants having dormancy, a low germination rate, and a short storage duration.

7.8 CONCLUSION

Medicinal plants are the source of several bioactive compounds with therapeutic properties that are used in indigenous, traditional, and modern medical systems. However, the population of these plant species in their natural habitat is declining alarmingly due to external factors such as over-exploitation, habitat destruction, and climate change, as well as internal factors such as dormancy, low germination, and the less durable nature of the seeds. Due to these factors, some of the commercially important medicinal plant species are either threatened or on the verge of extinction. Therefore, it is crucial to conserve such invaluable genetic resources that can serve as the sources of several noble drugs of pharmaceutical importance in the future. Different plant species respond to the environmental conditions of storage and germination in different ways due to the differences in their genetic, physiological, and morphological makeup. Therefore, the storage and germination conditions of most of the plants are species-specific. Temperature, moisture content of the seeds, relative humidity of the seeds, and the presence or absence of gases in the storage conditions play an important role in maintaining the quality of the seeds for a long period of time. Different scarification methods, such as heat treatment, chilling temperature, H_2SO_4 treatment, and so forth, are useful in releasing the dormancy of many plants and enhancing germination.

Priming seeds with different chemicals (KNO_3) and phytohormones (GA_3) is also useful to enhance seed germination in medicinal plants. The use of oxygen-absorbing substances (silica gel), nitrogen, and nanoparticles in storage conditions and the use of recent seed enhancement technologies such as magnetic field, ultrasound, UV rays, ozone, NTP, and so forth, can be implemented in medicinal plants to enhance seed germination, storage for longer durations, and conservation of commercially important but vanishing medicinal plants.

KEYWORDS

- **conservation**
- **genetic resources**
- **medicinal plants**
- **seed germination**
- **seed priming**
- **seed storage**
- **sustainable utilization**

REFERENCES

Abdalla, F. H., & Roberts, E. H. (1968). Effects of temperature, moisture, and oxygen on the induction of chromosome damage in seeds of barley, broad beans, and peas during storage. *Annals of Botany, 32*(1), 119–136. https://doi.org/10.1093/oxfordjournals.aob.a084187 (accessed on 12th August 2025).

Adetunji, A. E., Adetunji, T. L., Varghese, B., Sershen., & Pammenter, N. W. (2021). Oxidative stress, ageing and methods of seed invigoration: An overview and perspectives. *Agronomy, 11*(12), 2369. https://doi.org/10.3390/agronomy11122369 (accessed on 12th August 2025).

Afzal, I., Jaffar, I., Zahid, S., Ur Rehman, H., & Basr, S. M. A. B. (2020). Physiological and biochemical changes during hermetic storage of *Moringa oleifera* seeds. *South African Journal of Botany, 129*(3), 435–441. https://doi.org/10.1016/j.sajb.2019.11.011 (accessed on 12th August 2025).

Afzal, I., Mukhtar, K., Qasim, M., Basra, S. M. A., Shahid, M., & Haq, Z. U. (2012). Magnetic stimulation of marigold seed. *International Agrophysics, 26*(4), 335–339. https://doi.org/10.2478/v10247-012-0047-1 (accessed on 12th August 2025).

Ahmad, R., Sharma, V. K., Rai, A. K., Shivananda, R. D., & Shivananda, B. G. (2007). Production of lignans in callus culture of *Podophyllum hexandrum*. *Tropical Journal of Pharmaceutical Research, 6*(4), 803–808. https://doi.org/10.4314/tjpr.v6i4.14661 (accessed on 12th August 2025).

Airi, S., Rawal, R. S., Dhar, U., & Purohit, A. N. (1997). Population studies on *Podophyllum hexandrum* Royle – A dwindling, medicinal plant of the Himalaya. *Plant Genetic Resources Newsletter, 110*, 29–34.

Anand, U., Biswas, P., Kumar, V., Ray, D., Ray, P., Loake, V. I. P., Kandimalla, R., Chaudhary, A., Singh, B., Routhu, N. K., Chen, Z. S., Prockow, J., & Dey, A. (2022). *Podophyllum hexandrum* and its active constituents: Novel radioprotectants. *Biomedicine & Pharmacotherapy, 146*(4), 112555. https://doi.org/10.1016/j.biopha.2021.112555 (accessed on 12th August 2025).

Anand, U., Jacobo-Herrera, N., Altemimi, A., & Lakhssassi, N. (2019). A comprehensive review on medicinal plants as antimicrobial therapeutics: Potential avenues of biocompatible drug discovery. *Metabolites, 9*(11), 258. https://doi.org/10.3390/metabo9110258 (accessed on 12th August 2025).

Ara, H., Jaiswal, U., & Jaiswal, V. S. (2000). Synthetic seed: Prospects and limitations. *Current Science, 78*(12), 1438–1444.

Asci, Ö. A., Deveci, H., Erdeger, A., Özdemir, K. N., Demirci, T., & Baydar, N. G. (2019). Brassinosteroids promote growth and secondary metabolite production in lavandin (*Lavandula intermedia* Emeric ex Loisel.). *Journal of Essential Oil-Bearing Plants*. https://doi.org/10.1080/0972060X.2019.1585298 (accessed on 12th August 2025).

Atici, O., Agar, G., & Battal, P. (2005). Changes in phytohormone contents in chickpea seeds germinating under lead or zinc stress. *Biologia Plantarum, 49*(2), 215–222. https://doi.org/10.1007/s10535-005-5222-9 (accessed on 12th August 2025).

Badola, H. K., & Pal, M. (2002). Endangered medicinal plant species in Himachal Pradesh. *Current Science, 83*(7), 797–798.

Bano, H., Bhat, J. A. I., Siddique, M. A. A., Noor, F., Bhat, M. A., & Mir, S. A. (2018). Seed germination studies of *Saussurea costus* Clarke, a step towards conservation of a critically endangered medicinal plant species of North Western Himalaya. *Plant Archives, 18*(1), 963–968.

Baskey, S., Mahato, S. K., Thapa, B., Bantawa, P., Khawas, T., & Chhetri, B. (2016). Role of growth regulators in enhancement of germination of *Swertia chirayita* (Roxb. ex Fleming) H. Karst. *Indian Horticulture Journal, 6*(2), 183–186.

Baskin, C., Thompson, K., & Baskin, J. (2006). Mistakes in germination ecology and how to avoid them. *Seed Science Research, 16*(3), 165–168. https://doi.org/10.1079/SSR2006247 (accessed on 12th August 2025).

Baskin, J. M., & Baskin, C. C. (2004). A classification system for seed dormancy. *Seed Science Research, 14*(1), 1–16. https://doi.org/10.1079/SSR2003150 (accessed on 12th August 2025).

Belmehdi, O., El Harsal, A., Benmoussi, M., Laghmouchi, Y., Senhaji, N. S., & Abrini, J. (2018). Effect of light, temperature, salt stress and pH on seed germination of medicinal plant *Origanum elongatum* (Bonnet) Emb. and Maire. *Biocatalysis and Agricultural Biotechnology, 16*, 126–131. https://doi.org/10.1016/j.bcab.2018.07.032 (accessed on 12th August 2025).

Bentley, R. (2010). *Medicinal Plants*. London: Domville-Fife Press.

Bhardwaj, A. K., Kapoor, S., Naryal, A., Bhardwaj, P., Warghat, A. R., Kumar, B., & Chaurasia, O. P. (2016). Effect of various dormancy breaking treatments on seed germination, seedling growth and seed vigor of medicinal plants. *Tropical Plant Research, 3*(3), 508–516. https://doi.org/10.22271/tpr.2016.v3.i3.067 (accessed on 12th August 2025).

Bialek, K., Michalczuk, L., & Cohen, J. D. (1992). Auxin biosynthesis during seed germination in *Phaseolus vulgaris*. *Plant Physiology, 100*(1), 509–517. https://doi.org/10.1104/pp.100.1.509 (accessed on 12th August 2025).

Bodeker, G., Bhat, K. K. S., Burley, J., & Vantomme, P. (2014). *Medicinal plants for forest conservation and health care*. FAO Forestry Paper 142. http://www.fao.org/3/i3460e/i3460e.pdf (accessed on 12th August 2025).

Bonner, F. T., & Karrfalt, R. P. (2008). *The Woody Plant Seed Manual*. Agriculture Handbook No. 727. Washington, DC: United States Department of Agriculture, Forest Service.

Bose, B., Kumaria, S., Choudhury, H., & Tandon, P. (2016). Assessment of genetic homogeneity and analysis of phytomedicinal potential in micropropagated plants of *Nardostachys jatamansi*, a critically endangered medicinal plant of alpine Himalayas. *Plant Cell Tissue and Organ Culture, 124*(2), 331–349. https://doi.org/10.1007/s11240-015-0897-x (accessed on 12th August 2025).

Butola, J. S., & Samant, S. S. (2006). Seed viability of *Saussurea costus*. *Journal of Tropical Medicinal Plants, 7*(2), 197–203.

Chachalis, D., & Reddy, K. (2000). Factors affecting *Campsis radicans* seed germination and seedling emergence. *Weed Science, 48*(2), 212–216. https://doi.org/10.1614/0043-1745(2000)048[0212:FACRSG]2.0.CO;2 (accessed on 12th August 2025).

Chakraborty, A., Bhattacharya, D., Ghanta, S., & Chattopadhyay, S. (2010). An efficient protocol for *in vitro* regeneration of *Podophyllum hexandrum*, a critically endangered medicinal plant. *Indian Journal of Biotechnology, 9*(2), 217–220.

Chandra, B., Palni, L. M. S., & Nandi, S. K. (2006). Propagation and conservation of *Picrorhiza kurrooa* Royle ex Benth.: An endangered Himalayan medicinal herb of high commercial value. *Biodiversity and Conservation, 15*(7), 2325–2338. https://doi.org/10.1007/s10531-005-0770-z (accessed on 12th August 2025).

Chandra, K. K., Kumar, N., & Chand, G. (2010). Studies on mycorrhizal inoculation on dry matter yield and root colonization of some medicinal plants grown in different stress soils and forest soil. *Journal of Environmental Biology, 3*(6), 975–979.

Chandra, S., Kumar, V., Bandopadhyay, R., & Sharma, M. M. (2012). SEM and elemental studies of *Swertia chirayita*: A critically endangered medicinal herb of temperate Himalayas. *Current Trends in Biotechnology and Pharmacy, 6*(3), 381–388.

Chattopadhyay, S., Srivastava, A. K., Bhojwani, S. S., & Bisaria, V. S. (2001). Development of suspension culture of *Podophyllum hexandrum* for production of podophyllotoxin. *Biotechnology Letters, 23*(24), 2063–2066. https://doi.org/10.1023/A:1013704116860 (accessed on 12th August 2025).

Chattopadhyay, S., Srivastava, A. K., Bhojwani, S. S., & Bisaria, V. S. (2002). Production of podophyllotoxin by plant cell cultures of *Podophyllum hexandrum* in bioreactor. *Journal of Bioscience and Bioengineering, 93*(2), 215–220. https://doi.org/10.1263/jbb.93.215 (accessed on 12th August 2025).

Chauhan, B. S., Gill, G., & Preston, C. (2006). Influence of environmental factors on seed germination and seedling emergence of rigid ryegrass (*Lolium rigidum*). *Weed Science, 54*(6), 1004–1012. https://doi.org/10.1614/WS-06-087R.1 (accessed on 12th August 2025).

Chauhan, R. S., & Nautiyal, M. C. (2007). Seed germination and seed storage behavior of *Nardostachys jatamansi* DC., an endangered medicinal herb of high-altitude Himalaya. *Current Science, 92*(11), 1620–1624.

Chaurasia, O. P., Ballabh, B., Tayade, A., Kumar, R., Kumar, G. P., & Singh, S. B. (2012). *Podophyllum* L.: An endangered and anticancerous medicinal plant–An overview. *Indian Journal of Traditional Knowledge, 11*(2), 234–241.

Chen, C., Yang, L. Y., Lv, L. F., Zhao, Q., & Yuan, L. C. (2007). Study on seedling techniques of *Paris polyphylla* var. *yunnanensis*. *China Journal of Chinese Materia Medica, 32*(19), 1979–1983.

Chen, J., Shao, F., Igbokwe, C. J., Duan, Y., Cai, M., Ma, H., & Zhang, H. (2023). Ultrasound treatments improve germinability of soybean seeds: The key role of working frequency. *Ultrasonics Sonochemistry, 96*(1, 2), 106434. https://doi.org/10.1016/j.ultsonch.2023.106434 (accessed on 12th August 2025).

Chen, S. L., Yu, H., Luo, H. M., Wu, Q., Li, C. F., & Steinmetz, A. (2016). Conservation and sustainable use of medicinal plants: Problems, progress, and prospects. *Chinese Medicine, 11*(1), 37. https://doi.org/10.1186/s13020-016-0108-7 (accessed on 12th August 2025).

Chen, S. Y., Kuo, S. R., & Chien, C. T. (2007). Storage behavior of seeds of *Cinnamomum osmophloeum* and *Neolitsea aciculate* var. *variabillima* (Lauraceae). *Seed Science and Technology, 35*(1), 237–243. https://doi.org/10.15258/sst.2007.35.1.22 (accessed on 12th August 2025).

Chen, S. Y., Kuo, S. R., & Chien, C. T. (2008). Roles of gibberellins and abscisic acid in dormancy and germination of red bayberry (*Myrica rubra*) seeds. *Tree Physiology, 28*(9), 1431–1439. https://doi.org/10.1093/treephys/28.9.1431 (accessed on 12th August 2025).

Choudhary, D. K., Kaul, B. L., & Khan, S. (1998). Cultivation and conservation of *Podophyllum hexandrum*–An overview. *Journal of Medicinal and Aromatic Plant Sciences, 20*, 1071–1073.

Christensen, C. M. (1972). Microflora and seed deterioration. In R. H. Roberts (Ed.), *Viability of Seeds* (pp. 59–93). London: Chapman and Hall.

Copeland, L. O., & McDonald, M. B. (2001). *Principles of Seed Science and Technology* (4th ed.). Hingham, MA: Kluwer Academic.

Dadlani, M., Gupta, A., Sinha, S. N., & Kavali, R. (2023). Seed storage and packaging. In M. Dadlani & D. K. Yadava (Eds.), *Seed Science and Technology*. Singapore: Springer. https://doi.org/10.1007/978-981-19-5888-5_11 (accessed on 12th August 2025).

de Oliveira, R. S., Souza, F. V. D., dos Santos, I. L., Souza, S. O., Aona, L. Y. S., & de Souza, E. H. (2021). Cryopreservation and low-temperature storage of seeds of *Tillandsia* species (Bromeliaceae) with ornamental potential. *3 Biotech, 11*(4), 186. https://doi.org/10.1007/s13205-021-02730-x (accessed on 12th August 2025).

Dhiman, N., Patial, V., & Bhattacharya, A. (2016). *In vitro* approaches for conservation and sustainable utilization of *Podophyllum hexandrum* and *Picrorhiza kurroa*: Endangered medicinal herbs of Western Himalaya. In M. Anis & N. Ahmad (Eds.), *Plant Tissue*

Culture: Propagation, Conservation and Crop Improvement (pp. 45–69). Singapore: Springer. https://doi.org/10.1007/978-981-10-1917-3_3 (accessed on 12th August 2025).

Dinato, N. B., Santos, I. R. I., Vigna, B. B. Z., Paula, A. F., & Fávero, A. P. (2020). Pollen cryopreservation for plant breeding and genetic resources conservation. *CryoLetters, 41*(3), 115–127.

Doijode, S. D. (1997). Effect of storage temperature and packaging on longevity of tomato seeds. *Vegetable Science, 24*(1), 70–72.

Doijode, S. D. (2007). Low cost seed storage techniques for certain hardy indigenous pod vegetables. *Acta Horticulturae, 752*, 589–591. https://doi.org/10.17660/ActaHortic.2007.752.111 (accessed on 12th August 2025).

Elhindi, K. M., Dewir, Y. H., Asrar, A., Abdel-Salam, E., El-Din, A. S., & Ali, M. (2016). Improvement of seed germination in three medicinal plant species by plant growth regulators. *HortScience, 51*(7), 887–891. https://doi.org/10.21273/HORTSCI.51.7.887 (accessed on 12th August 2025).

Ellis, R. H., Hong, T. D., & Roberts, E. H. (1990). An intermediate category of seed storage behavior? I. Coffee. *Journal of Experimental Botany, 41*(9), 1167–1174. https://doi.org/10.1093/jxb/41.9.1167 (accessed on 12th August 2025).

Fabricant, D. S., & Farnsworth, N. R. (2001). The value of plants used in traditional medicine for drug discovery. *Environmental Health Perspectives, 109*(1), 69–75. https://doi.org/10.1289/ehp.01109s169 (accessed on 12th August 2025).

Finch-Savage, W. E. (1995). Influence of seed quality on crop establishment, growth and yield. In A. S. Basra (Ed.), *Seed Quality: Basic Mechanisms and Agricultural Implications* (p. 470). Binghamton, NY: Food Products Press.

Finkelstein, R., Reeves, W., Ariizumi, T., & Steber, C. (2008). Molecular aspects of seed dormancy. *Annual Review of Plant Biology, 59*(1), 387–415. https://doi.org/10.1146/annurev.arplant.59.032607.092740 (accessed on 12th August 2025).

Fleming, M. B., Hill, L. M., & Walters, C. (2019). The kinetics of ageing in dry-stored seeds: A comparison of viability loss and RNA degradation in unique legacy seed collections. *Annals of Botany, 123*(7), 1133–1146. https://doi.org/10.1093/aob/mcy217 (accessed on 12th August 2025).

Gantait, S., Kundu, S., Ali, N., & Sahu, N. C. (2015). Synthetic seed production of medicinal plants: A review on influence of explants, encapsulation agent and matrix. *Acta Physiologiae Plantarum, 37*(98). https://doi.org/10.1007/s11738-015-1847-2 (accessed on 12th August 2025).

Gayathri, P., & Kanakamany, M. T. (2018). Influence of storage environment and packing materials on seed germination and viability of *Desmodium gangeticum* (L.) DC. *Open Access Journal of Medicinal and Aromatic Plants, 9*(1), 23–33.

Generoso, A. L., Carvalho, V. S., Walter, R., Campbell, G., Araújo, L. S., Santana, J. G. S., & Cunha, M. (2019). Mature-embryo culture in the cryopreservation of passion fruit (*Passiflora edulis* Sims) seeds. *Scientia Horticulturae, 256*(4), 108638. https://doi.org/10.1016/j.scienta.2019.108638 (accessed on 12th August 2025).

Ghimire, S. K., McKey, D., & Aumeeruddy-Thomas, Y. (2005). Conservation of Himalayan medicinal plants: Harvesting patterns and ecology of two threatened species, *Nardostachys grandiflora* DC. and *Neopicrorhiza scrophulariiflora* (Pennell) Hong. *Biological Conservation, 124*(4), 463–475. https://doi.org/10.1016/j.biocon.2005.02.005 (accessed on 12th August 2025).

Giri, A., & Lakshmi Narasu, M. (2000). Production of podophyllotoxin from *Podophyllum hexandrum*: A potential natural product for clinically useful anticancer drugs. *Cytotechnology, 34*(1), 17–26. https://doi.org/10.1023/A:1008138230896 (accessed on 12th August 2025).

Goel, A., Goel, A. K., & Sheoran, I. S. (2003). Changes in oxidative stress enzymes during artificial ageing in cotton (*Gossypium hirsutum* L.) seeds. *Journal of Plant Physiology, 160*(9), 1093–1100. https://doi.org/10.1078/0176-1617-00881 (accessed on 12th August 2025).

Gowdey, G., Lee, R. K., & Carpenter, W. M. (1995). Treatment of HIV-related hairy leukoplakia with *Podophyllum* resin 25% solution. *Oral Surgery Oral Medicine Oral Pathology Oral Radiology and Endodontology, 79*(1), 64–67. https://doi.org/10.1016/S1079-2104(05)80076-9 (accessed on 12th August 2025).

Groot, S. P. C., de Groot, L., Kodde, J., & van Treuren, R. (2014). Prolonging the longevity of *ex situ* conserved seeds by storage under anoxia. *Plant Genetic Resources: Characterization and Utilization, 13*(1), 18–26. https://doi.org/10.1017/S1479262114000586 (accessed on 12th August 2025).

Groot, S. P. C., Surki, A. A., de Vos, R. C. H., & Kodde, J. (2012). Seed storage at elevated partial pressure of oxygen, a fast method for analyzing seed ageing under dry conditions. *Annals of Botany, 110*(6), 1149–1159. https://doi.org/10.1093/aob/mcs198 (accessed on 12th August 2025).

Guo, Q., Zhou, J., Wang, Z., & Yang, H. (2012). *In vitro* rooting of *Podophyllum hexandrum* and transplanting technique. *Engineering, 4*(10), 142–145. https://doi.org/10.4236/eng.2012.410B037 (accessed on 12th August 2025).

Gurmani, A. R., Bano, A., Ullah, N., Khan, H., Jahangir, M., & Flowers, T. J. (2013). Exogenous abscisic acid (ABA) and silicon (Si) promote salinity tolerance by reducing sodium (Na^+) transport and bypass flow in rice (*Oryza sativa* indica). *Australian Journal of Crop Science, 7*(9), 1219–1226.

Habibi, A., & Abdoli, M. (2013). Influence of salicylic acid pre-treatment on germination, vigor and growth parameters of garden cress (*Lepidium sativum*) seedlings under water potential loss at salinity stress. *International Journal of Sciences Basic and Applied Research, 4*(6), 1393–1399.

Hamilton, A. C. (2008). Medicinal plants in conservation and development: Case studies and lessons learned. In C. P. Kala (Ed.), *Medicinal plants in conservation and development* (pp. 1–43). Salisbury: Plantlife International Publisher.

Han, B., Fernandez, V., Pritchard, H. W., & Colville, L. (2021). Gaseous environment modulates volatile emission and viability loss during seed artificial ageing. *Planta, 253*, 106. https://doi.org/10.1007/s00425-021-03620-5 (accessed on 12th August 2025).

Harman, G., & Mattick, L. (1976). Association of lipid oxidation with seed ageing and death. *Nature, 260*, 323–324. https://doi.org/10.1038/260323a0 (accessed on 12th August 2025).

Harrington, J. F. (1960). Thumb rules of drying seeds. *Crops Soils, 13*, 16–17.

Harrington, J. F. (1972). Seed storage and longevity. In T. T. Kozlowski (Ed.), *Seed Biology, Insects, and Seed Collection, Storage, Testing and Certification* (pp. 145–245). New York: Academic Press. https://doi.org/10.1016/B978-0-12-395605-7.50009-0 (accessed on 12th August 2025).

Hentrich, M., Boettcher, C., & Duchting, P. (2013). The jasmonic acid signaling pathway is linked to auxin homeostasis through the modulation of *YUCCA8* and *YUCCA9* gene expression. *The Plant Journal, 74*, 626–637. https://doi.org/10.1111/tpj.12152 (accessed on 12th August 2025).

Heywood, V. H., & Iriondo, J. M. (2003). Plant conservation: Old problems, new perspectives. *Biological Conservation, 113*(3), 321–335. https://doi.org/10.1016/S0006-3207(03)00121-6 (accessed on 12th August 2025).

Hong, T. D., & Ellis, R. H. (1996). *A Protocol to Determine Seed Storage Behavior*. Rome: International Plant Genetic Resources Institute.

Hong, T. D., Linington, S., & Ellis, R. H. (1996). *Seed Storage Behavior: A Compendium*. Handbooks for Genebanks No. 4. Rome: International Plant Genetic Resources Institute.

Hu, Q. J., Chen, M. X., Song, T., Cheng, C. L., Tian, Y., Hu, J., & Zhang, J. H. (2020). Spermidine enhanced the antioxidant capacity of rice seeds during seed aging. *Plant Growth Regulation, 91*, 397–406. https://doi.org/10.1007/s10725-020-00613-4 (accessed on 12th August 2025).

Jan, A., Thomas, G., Shawl, A. S., Jabeen, N., & Kozgar, M. I. (2010). Improved micropropagation protocol of an endangered medicinal plant–*Picrorhiza kurroa* Royle ex Benth. promptly through auxin treatments. *Chiang Mai Journal of Science, 37*(2), 304–313.

Jenderek, M. M., & Reed, B. M. (2017). Cryopreserved storage of clonal germplasm in the USDA national plant germplasm system. *In Vitro Cellular and Developmental Biology–Plant, 53*(4), 299–308. https://doi.org/10.1007/s11627-017-9828-3 (accessed on 12th August 2025).

Johnson, T. S., Badari, S., Naryan, B., & Narayana, D. B. A. (1997). Rapid *in vitro* propagation of *Saussurea lappa*, an endangered medicinal plant, through multiple shoot cultures. *In Vitro Cellular and Developmental Biology–Plant, 33*(2), 128–130. https://doi.org/10.1007/s11627-997-0010-1 (accessed on 12th August 2025).

Joshi, P., & Dhawan, V. (2005). *Swertia chirayita*–An overview. *Current Science, 89*(4), 635–640.

Kala, C. P. (2005). Indigenous uses, population density, and conservation of threatened medicinal plants in protected areas of the Indian Himalayas. *Conservation Biology, 19*(2), 368–378. https://doi.org/10.1111/j.1523-1739.2005.00602.x (accessed on 12th August 2025).

Kaur, R., Arora, S., & Thukral, A. K. (2009). Enhancing seed germination of *Chlorophytum borivilianum* Sant. et Fernand. with PGRs, steroidal hormones and zinc. *Research Journal of Seed Science, 2*(2), 32–39. https://doi.org/10.3923/rjss.2009.32.39 (accessed on 12th August 2025).

Khalaki, M. A., Moameri, M., Asgari Lajayer, B., & Astatkie, T. (2021). Influence of nano-priming on seed germination and plant growth of forage and medicinal plants. *Plant Growth Regulation, 93*, 13–28. https://doi.org/10.1007/s10725-020-00670-9 (accessed on 12th August 2025).

Khalequzzaman, K. M., Rashid, M. M., Hasan, M. A., & Reza, M. M. A. (2012). Effect of storage containers and storage periods on the seed quality of French bean (*Phaseolus vulgaris*). *Bangladesh Journal of Agricultural Research, 37*(2), 195–205. https://doi.org/10.3329/bjar.v37i2.11221 (accessed on 12th August 2025).

Khan, A., Shah, A. H., & Ali, N. (2021). *In vitro* propagation and phytochemical profiling of a highly medicinal and endemic plant species of the Himalayan region (*Saussurea costus*). *Scientific Reports, 11*, 23575. https://doi.org/10.1038/s41598-021-03032-1 (accessed on 12th August 2025).

Khan, M. A., & Ungar, I. A. (1997). Alleviation of seed dormancy in the desert forb *Zygophyllum simplex* L. from Pakistan. *Annals of Botany, 80*(4), 395–400. https://doi.org/10.1006/anbo.1997.0449 (accessed on 12th August 2025).

Kharkwal, A. C., Kushwaha, R., Prakash, O., Ogra, R. K., Bhattacharya, A., Nagar, P. K., & Ahuja, P. S. (2008). An efficient method of propagation of *Podophyllum hexandrum*: An endangered medicinal plant of the Western Himalayas under *ex situ* conditions. *Journal of Natural Medicine, 62*(2), 211–216. https://doi.org/10.1007/s11418-007-0217-9 (accessed on 12th August 2025).

Koger, C. H., Reddy, K. N., & Poston, D. H. (2004). Factors affecting seed germination, seedling emergence, and survival of Texasweed (*Caperonia palustris*). *Weed Science, 52*(6), 989–995. https://doi.org/10.1614/WS-03-139R2 (accessed on 12th August 2025).

Kranner, I. (2013). Mechanisms of seed ageing. *South African Journal of Botany, 86*, 140. https://doi.org/10.1016/j.sajb.2013.02.010 (accessed on 12th August 2025).

Kulus, D. (2019). Managing plant genetic resources using low and ultra-low temperature storage: A case study of tomato. *Biodiversity and Conservation, 28*(3), 1003–1027. https://doi.org/10.1007/s10531-019-01710-1 (accessed on 12th August 2025).

Kumar, R., Swapnil, P., Meena, M., Selpair, S., & Yadav, B. G. (2022). Plant growth-promoting rhizobacteria (PGPR): Approaches to alleviate abiotic stresses for enhancement of growth and development of medicinal plants. *Sustainability, 14*(23), 15514. https://doi.org/10.3390/su142315514 (accessed on 12th August 2025).

Kumar, V., Kumar, K., Kumari, P., Kumar, R., Saxena, S., & Chaurasia, O. P. (2017). Effect of plant growth regulators on germination of seed of *Podophyllum hexandrum* in high altitude region of Ladakh, India. *International Journal of Bioassays, 6*(9), 5470–5473.

Kumar, V., Singh, S. K., Bandopadhyay, R., Sharma, M. M., & Chandra, S. (2014). *In vitro* organogenesis, secondary metabolite production and heavy metal analysis in *Swertia chirayita*. *Central European Journal of Biology, 9*(7), 686–698. https://doi.org/10.2478/s11535-014-0300-7 (accessed on 12th August 2025).

Kumari, R., Negi, M., Thakur, P., Mahajan, H., Raina, K., Sharma, R., Singh, R., Anand, V., Ming, L. C., Goh, K. W., Calina, D., Sharifi Rad, J., & Chaudhary, A. (2023). *Saussurea costus* (Falc.) Lipsch.: A comprehensive review of its pharmacology, phytochemicals, ethnobotanical uses, and therapeutic potential. *Naunyn-Schmiedeberg's Archives of Pharmacology*. https://doi.org/10.1007/s00210-023-02694-0 (accessed on 12th August 2025).

Kurek, K., Plitta-Michalak, B., & Ratajczak, E. (2019). Reactive oxygen species as potential drivers of the seed aging process. *Plants, 8*(6), 174. https://doi.org/10.3390/plants8060174 (accessed on 12th August 2025).

Leubner-Metzger, G. (2001). Brassinosteroids and gibberellins promote tobacco seed germination by distinct pathways. *Planta, 213*(5), 758–763. https://doi.org/10.1007/s004250100542 (accessed on 12th August 2025).

Li, S., Yan, N., Tanveer, M., Zhao, Z., Jiang, L., & Wang, H. (2023). Seed germination ecology of the medicinal plant *Peganum harmala* (Zygophyllaceae). *Plants, 12*(14), 2660. https://doi.org/10.3390/plants12142660 (accessed on 12th August 2025).

Li, X., Baskin, J. M., & Baskin, C. C. (1999). Physiological dormancy and germination requirements of seeds of several North American *Rhus* species (Anacardiaceae). *Seed Science Research, 9*(3), 237–245.

Lin, H. W., Kwok, K. H., & Doran, P. M. (2003). Production of podophyllotoxin using cross-species coculture of *Linum flavum* hairy roots and *Podophyllum hexandrum* cell suspensions. *Biotechnology Progress, 19*(5), 1417–1426. https://doi.org/10.1021/bp034036h (accessed on 12th August 2025).

Liu, X. C., & Liu, Z. L. (2014). Evaluation of insecticidal activity of *Nardostachys jatamansi* essential oil against some grain storage insects. *Journal of Entomology and Zoology Studies, 2*(4), 335–340.

Mathur, J. (1992). *In vitro* morphogenesis in *Nardostachys jatamansi* DC.: Shoot regeneration from callus derived roots. *Annals of Botany, 70*(5), 419–422. https://doi.org/10.1093/oxfordjournals.aob.a088497 (accessed on 12th August 2025).

Mathur, J. (1993). Somatic embryogenesis from callus cultures of *Nardostachys jatamansi*. *Plant Cell Tissue and Organ Culture, 33*(2), 163–169. https://doi.org/10.1007/BF01983230 (accessed on 12th August 2025).

Mehta, P., Sekar, K. C., Bhatt, D., Tewari, A., Bisht, K., Upadhyay, S., Negi, V. S., & Soragi, B. (2020). Conservation and prioritization of threatened plants in Indian Himalayan region. *Biodiversity and Conservation, 29*(6), 1723–1745. https://doi.org/10.1007/s10531-020-01959-x (accessed on 12th August 2025).

Miransari, M., & Smith, D. L. (2014). Plant hormones and seed germination. *Environmental and Experimental Botany, 99*, 110–121. http://dx.doi.org/10.1016/j.envexpbot.2013.11.005 (accessed on 12th August 2025).

Mishra, D., Chaturvedi, R. V., & Tripathi, S. C. (1995). The fungitoxic effect of the essential oil of the herb *Nardostachys jatamansi* DC. *Tropical Agriculture, 72*(1), 48–52.

Mohammadi, M., Pouryousef, M., & Farhang, N. (2023). Study on germination and seedling growth of various ecotypes of fennel (*Foeniculum vulgare* Mill.) under salinity stress. *Journal of Applied Research on Medicinal and Aromatic Plants, 34*(11), 100481. https://doi.org/10.1016/j.jarmap.2023.100481 (accessed on 12th August 2025).

Mohi-ud-din, G., Nawchoo, I. A., & Wafai, B. A. (2005). Seed germination studies on *Saussurea costus* (Falc.) Lipsch.: A threatened medicinal plant of N.W. Himalaya. *Indian Journal of Plant Physiology, 10*, 394–396.

Moncini, L., Simone, G., Romi, M., Cai, G., Guerriero, G., Whittaker, A., Benedettelli, S., & Berni, R. (2020). Controlled nitrogen atmosphere for the preservation of functional molecules during silos storage: A case study using old Italian wheat cultivars. *Journal of Stored Product Research, 88*, 101638. https://doi.org/10.1016/j.jspr.2020.101638 (accessed on 12th August 2025).

Morrison, D. A., McClay, K., Porter, C., & Rish, S. (1998). The role of the lens in controlling heat-induced breakdown of testa-imposed dormancy in native Australian legumes. *Annals of Botany, 82*(1), 35–40. https://doi.org/10.1006/anbo.1998.0640 (accessed on 12th August 2025).

Nadeem, M., Palni, L. M. S., Purohit, A. N., Pandey, H., & Nandi, S. K. (2000). Propagation and conservation of *Podophyllum hexandrum* Royle: An important medicinal herb. *Biological Conservation, 92*(1), 121–129. https://doi.org/10.1016/S0006-3207(99)00059-2 (accessed on 12th August 2025).

Nag, A., Choudhary, S., Masand, M., Parmar, R., Bhandawat, A., Seth, R., Singh, G., Dhyani, P., & Sharma, R. K. (2020). Spatial transcriptional dynamics of geographically separated genotypes revealed key regulators of podophyllotoxin biosynthesis in *Podophyllum hexandrum*. *Industrial Crops and Products, 147*, 112247. https://doi.org/10.1016/j.indcrop.2020.112247 (accessed on 12th August 2025).

Nalawade, S. M., Sagare, A. P., Lee, C. Y., Kao, C. L., & Tsay, H. S. (2003). Studies on tissue culture of Chinese medicinal plant resources in Taiwan and their sustainable utilization. *Botanical Bulletin - Academia Sinica Taipei, 44*(2), 79–98.

Nautiyal, O. H. (2013). Anti-fungal activity of *Nardostachys jatamansi* essential oil beneficial for treating (dermatophytosis) ringworm. *Natural Products, 9*(6), 241–244.

Nautiyal, P. C., Sivasubramaniam, K., & Dadlani, M. (2023). Seed dormancy and regulation of germination. In M. Dadlani & D. K. Yadava (Eds.), *Seed Science and Technology*. Singapore: Springer. https://doi.org/10.1007/978-981-19-5888-5_3 (accessed on 12th August 2025).

Neelamegam, R., & Sutha, T. (2015). UV-C irradiation effect on seed germination, seedling growth and productivity of groundnut (*Arachis hypogaea* L.). *International Journal of Current Microbiology and Applied Sciences, 4*(8), 430–443.

Netra, N., Rani, K. U., Gowda, R., Siddegowda, R. P., & Narayanaswamy, S. (2015). Effect of packaging material and desiccant on storability of soybean seeds. *Seed Science and Technology, 44*(1), 207–211. https://doi.org/10.15258/sst.2016.44.1.05 (accessed on 12th August 2025).

Nikolic, R., Mitic, N., Miletic, R., & Neskovic, M. (2006). Effects of cytokinins on *in vitro* seed germination and early seedling morphogenesis in *Lotus corniculatus* L. *Journal of Plant Growth Regulation, 25*(3), 187–194. https://doi.org/10.1007/s00344-005-0129-4 (accessed on 12th August 2025).

Pandey, H., Nandi, S. K., Mohammad, N., & Palni, L. M. S. (2000). Chemical stimulation of seed germination in *Aconitum heterophyllum* Wall. and *A. balfourii* Stapf.: Important Himalayan species of medicinal value. *Seed Science and Technology, 28*(1), 39–48.

Panis, B., & Lambardi, M. (2006). Status of cryopreservation technologies in plants (crops and forest trees). *The Role of Biotechnology, 5*(7), 43–54.

Parimala, K., Subramanian, K., Kannan, S. M., & Vijayalakshmi, K. (2013). *Seed Storage Techniques–A Primer*. Centre for Indian Knowledge Systems (CIKS), Seed Node of the Revitalizing Rainfed Agriculture Network.

Peleg, Z., & Blumwald, E. (2011). Hormone balance and abiotic stress tolerance in crop plants. *Current Opinion in Plant Biology, 14*(3), 290–295. https://doi.org/10.1016/j.pbi.2011.02.001 (accessed on 12th August 2025).

Peng, P. L., Geng, B. Y., Qin, Y. B., Yang, L. E., Baskin, C. C., & Baskin, J. M. (2023). Ecophysiology of seed dormancy and germination in the alpine-subalpine medicinal plant species *Sinopodophyllum hexandrum* (Royle) T. S. Ying. *Journal of Applied Research on Medicinal and Aromatic Plants, 32*, 100448. https://doi.org/10.1016/j.jarmap.2022.100448 (accessed on 12th August 2025).

Pennazio, S., & Roggero, P. (1991). Effects of exogenous salicylate on basal and stress-induced ethylene formation in soybean. *Biologia Plantarum, 33*(1), 58–65. https://doi.org/10.1007/BF02873789 (accessed on 12th August 2025).

Perveen, S., & Anis, M. (2014). Encapsulation of internode regenerated adventitious shoot buds of Indian siris in alginate beads for temporary storage and twofold clonal plant production. *Acta Physiologiae Plantarum, 36*(8), 2067–2077. https://doi.org/10.1007/s11738-014-1584-y (accessed on 12th August 2025).

Pradhan, B. K., & Badola, H. K. (2008). Seed germination response of populations of *Swertia chirayita* following periodical storage. *Seed Technology, 30*(1), 63–69. http://www.jstor.org/stable/23433371 (accessed on 12th August 2025).

Pradhan, B. K., & Badola, H. K. (2012). Effect of storage conditions and storage periods on seed germination in eleven populations of *Swertia chirayita*: A critically endangered medicinal herb in Himalaya. *The Scientific World Journal, 2012*(1), 128105. https://doi.org/10.1100/2012/128105 (accessed on 12th August 2025).

Probert, R. J. (2000). The role of temperature in the regulation of seed dormancy and germination. In M. Fenner (Ed.), *Seeds: The Ecology of Regeneration in Plant Communities* (pp. 261–292). https://doi.org/10.1079/9780851994321.0261 (accessed on 12th August 2025).

Puligundla, P., Kim, J. W., & Mok, C. (2017). Effect of corona discharge plasma jet treatment on decontamination and sprouting of rapeseed (*Brassica napus* L.) seeds. *Food Control, 71*, 376–382. https://doi.org/10.1016/j.foodcont.2016.07.021 (accessed on 12th August 2025).

Rahman, M. M. K., & Rahman, G. M. M. (1997). Effect of containers and length of storage on germination and seed-borne fungi associated with jute seed. *Bangladesh Journal of Plant Pathology, 13*, 13–16.

Rajesh, M., Sivanandhan, G., Jeyaraj, M., Chackravarthy, R., Manickavasagam, M., Selvaraj, N., & Ganapathi, A. (2014a). An efficient *in vitro* system for somatic embryogenesis and podophyllotoxin production in *Podophyllum hexandrum* Royle. *Protoplasma, 251*(5), 1231–1243. https://doi.org/10.1007/s00709-014-0632-1 (accessed on 12th August 2025).

Rajesh, M., Sivanandhan, G., Subramanyam, K., Kapildev, G., Jaganath, B., Kasthurirengan, S., Manickavasagam, M., & Ganapathi, A. (2014b). Establishment of somatic embryogenesis and podophyllotoxin production in liquid shake cultures of *Podophyllum hexandrum* Royle. *Industrial Crops and Products, 60*(3), 66–74. https://doi.org/10.1016/j.indcrop.2014.05.046 (accessed on 12th August 2025).

Rathore, S., Debnath, P., & Kumar, R. (2021). Kuth (*Saussurea costus* (Falc.) Lipsch.): A critically endangered medicinal plant from Himalaya. *Journal of Applied Research on Medicinal and Aromatic Plants, 20*(4), 100277. https://doi.org/10.1016/j.jarmap.2020.100277 (accessed on 12th August 2025).

Rawat, J. M., Bhandari, A., Rajneesh, K., Raturi, M., Rawat, B., Dhakad, A. K., Negi, R. K., Thakur, A., & Chandra, A. (2018). RAPD-based genetic fidelity analysis of *in vitro* regenerated plants of *Nardostachys jatamansi*. *International Journal on Biological Sciences, 8*(2), 141–147.

Reed, R. C., Bradford, K. J., & Khanday, I. (2022). Seed germination and vigor: Ensuring crop sustainability in a changing climate. *Heredity, 128*(6), 450–459. https://doi.org/10.1038/s41437-022-00497-2 (accessed on 12th August 2025).

Rekha, K., Rao, R. R., Pandey, R., Prasad, K. R., Babu, K. S., Vangala, J. R., Kalivendi, S. V., & Rao, J. M. (2013). Two new sesquiterpenoids from the rhizomes of *Nardostachys jatamansi*. *Journal of Asian Natural Product Research, 15*(2), 111–116. https://doi.org/10.1080/10286020.2012.738673 (accessed on 12th August 2025).

Richards, D. E., King, K. E., Ait-ali, T., & Harberd, N. P. (2001). How gibberellin regulates plant growth and development: A molecular genetic analysis of gibberellin signaling. *Annual Review of Plant Physiology and Plant Molecular Biology, 52*, 67–88. https://doi.org/10.1146/annurev.arplant.52.1.67 (accessed on 12th August 2025).

Roberts, E. H. (1972). Storage environment and the control of viability. In E. H. Roberts (Ed.), *Viability of seeds* (pp. 14–58). London: Chapman and Hall.

Roberts, E. H. (1973). Predicting the storage life of seeds. *Seed Science and Technology, 1*(3), 499–514.

Roberts, E., & Ellis, R. (1989). Water and seed survival. *Annals of Botany, 63*(1), 39–52. https://doi.org/10.1093/oxfordjournals.aob.a087727 (accessed on 12th August 2025).

Robertson, L. D., & Labate, J. A. (2006). Genetic resources of tomato (*Lycopersicon esculentum* Mill.) and wild relatives. In M. K. Razdan & A. K. Mattoo (Eds.), *Genetic improvement of Solanaceae crops, vol. 2: Tomato* (pp. 285–313). Boca Raton: Taylor and Francis. https://doi.org/10.1201/b10744-3 (accessed on 12th August 2025).

Ross, I. A. (2005). *Medicinal plants of the world* (Vol. 3, pp. 110–132): Chemical constituents, traditional and modern medicinal uses. New Jersey: Humana Press Inc.

Saisanthosh, K., & Patil, N. K. B. (2018). Effect of packaging materials and moisture content on seed storability of onion. *Journal of Pharmacognosy and Phytochemistry, 7*(4), 1745–1750.

Samant, S. S., Dhar, U., & Palni, L. M. S. (1998). *Medicinal plants of Indian Himalaya: Diversity, distribution, potential values*. G.B. Pant Institute of Himalayan Environment and Development; Almora, India.

Schmidt, L. (2000). *Guide to handling of tropical and subtropical forest seed* (pp. 511). Humlebaek: Danida Forest Seed Centre.

Sewell, R. D. E., & Rafieian-Kopaei, M. (2014). The history and up and downs of herbal medicinal usage. *Journal of Herbmed Pharmacology, 3*(1), 1–3.

Shah, N. C. (2019). Kuṣṭha, *Saussurea costus* (*Saussurea lappa*): Its unexplored history from the Atharvaveda. *Indian Journal of History of Science, 54*(4), 420–434. https://doi.org/10.16943/ijhs/2019/v54i4/49769 (accessed on 12th August 2025).

Shankar, B. S. (2015). Efficacy of temperature regime on seed germination and vigor in kuth (*Saussurea costus* (Falc.) Lipsch.). *Indian Horticulture Journal, 5*(3, 4), 70–73. https://doi.org/174-15-IHJ-2206-2015-15 (accessed on 12th August 2025).

Sharma, N., Varshney, V. K., Kala, R. P., Bisht, B., & Sharma, M. (2013). Antioxidant capacity and total phenolic content of *Swertia chirayita* (Roxb. ex Fleming) H. Karst. in Uttarakhand. *International Journal of Pharmaceutical Sciences Review and Research, 23*(1), 259–261.

Sharma, R. K., Sharma, S., & Sharma, S. S. (2006). Seed germination behavior of some medicinal plants of Lahaul and Spiti cold desert (Himachal Pradesh): Implications for conservation and cultivation. *Current Science, 90*(8), 1113–1118.

Sharma, R. K., Sharma, S., & Sharma, S. S. (2010). Storage-dependent changes in dormancy and germination of Himalayan mayapple (*Podophyllum hexandrum* Royle) seeds and their response to gibberellic acid. *Journal of Herbs, Spices and Medicinal Plants, 16*(1), 69–82. https://doi.org/10.1080/10496475.2010.481944 (accessed on 12th August 2025).

Sharma, R. K., Sharma, S., & Sharma, S. S. (2014). The long-term ambient storage-induced alterations in seed viability, germination and seedling vigor of *Saussurea costus*, a critically endangered medicinal herb of North West Himalaya. *Journal of Applied Research on Medicinal and Aromatic Plants, 1*(3), 92–97. http://dx.doi.org/10.1016/j.jarmap.2014.07.002 (accessed on 12th August 2025).

Silva, S. S. S., Souza, E. H., Souza, F. V. D., Max, D. A. S., Rossi, M. L., & Costa, M. A. P. C. (2021). Post-seminal development and cryopreservation of endemic or endangered bromeliads. *Anais da Academia Brasileira de Ciências*.

Singh, D. K., & Hajra, P. K. (1996). Floristic diversity. In G. S. Gujral & V. Sharma (Eds.), *Changing Perspective of Biodiversity Status in the Himalaya* (pp. 23–38). New Delhi: British Council Division, British High Commission Publication, Wildlife Youth Services.

Singh, H., Singh, T., & Handa, K. (1957). A note on costus oil from Kashmir costus roots. *Indian Forester, 83*(10), 606, 607.

Singh, M., Khan, M. A., Khan, M. S., Ansari, S. H., & Ahmad, S. (2015). Standardization and *in vitro* antioxidant activity of jatamansi rhizome. *Journal of Pharmacy Bioallied Sciences, 7*(4), 275–279. https://doi.org/10.4103/0975-7406.168025 (accessed on 12th August 2025).

Songnuan, W., & Kirawanich, P. (2012). Early growth effects on *Arabidopsis thaliana* by seed exposure of nanosecond pulsed electric field. *Journal of Electrostatics, 70*(5), 445–450.

Souza, F. V. D., Kaya, E., Vieira, L. J., Souza, E. H., Amorim, V. B. O., Skogerboe, D., Matsumoto, T., Alves, A. A. C., Ledo, C. A. S., & Jenderek, M. M. (2016). Droplet-vitrification and morphohistological studies of cryopreserved shoot tips of cultivated and wild pineapple genotypes. *Plant Cell Tissue and Organ Culture, 124*, 351–360. https://doi.org/10.1007/s11240-015-0899-8 (accessed on 12th August 2025).

Srivastava, N., Sharma, V., Dobriyal, A. K., Kamal, B., Gupta, S., & Jadon, V. S. (2011). Influence of pre-sowing treatments on *in vitro* seed germination of ativisha (*Aconitum heterophyllum* Wall) of Uttarakhand. *Biotechnology, 10*(2), 215–219. https://doi.org/10.3923/biotech.2011.215.219 (accessed on 12th August 2025).

Sudhakar, N., Nagendra-Prasad, D., Mohan, N., & Murugesan, K. (2008). A preliminary study on the effects of ozone exposure on growth of the tomato seedlings. *Australian Journal of Crop Science, 2*(1), 33–39.

Sultan, P., Shawl, A. S., Fayaz, A. S., & Ramteke, P. W. (2009). *In vitro* embryogenesis and marker guided analysis of *Podophyllum hexandrum*: An endangered medicinal plant. *International Journal of Biodiversity and Conservation, 1*(7), 215–223. http://www.academicjournals.org/ijbc (accessed on 12th August 2025).

Sultan, P., Shawl, A. S., Ramteke, P. W., Jan, A., Chisti, N., Jabeen, N., & Shabir, S. (2006). *In vitro* propagation for mass multiplication of *Podophyllum hexandrum*: A high value medicinal herb. *Asian Journal of Plant Sciences, 5*(2), 179–184. https://doi.org/10.3923/ajps.2006.179.184 (accessed on 12th August 2025).

Tiwari, R. K. S., & Das, K. (2014). Impact of different storage conditions on seed germination and viability of some medicinal plants. *African Journal of Agricultural Research, 9*(20), 1578–1585. https://doi.org/10.5897/AJAR2012.1758 (accessed on 12th August 2025).

Tiwari, R. S., Chandra, K. K., Dubey, S., & Tripathi, S. (2022). Influence of packaging materials and storage conditions on seed germination ability and biochemical changes in some medicinal plants of Indian forest. *Frontiers in Forests and Global Change, 5*, 868237. https://doi.org/10.3389/ffgc.2022.868237 (accessed on 12th August 2025).

Trusiak, M., Patrycja, B., Plitta-Michalak, & Michalak, M. (2023). Choosing the right path for the successful storage of seeds. *Plants, 12*(1), 72. https://doi.org/10.3390/plants12010072 (accessed on 12th August 2025).

United Nations Environment Programme (UNEP). (2015). *The State of the World's Biodiversity for Food and Agriculture*. http://www.fao.org/3/i9022en/i9022en.pdf (accessed on 12th August 2025).

Ved, D., Saha, D., Ravikumar, K., & Haridasan, K. (2015). *Nardostachys jatamansi* (D. Don) DC.; The IUCN red list of threatened species 2015. http://www.iucnredlist.org/details/50126627 (accessed on 12th August 2025).

Verma, S., & Gupta, M. L. (2015). Radiation-induced hematopoietic myelosuppression and genotoxicity get significantly countered by active principles of *Podophyllum hexandrum*: A study in strain 'A' mice. *International Journal of Radiation Biology, 91*(9), 757–770. https://doi.org/10.3109/09553002.2015.1062576 (accessed on 12th August 2025).

Vertucci, C. W., & Roos, E. E. (1990). Theoretical basis of protocols for seed storage. *Plant Physiology, 94*(3), 1019–1023. https://doi.org/10.1104/pp.94.3.1019 (accessed on 12th August 2025).

Vertucci, C. W., Roos, E. E., & Crane, J. (1994). Theoretical basis of protocols for seed storage III. Optimum moisture contents for pea seeds stored at different temperatures. *Annals of Botany, 74*(5), 531–540. https://doi.org/10.1006/anbo.1994.1151 (accessed on 12th August 2025).

Vettorazzi, R. G., Carvalho, V. S., Teixeira, M. C., Campostrini, E., Cunha, M., Matos, E. M., & Viccini, L. F. (2019). Cryopreservation of immature and mature seeds of Brazilian orchids of the genus *Cattleya*. *Scientia Horticulturae, 256*(15), 108603. https://doi.org/10.1016/j.scienta.2019.108603 (accessed on 12th August 2025).

Vishvamitera, S., Dhiman, D., Baghla, S., Singh, S., Kumar, M., Kumar, A., Kumar, D., Singh, S., & Chauhan, R. (2023). Sustainable production of *Saussurea costus* under different levels of nitrogen, phosphorus and potassium fertilizers in cold desert region of Western Himalaya. *Frontiers in Plant Science, 14*, 1179183. https://doi.org/10.3389/fpls.2023.1179183 (accessed on 12th August 2025).

Vitis, M. D., Hay, F. R., Dickie, J. B., Trivedi, C., Choi, J., & Fiegener, R. (2020). Seed storage: Maintaining seed viability and vigor for restoration use. *Restoration Ecology*. https://doi.org/10.1111/rec.13174 (accessed on 12th August 2025).

Walters, C., Ballesteros, D., & Vertucci, V. A. (2010). Structural mechanics of seed deterioration: Standing the test of time. *Plant Science, 179*(6), 565–573. https://doi.org/10.1016/j.plantsci.2010.06.016 (accessed on 12th August 2025).

Wang, M., Lambardi, M., Engelmann, F., Pathirana, R., Panis, B., Volk, G. M., & Wang, Q. C. (2020). Advances in cryopreservation of *in vitro*-derived propagules: Technologies and explant sources. *Plant Cell Tissue and Organ Culture, 144*(1), 7–20. https://doi.org/10.1007/s11240-020-01770-0 (accessed on 12th August 2025).

Warghat, A. R., Bajpai, P. K., Rewang, R., Kapoor, S., Kumar, J., Chaurasia, O. P., & Srivastava, R. B. (2016). *In vitro* callus induction and plantlet regeneration of *Saussurea lappa* (Clarke.) from Ladakh region of India. *Proceedings of the National Academy of Sciences, India Section B: Biological Sciences, 86*(3), 651–660. https://doi.org/10.1007/s40011-015-0496-y (accessed on 12th August 2025).

Wcislińska, B. (1977). The role of gibberellic acid (GA_3) in the removal of dormancy in *Fraxinus excelsior* L. seeds. *Biologia Plantarum, 19*(5), 370–376.

Wong, S. K., Tsui, S. K., Kwan, S. Y., Su, X. L., & Lin, R. C. (2000). Identification and characterization of *Podophyllum emodi* by API-LC/MS/MS. *Journal of Mass Spectrometry, 35*(11), 1246–1251. https://doi.org/10.1002/1096-9888(200011)35:11<1246::AID-JMS55>3.0.CO;2-L (accessed on 12th August 2025).

Xia, F., Cheng, H., Chen, L., Zhu, H., Mao, P., & Wang, M. (2020). Influence of exogenous ascorbic acid and glutathione priming on mitochondrial structural and functional systems to alleviate aging damage in oat seeds. *BMC Plant Biology, 20*, 104. https://doi.org/10.1186/s12870-020-2321-x (accessed on 12th August 2025).

Yamaguchi, S. (2008). Gibberellin metabolism and its regulation. *Annual Review of Plant Biology, 59*(1), 225–251. https://doi.org/10.1146/annurev.arplant.59.032607.092804 (accessed on 12th August 2025).

Young, A., Boyle, T., & Brown, T. (1996). The population genetic consequences of habitat fragmentation for plants. *Trends in Ecology and Evolution, 11*(10), 413–418. https://doi.org/10.1016/0169-5347(96)10045-8 (accessed on 12th August 2025).

Zapata, P. J., Serrano, M., Pretel, M. T., Amorós, A., & Botella, M. A. (2004). Polyamines and ethylene changes during germination of different plant species under salinity. *Plant Science, 167*(4), 781–788. https://doi.org/10.1016/j.plantsci.2004.05.014 (accessed on 12th August 2025).

Zerabruk, S., & Yirga, G. (2012). Traditional knowledge of medicinal plants in Gindeberet district, Western Ethiopia. *South African Journal of Botany, 78*, 165–169. https://doi.org/10.1016/j.sajb.2011.06.006 (accessed on 12th August 2025).

Zhang, L., Tian, L. H., Zhao, J., Song, Y., Zhang, C., & Guo, Y. (2009b). Identification of an apoplastic protein involved in the initial phase of salt stress response in rice root by two-dimensional electrophoresis. *Plant Physiology, 149*(2), 916–928. https://doi.org/10.1104/pp.108.131144 (accessed on 12th August 2025).

Zhang, S., Cai, Z., & Wang, X. (2009a). The primary signaling outputs of brassinosteroids are regulated by abscisic acid signaling. *Proceedings of the National Academy of Sciences, 106*(11), 4543–4548. https://doi.org/10.1073/pnas.0900349106 (accessed on 12th August 2025).

Zhou, L., Wu, J., & Wang, S. (2003). Low-temperature stratification strategies and growth regulators for rapid induction of *Paris polyphylla* var. *yunnanensis* seed germination. *Plant Growth Regulation, 41*(2), 179–183. https://doi.org/10.1023/A:1027310507257 (accessed on 12th August 2025).

CHAPTER 8

Conservation of Ethnomedicinal Endangered Plant Species of the Eastern Sub-Himalayan Regions of India

PRIYANKA SHARMA,[1,*] MONISH ROY,[2] KISHORE KUMAR SHARMA,[3] BIDHAN ROY,[2] and GOPAL SHUKLA[4]

[1]*Department of Plant Breeding and Genetics, AICRP on Crops, Faculty of Agriculture, Assam Agricultural University, Jorhat, Assam, India*

[2]*Department of Seed Science and Technology, Uttar Banga Krishi Viswavidyalaya, Pundibari, Cooch Behar, West Bengal, India*

[3]*Department of Plant Breeding and Genetics, Faculty of Agriculture, Assam Agricultural University, Jorhat, Assam, India*

[4]*Department of Forestry, North-Eastern Hill University, Tura Campus, Tura, Meghalaya, India*

*Corresponding author

ABSTRACT

The distribution of medicinal plants varies according to the climatic zones on various hills in North Eastern India. While some therapeutic plants have wide distribution and tremendous promise, others continue to disappear from their natural habitats. Due to overexploitation and habitat fragmentation, the plant species population stock in its native habitat has been rapidly diminishing, thereby putting it on the verge of either extinction or endangered status. About 427 Indian medicinal plant entries are identified as endangered species in Red Data Book records, of which 28 are believed to be extinct, 124 are threatened with extinction, 81 are uncommon, and

Seed Technology for Medicinal and Aromatic Plants: Production, Testing, Storage, and Marketing.
Priyanka Sharma, Kishore Kumar Sharma, and Gopal Shukla (Eds.)

DOI: 10.1201/9781779641199-8

34 are not sufficiently known. This depletion is a result of intense pressure to overuse them through shifting cultivation, urbanization, road construction, and agricultural land use, as well as some natural disasters such as landslides. Additionally, portions of plants, including wild sources, are indiscriminately raided for materials such as leaves, bark, roots, fruits, seeds, or even the entire plant, with no concern for the species' survival. Because more than 95% of medicinal plants are foraged from the wild, many are currently endangered in their natural habitats. Therefore, several conservational considerations for *in situ* conservation and simple agriculture techniques that are easily acceptable by locals in close proximity to forested areas have been eliminated. The cultivation and maintenance of plants in botanical gardens, parks, and other suitable locations, as well as the long-term storage of plant propagules in gene banks, seed banks, pollen banks, DNA libraries, and plant tissue culture repositories, in addition to cryopreservation, can help to preserve medicinal plants *ex situ* or outside of their natural habitat.

8.1 AN OVERVIEW OF NORTH EASTERN INDIA

In northeastern India, there are two biodiversity hotspots: the Himalayas and Indo-Burma. With the largest reservoir of plant diversity, it sustains around 50% of India's biodiversity and is one of the world's richest biodiversity hotspots (Mao and Hynniewta, 2000). In northeastern India, there are many Assamese plains and a hilly forest region. Its coverage extends to Assam, Arunachal Pradesh, Manipur, Meghalaya, Mizoram, Nagaland, Sikkim, and Tripura. The Himalayan hills are found in Sikkim and Arunachal Pradesh, whereas the Naga hills are found in Manipur and Nagaland, and the Garo, Jayantiya, and Khasi hills are found in Meghalaya. Lusai's hills are in Mizoram. According to their names, each of the hills has a distinctive range of topographic and altitudinal variations that regulate the presence and distribution of biodiversity, including the distribution of medicinal plants. Due to nature's abundance of medicinal plants, India is commonly referred to as the "Medicinal Garden" or "Botanical Garden" of the world (Ahmadullah and Nayar, 1999). About 7500 plant species are used medicinally in both conventional and contemporary medical systems, and 900 and 50 species are noteworthy from the viewpoint of ethnobotany out of the 48,655 plant species that are currently known (including viruses, bacteria, algae, fungus, and lichens; Sharma and Pandey, 2013). More than 90% of medicinal plants in India are at risk because of excessive and unsustainable collection, usage, overexploitation, or inexperienced harvesting (Kumari et al., 2011). According to the worldwide rates of plant species facing extinction, there may be 1000 medicinal plant species under peril in diverse ecosystems across India (Frlhtenvis, 2016a). The supply and demand of medicinal plants have increased dramatically during the past decade (2005–2014) (Goraya and Ved, 2017). According to Ved and Goraya (2008), the yearly demand for raw herbal pharmaceuticals increased by 62% in 2014–15 compared with the demand predicted for 2005–06. The rapid extinction of species is due to population growth and related human-driven substantial changes in land use. Additionally, it has raised the likelihood that many essential services that people rely on would be lost (Khandel et al., 2012). According to Pao and Upadhaya (2017), anthropogenic disturbances and the resulting habitat fragmentation are major contributors to the decline in biodiversity. Particular attention should be paid to areas with unique biological characteristics that include high levels of endemism and diversity. Scientists, practitioners, and planners interested in conservation have begun to pay attention to these areas (Margules and Pressey, 2000). The degree of endemism for a location must be considered as a barometer of the uniqueness of the flora when selecting sites for conservation (Young et al., 2002). Endangered species with restricted geographic ranges are more likely to become extinct than species with vast geographic ranges because of their extreme susceptibility to anthropogenic disturbances and environmental change (Myers, 1988). Given the intimate relationship between endemism and extinction threats, efforts to reduce global extinction must concentrate on endemism patterns and species with restricted ranges (Pimm and Brooks, 2000). Exploration of medicinal plants has been conducted in North Eastern India in a variety of forests, from crude drug market places up to the point where marketable medicinal plant items are sold. In addition, numerous attempts have been made to acclimatize people in the region by cultivating therapeutic plants in different zones. The regions' illicit drug production is offered in Assam, West Bengal, Bihar, and even Delhi's Central Market marketplaces. *Aquilaria malaccensis*, *Acorus calamus*, *Aconitum heterophyllum*, *Bacopa monnieri*, *Berberis aristata*, *Coptis teeta*, *Costus speciosus*, *Curcuma caesia*, *Elaeocarpus sphericus*, *Embelia ribes*, Gmelina arborea, Gynocordia odorata (Figure 8.1).

FIGURE 8.1 Essential plant species required for pharmaceutical uses.

8.2 STATUS OF CONSERVATION OF BIODIVERSITY

The preservation of biodiversity is a difficult endeavor that involves social, political, and scientific difficulties (Pelletier et al., 2018). Prioritizing species for conservation, determining the level of threat, and determining the best course of action to follow are essential prerequisites due to the large variety of species and the limited funding for these conservation projects. The International Union for Conservation of Nature (IUCN), a membership union comprising both governmental and non-governmental organizations, was founded to further this goal. The IUCN established a "Red List of Threatened Species" to provide data on extinction risk and species distribution (Brooks et al., 2004). The IUCN classification categorizes species into ranked threat categories such as extinct (EX), extinct in the wild (EW), critically endangered (CR), endangered (EN), vulnerable (VU), near threatened (NT), least concern (LC), and data deficient (DD) based on quantitative criteria on indicators of extinction risk (Collen et al., 2016) (Figure 8.2). Nearly 2.4% of the planet's geographical area and 8% of its biodiversity are found in India, one of the world's 12 mega-diversity countries (Bapat et al., 2008). Four of the 35 global biodiversity hotspots (GBH) that are situated inside the Indian geopolitical region are the Himalayas, Indo-Burma, Sundaland, and Western Ghats (Jaisankar et al., 2018). India has a significant phyto-diversity because it has 15 agro-climatic zones with different biological environments. India is home to 10% of the world's floral variety. More than 50% of the world's plant species are found in the 35 GBH, which once made up 15.9% of the planet's land area but are currently only 2.3% (Mitter et al., 2011). These hotspots are home to several endemic species that are in danger of becoming extinct (Hazarika et al., 2016). According to the IUCN, 40,468 species, or 13.49% of the world's vascular plants (3,000 species), are either threatened to varying degrees or are in danger of going extinct. These threats include population decline, pollinator loss, decreased reproductive and/or seed germination capacity, habitat degradation both naturally occurring and caused by humans, overexploitation, and genetic instability. Of India's 18,532 vascular plant species, approximately 2142 species, or 11.53%, are red-listed. Approximately 54 species are listed as near endangered, whereas IUCN 2020 classifies 432 species as threatened (CR, EN, and VU), 8 of which are extinct, indicated through a schematic diagram in Figure 8.3.

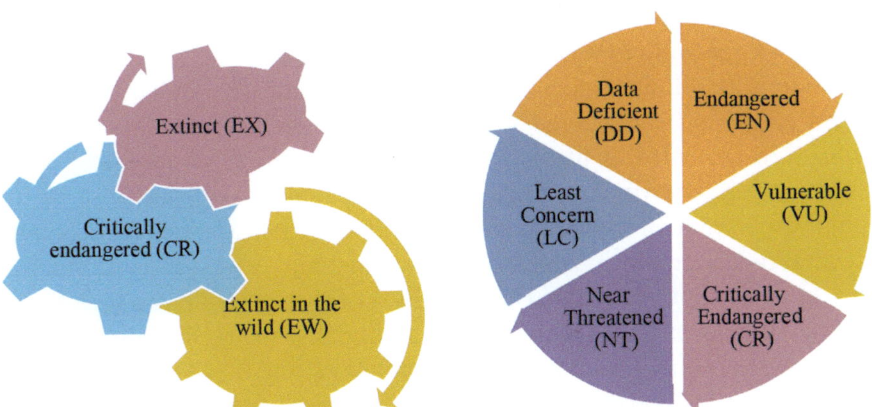

FIGURE 8.2 Schematic representation of classification of species based on threat categories.

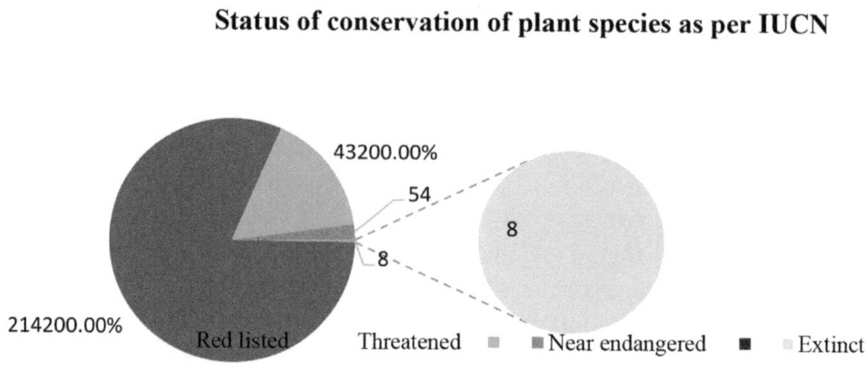

FIGURE 8.3 Graphical representation of status of conservation of plant species as per IUCN.

8.3 GEOGRAPHIC DISTRIBUTION OF MEDICINAL PLANT SPECIES

India, which has 10 biogeographic areas, is one of the top 12 countries in the world for mega diversity. Two of the eight biodiversity hotspots in the world are located in India alone. The country's diverse ecological habitats, along with the country's climatic and altitudinal variations, have helped to develop an incredibly rich vegetation, including a rare diversity of medicinal plants. These plants serve as an important source of medicinal raw materials for both domestic and international pharmaceutical industries and traditional medical systems. Over 21,000 plant species used for therapeutic purposes are classified by the World Health Organization (WHO). Approximately 2500 plant species are used in India's traditional medical practices. In total, 427 Indian medicinal plant entries on endangered species are listed in the Red Data Book, of which 28 are listed as extinct, 124 as endangered, 81 as rare, and 34 as inadequately known. Another aspect of relevance in this situation is the natural presence of traded species. The distribution of these species over several biogeographical zones, diversified habitats, and landscape features has been discovered. Approximately 18% of these species are found only in the Himalayan and Trans-Himalayan regions, which include northeast India; 4% are only found in the Western Ghats; and 0.5% are found in the Desert region. Approximately 77% are dispersed widely over the different biogeographic regions of the nation.

8.4 LOSS OF MEDICINAL PLANT BIODIVERSITY

8.4.1 EXTERNAL VARIABLES

- **Rainfall:** Over the past few years, the yearly rainfall has decreased, which has improved the summer health of many herbaceous plants.
- **Deforestation:** Over the past 20 years, deforestation has been documented. The main causes of the decline in the area covered by important woods include the expansion of agriculture, logging, firewood collection, heavy wood collection, and heavy grazing. Due to deforestation efforts, numerous valuable wild medicinal plant species are eliminated or reduced every year. Water body siltation: The water holding capacity has decreased due to the situations of water bodies in forests, which will eventually deplete subsurface water.
- **Development Oriented Actions:** Because of submersion, several species of medicinal plants have disappeared from the forests. For instance, the Maradavally forest serves as a catchment for the Linganamakki Dam, which is the largest reservoir used for irrigation and power generation in Karnataka. Nearly 10 sq. km of forest was submerged during the monsoon, which led to the extinction of important medicinal plant species.
- **Infrastructure:** The Devanarayanadurga forest in Karnataka has suffered considerable damage because of road expansion, installation of power lines, and construction of buildings.

8.5 TECHNIQUES IN FORESTRY AND AGRICULTURE

- **Monoculture:** Monoculture plantations of commercially significant indigenous and foreign plants have been steadily expanding throughout the forest. Monoculture plantations, such as the eucalyptus and acacia species found in many woods, completely disrupt organic productivity and impair the natural stability and complexity, leading to the loss of medicinal plants.

- **Encroachments:** Forestland encroachments have reached frightening heights. The agricultural methods utilized on high sloppy lands have also resulted in soil erosion and a loss of the richness of medicinal plants, in addition to the fall of trees and clearance of vegetation.
- **Over Exploitation:** There is extensive gathering of medicinal plants from the forests. Unorganised forest collectors were responsible for the collection, and they then sold the product to a contractor at a price set by the latter.

8.6 MEDICINAL PLANT CONSERVATION STRATEGIES AND ORGANIZATION IN INDIA

Since the early 1990s, the Government of India (GoI) has given medicinal plants the attention they deserve, and a number of initiatives have been made to protect and preserve these species both *in situ* (biosphere reserves, national parks, wildlife sanctuaries, sacred groves, etc.) and *ex situ* (botanical gardens, field gene banks, seed gene banks, *in vitro* gene banks, and cryo gene-banks).

Approximately 16.5 million ha (5.02%) of India's total geographic area is covered by protected areas, while 70.8 million ha (21.54%) is covered by forests, allowing for *in situ* conservation. About 870 protected areas, including 104 national parks, 551 animal sanctuaries, 127 community reserves, and 88 conservation reserves, are designated in India (Wiienvis, 2019). Several institutions and organizations, including the Central Council for Research in Ayurvedic Sciences (CCRAS), Central Council for Research in Homoeopathy (CCRH), Central Council for Research in Siddha (CCRS), and Central Council for Research in Unani Medicine (CCRUM) under the Ministry of Ayurveda, have been established by the Government of India specifically to conduct research on medicinal plants for *ex situ* conservation. The Jawaharlal Nehru Tropical Botanic Garden and Research Institute (KSCSTE-JNTBGRI, formerly TBGRI), the Indian Council of Forestry Research and Education (ICFRE), the Central Drug Research Institute (CDRI), the Central Institute of Medicinal and Aromatic Plants (CIMAP), the National Botanical Research Institute (NBRI), and the Regional Research Laboratories are all CSIR institutions. Additionally, there are a number of other businesses, ayurveda practitioners, and government and non-government organizations. In 1993, the G-15 GEBMAP program established a network of four National Gene Banks for Medicinal and Aromatic Plants (GEBMAP) at (i) ICAR-NBPGR; (ii) KSCSTE-JNTBGRI; (iii) CSIR-CIMAP; and (iv) Regional Research Laboratory (RRL) Jammu. The Department of Biotechnology (DBT), Ministry of Science and Technology, GoI, served as the programs' nodal agency. This has not only improved the focus and impetus, particularly on the collection and conservation of medicinally important vulnerable species, but it has also assisted in coordinating ongoing activities in the nation (Sharma et al., 2020). The 10 acres of the JNTBGRI herbal garden contain a collection of 1,200 taxa of herbs (Kscste-Jntbgri, 2019). The national coordinator for all activities involving plant genetic resources (PGR) introduction, collection, conservation, documentation, assessment, and distribution is the ICAR-NBPGR. It has 59 National Active Germplasm Sites (NAGS) based at ICAR institutes (specific crops groups), which are tasked with the distribution of active germplasm collections to users at the national and international levels. It also has 10 regional stations spread across the country's various agroclimatic zones. The network is also connected to some ICAR institutes, State Agricultural Universities, and other stakeholders (Singh et al., 2016). Four different types of conservation facilities are available at the National Gene Bank (NGB) of the ICAR-NBPGR: seed gene banks, cryo gene banks, *in vitro* gene banks, and field gene banks. About 8,071 MAP accessions are being conserved at ICAR-NBPGR in the seed gene bank, 178 in the *in vitro* gene bank, and 1,041 in the cryo gene bank (Sharma et al., 2020). Along with creating NMPB under the Ministry of AYUSH in 2000, the GoI also provided funding for studies on medicinal plants. Along with NMPB, DBT and the Department of Science and Technology (DST) have offered funding to researchers in studying medicinal plants. *Asparagus racemosus* (Burm.), *Aegle marmelos* (L.), *Andrographis paniculata, Bacopa monnieri* (L.), *Berberis aristata, Cassia angustifolia, Senna alexandrina, Chlorophytum borivilianum, Commiphora wightii* (Arn.), *Coleus barbatus, Phyllanthus emblica* L., *Garcinia indica, Emblica officinalis, Embelia ribes Burm., Crocus sativus, Gymnema sylvestre* (Retz.) R. Br., *Nardostachys jatamansi* (Retz.) R. Br., *Gloriosa superba* L., *Glycyrrhiza glabra* L., *Ocimum sanctum* L., *Phyllanthus amarus, Santalum album* L., *Rauvolfia serpentina Benth.* ex Kurz., *Plantago ovata* Forsk.,

Picrorhiza kurroa Royle ex Benth., *Piper longum* L., (NMPB 2020b) *Swertia chirata* Roxb. ex Flem., *Withania somnifera, Saraca asoca* (Roxb.) De Wilde, *Saussurea costus* (Falc.) Lipsch, *S. nigrum* L. In addition, the NMPB has established 36 State Medicinal Plant Boards in State/UT and 72 Medicinal Plant Conservation and Development Areas (MPCDAs) across 13 states of the nation, totaling 10,935 ha in forested areas (Biswas et al., 2017). These initiatives are aimed at the *in situ* conservation of medicinal plants. FRLHT have established 108 medicinal plant conservation areas (MPCAs) across 13 states and 18 medicinal plants conservation parks (MPCPs) in southern India to focus on the conservation of prioritized wild medicinal plants that occur in various regions of the country (Frlhtenvis, 2016a). These projects were funded by the Danish International Development Aid (DANIDA), The Netherlands, and the United Nations Development Program (UNDP). DBT launched a massive network effort in 2012 with the goal of "conserving 100 threatened species of India" and "preventing extinction and improving the conservation status of threatened plants through application of biotechnological tools" (Barik et al., 2018). About 115 species were reintroduced as part of this effort, and 106 species' macro and 76 species' micro-propagation methods were standardized (Barik et al., 2018). In India, there are no specific laws or rules protecting medicinal plants that thrive in forests. The forestry rules currently in effect cover their conservation.

Only a very small percentage of the medicinal plant species traded (both domestically and internationally) are grown; the rest are acquired from the country's many distinct agroclimatic zones (Sharma and Pandey, 2013). It is significant to remember that, of the 242 species with the highest commercial demand (100 MT annually), 72% are mostly supplied from wild plants in the form of roots, bark, wood stems, or the entire plant. This poses a serious problem because it concerns the continued existence of these species. The availability of many medicinal plant species in the nation has rapidly decreased over the past three decades because of unsustainable collecting practices, little to no replenishment/cultivation efforts, long plant life cycles, climate change, and development activities. This has led to a large number of species becoming threatened both nationally and internationally. The GoI has created a few laws and policies for the conservation of forests that either directly or indirectly conserve the wild herbal flora to reduce this, as indicated in the flow chart (Figure 8.4).

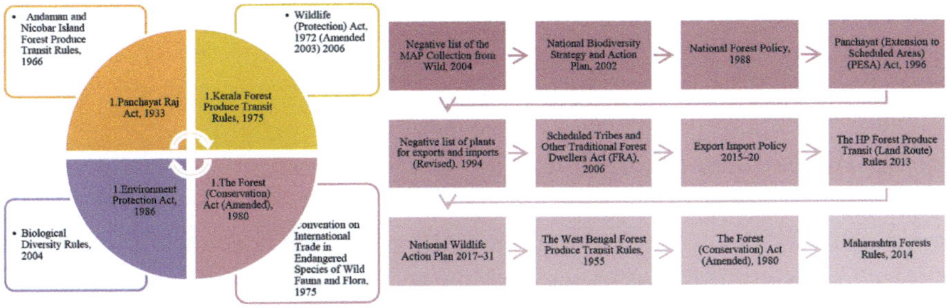

FIGURE 8.4 Laws and policies enacted by the government of India for conservation of forests.

The Central Government announces plants that are close to extinction in accordance with Section 38 of the Biological Diversity Act, 2002, in conjunction with the Government of States. To promote and expand the market for medicinal plants, the NMPB has created many policies, strategies, and programs for conservation, ethical harvesting, economical cultivation, research and development, processing, and marketing of raw materials. The National Gene Bank of ICAR-NBPGR (ICAR-NBPGR, 2020) and NMPB recently signed a Memorandum of Understanding (MoU) for the safe conservation of medicinal and aromatic PGR for long-term storage. India has signed some international accords in addition to its national policy framework: (i) The Convention on International Trading in Endangered Species of Wild Fauna and Flora (CITES), a treaty between nations that attempts to ensure that trading in endangered wild plants and animals does not endanger their survival. About 5000 animal species and 28,000 plant species are protected from overexploitation through international trade by CITES. A list of 113 species (CITES, 2017) that the Ministry of Environment, Forests, and Climate Change of the Government of India has made public requires a "Certificate of Cultivation" or "Legal Procurement Certificate" from the designated officials of the Forest Departments to be exported.

8.6.1 *IN SITU CONSERVATION*

Conservation of a species in its natural environment or where it grows naturally is referred to as *in situ* conservation. (i) It entails things like gene banks and gene sanctuaries, biosphere reserves, national parks, holy locations, and sacred groves, among other things; (ii). Plant variety at the genetic, species, and ecosystem levels can only be preserved over time in nature; (iii). It is important to preserve interspecific and intraspecific genetic variation in separate, representative biogeographic zones. The most efficient method of preserving the genetic and biological diversity that already exists is through "*in situ*" or "on-site" conservation, which involves preserving a wild population of a particular species or a stock of a biological community in its native habitat. Examples of "*in situ*" methods of conservation include the creation of biosphere reserves, national parks, wildlife sanctuaries, sacred groves, and other protected areas. At the national, international, and global levels, the concept of creating a network of protected areas has become important to all policy decision-making processes related to biodiversity conservation. Below, we have highlighted the *in situ* conservation techniques that are most frequently used:

- **National Parks:** Pin Valley National Park and Great Himalayan Park, located in the regions of Lahaul Spiti and Kulu, respectively, are two of the 91 national parks that have been established in the nation. Wetlands (19, including 6 Ramsagar sites), mangroves (15), coral reefs (4), and other locations such as holy groves, natural monuments, ethno-biological reserves, and so forth, are all protected areas in addition to the 448 wildlife sanctuaries mentioned above. These will undoubtedly help preserve biodiversity in the aforementioned places. However, experiences have adequately shown this in a populous, growing nation like India, where a large portion of the population lives close to the woods.

8.6.2 ADDITIONAL IN SITU CONSERVATION TECHNIQUES

- **On Farm Conservation:** Within the confines of the conventional agricultural system, producers maintain traditional crop cultivars (sometimes referred to as "land races") or farming methods. Land races, which the farmer develops and are well fitted to the local environment, are used by traditional farmers. Although farmers have been using it for millennia, this conservation technique has recently gained popularity. The consequences of growers' practices are crucial for agrobiodiversity. It is necessary to systematically document farmers' understanding of diversity and its applications.

- **Gardens at Home:** Although on-farm conservation is on a much larger scale, conservation in gardens at home is comparable. Home gardens in rural areas frequently include a wide variety of species, including fruits, vegetables, medicinal and spice plants.

Para-frost conservation in the Himalayan region, which is still covered in snow, is based on a concept with zero energy input.

8.6.3 EX SITU CONSERVATION

By cultivating and maintaining plants in botanic gardens, parks, and other suitable locations, as well as by long-term preservation of plant propagules in gene banks (seed banks, pollen banks, DNA libraries, etc.), plant tissue culture repositories, and by cryopreservation, it is possible to conserve medicinal plants *ex situ*, or outside of their natural habitat.

8.7.1 FIELD GENE BANK (FIELD REPOSITORY/CLONAL REPOSITORY)

8.7.1.1 GENE BANK

Seed storage (Active collection at 4°C to 10°C; Base collection at –20°C). For the *ex situ* preservation of medicinal and aromatic plants, India has created three NGB.

- The ICAR-affiliated National Bureau of Plant Genetic Resources (NBPGR), New Delhi.
- CIMAP, Lucknow, Uttar Pradesh; CSIR; Ministry of Science and Technology; Government of India.
- Thiruvananthapuram (Kerala)'s Tropical Botanical Gardens Research Institute (TBGRI), Palode.
- The National Bureau of Plant Genetic Resources (NBPGR), which is housed within the Department of Agricultural Research and Education (DARE) of the Indian Council of Agricultural Research (ICAR), is solely responsible for the conservation of genetic variability in domesticated plants and their wild relatives.

8.7.1.2 SEED GENE BANK

In the Seed Gene Bank, germplasm preservation is more cost-effective. The NGB, which is housed at the NBPGR in New Delhi, is largely in charge of conserving the germplasm of agri-horticultural crops and their wild cousins for long-term seed storage for future generations. The "base collection" is kept in modules at –20°C and is known as such. The seeds are hermetically packed in packets made of moisture-proof aluminum foil after drying to a moisture content of 4–6%. For 50 to 100 years after storage, these seeds are still useful. Only seed samples with a seed viability of more than 85% are treated in crops. Preferably, the seeds in gene banks are preserved in accordance with the gene bank criteria advised by the FAO/IPGRI.

8.7.2 PLACES FOR NATIONALLY ACTIVE GERMPLASM

The network's core element is the NAGS. There are currently 40 NAGS, which are housed in SAUs and ICAR institutes (institutes based on crops for a particular crop or a range of crops). These are essential components of the national network for conserving plant biodiversity. The NAGS is in charge of dissemination to legitimate users at the national and international levels, as well as the multiplication, evaluation, maintenance, and conservation of the active collection. The modules housing these active/working collections are maintained at 4°C and 35–40% relative humidity (RH). Seeds are predicted to be viable at these temperatures for 15 to 50 years. Seed moisture content is reduced to 8–10% for medium-term storage. To support the active germplasm conservation activities of the regions, the NBPGR maintains a network of II regional stations spread across several agroclimatic zones of the nation.

8.7.2.1 CRYOPRESERVATION IN LIQUID NITROGEN (–165°C TO 196°C)

- **Preservation of Seeds:** Based on how they responded to dehydration, the seeds have been roughly divided into two types. Since the bulk of them are "Orthodox," or desiccation-tolerant, they can be kept for longer periods of time. The second category of plant species is referred to as "Recalcitrant," and their seeds suffer damage during drying and cannot be kept at extremely low temperatures.

- **Pollen Preservation:** The primary purpose of pollen storage is to facilitate the regulated pollination of plants that flower at the same time, particularly fruit tree species. Pollen preservation has also been regarded as a cutting-edge technique for genetic preservation.4–6 Large amounts of pollen can be conveniently harvested and cryopreserved in comparatively little space.

8.8 PLANT QUARANTINE ISSUES ARE REDUCED WHEN GERMPLASM IS TRANSFERRED BY POLLEN

Different types of pollen have recently undergone cryopreservation techniques (Towill, 1985), and cryobanks of pollen have been established for various fruit-tree species in various countries (Alexander et al., 1993). Preservation of *in vitro* (on tissue culture): The establishment of specialized facilities (culture rooms with regulated environments, artificial lights, laminar airflow cabinets, autoclave, etc.), as well as qualified scientists and technicians, are needed for an *in vitro* conservation strategy. Also desirable is information on plant species' *in vitro* multiplication and/or conservation techniques. *In vitro* multiplication to create many plants and *in vitro* storage are the two main stages of any *in vitro* conservation strategy. Meristems, shoot tips, axillary buds, embryos, calluses, and cell suspensions are just a few examples of materials that can be preserved *in vitro*. *In vitro* gene banks can be economical and simple to maintain if efficient storage solutions are created. DNA preservation (at –20°C): In theory, storing DNA is a straightforward and frequently used process. DNA appears to be relatively simple and inexpensive to store. In recent years, gene transfer has become possible across species and genera thanks to genetic engineering. With the help of genes taken from viruses, bacteria, fungi, and even mice, transgenic plants have been created. DNA libraries have been established as a result of these efforts. To take the advantage of the data stored in DNA, necessary methods and techniques must be created.

8.8.1 ARBORETA AND BOTANICAL GARDENS

For scientific research, conservation, display, and education, a botanical garden is an establishment that maintains a documented collection of living plants (Jackson, 1999). They act as storage facilities for collections of rare and endangered genetic materials from both domestic and foreign origins (Sharma and Goel, 1994). To foster worldwide cooperation and keep track of botanical gardens' conservation initiatives, Botanic Garden Conservation Worldwide (BGCI), a global association with its headquarters in London (UK), was founded in 1987. Worldwide, there are 500 botanical gardens that are members of the BGCI (Jackson, 1999). According to the BGCI database, there are approximately 1846 botanical gardens worldwide. India has a network of over 140 botanical gardens, 33 of which are connected to botany departments at 33 institutions. However, there are only about 30 active conservation programs in botanical gardens. Kerala's Tropical Botanical Garden and Research Institute (TBGRI), situated in a degraded forest area of the Western Ghat Mountains, is a prime example of *ex situ* plant variety conservation (Figure 8.5).

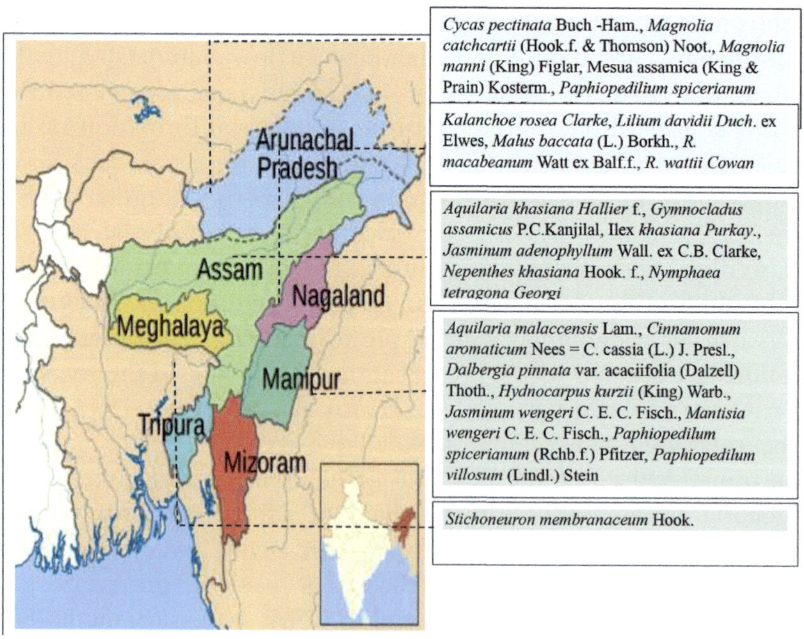

FIGURE 8.5

Source: Modified concept from Gowthami et al. (2022).

8.8.2 SIGNIFICANCE OF MEDICINAL PLANTS

Since ancient times, medicinal plants have been used to manage human health, and through several scientific analyses, they continue to do so with stability in accordance with the demand for the treatment of disorders. In addition, it has opened up opportunities for plant research and identification in diverse climate zones, regions, and states (Kirtikar and Basu, 1935). The value of medicinal plants and the conventional healthcare system is becoming increasingly important in addressing global health problems. During the survey study in different northeastern states, it was found that medicinal plants are classified in several categories having a well-established system of medicines such as Ayurveda, Unani, Siddha, Homoeopathy, and modern remedy. In addition, the plants used by the locals provide health administration by obtaining locally accessible herbs from nature. Thus, northeast India has an incredibly unique ecological ecosystem rich in plant species having ethnomedicinal properties (Tynsong et al., 2020). Among these many varieties of medicinal plants, the northeast origin has changed allopathic medical paradigms. The importance of therapeutic plant species has also generally been linked to an important issue in the sociocultural, spiritual, and therapeutic lives of local residents. According to the WHO, 80% of rural populations in Asia and Africa use naturally available medicinal herbs to meet their basic medical needs. Only 10% of medicinal plants in India are distributed among open fields, horticultural fields, squatter land, and in and near freshwater bodies (Mazid et al., 2012). This is because approximately 90% of India's therapeutic plants having folk medicinal purpose are found in woodland regions. A significant aspect of northeast Indian culture is conventional therapeutic practice. Despite this circumstance, traditional medical services have faced challenges due to modern human advancement, industrialization, and a decline in support, particularly in urban areas.

8.9 CONCLUSION AND FUTURE PROSPECTIVES

Although northeastern India is endowed with wild natives of MAP, overexploitation of these naturally grown resources for human needs has caused several medicinal plant species to be on the verge of extinction and endangered, according to information collected from IUCN lists. Therefore, scientific research has been carried out by several researchers worldwide for further conservation and preservation of these highly valuable medicinal plant species. In this chapter, we have discussed the strategies and methods of conservation based on geographical locations and status of conservation. In addition, various acts related to the conservation of biodiversity have also been discussed. In terms of health benefits, the significance of medicinal plants is explained in detail in this manuscript, since the value of medicinal plants and the conventional healthcare system is becoming increasingly important in addressing global health problems. Further research could be conducted for the long-term conservation and utilization of these plant species in order to maintain ecological balance and biodiversity conservation in northeastern India.

KEYWORDS

- **botanical gardens**
- **conservation**
- **ethnomedicinal values**
- **medicinal plants**
- **overexploitation**
- **therapeutic plants**

REFERENCES

Ahmadullah, M., & Nayar, M. P. (1999). *Red Data Book of Indian Plants (Peninsular India)*. Botanical Survey of India, Calcutta.

Alexander, M. P., & Ganeshan, S. (1993). Pollen storage. In K. L. Chadha et al. (Eds.), *Advances in Horticulture: Fruit Crops–Part I* (pp. 481–496). New Delhi: Malhotra Publishing House.

Bapat, V. A., Yadav, S. R., & Dixit, G. B. (2008). Rescue of endangered plants through biotechnological applications. *National Academy Science Letters, 31*, 201–210.

Barik, S. K., Tiwari, O. N., Adhikari, D., Singh, P. P., Tiwary, R., & Barua, S. (2018). Geographic distribution pattern of threatened plants of India and steps taken for their conservation. *Current Science, 114*, 470–503.

Biswas, S., Rawat, M. S., Tantray, F. A., & Sharma, S. (2017). Medicinal plants conservation and development areas (MPCDAs)–An initiative towards conservation of medicinal plants. *Medicinal Plants–International Journal of Phytomedicines and Related Industries, 9*, 143–149.

Brooks, T. M., Bakarr, M. I., Boucher, T., Da Fonseca, G. A., Hilton-Taylor, C., Hoekstra, J. M., Moritz, T., Olivieri, S., Parrish, J., Pressey, R. L., & Rodrigues, A. S. (2004). Coverage provided by the global protected-area system: Is it enough? *Bioscience, 54*, 1081–1091.

CITES. (2017). CITES listed species. https://www.cites.org/eng/disc/species.php (accessed on 12th August 2025).

Collen, B., Dulvy, N. K., Gaston, K. J., Gärdenfors, U., Keith, D. A., Punt, A. E., Regan, H. M., Böhm, M., Hedges, S., Seddon, M., & Butchart, S. H. (2016). Clarifying misconceptions of extinction risk assessment with the IUCN Red List. *Biology Letters, 12*, 20150843.

FRLHTENVIS. (2016a). Centre on medicinal plants–Medicinal plants under threat. http://envis.frlht.org/overview.html (accessed on 12th August 2025).

Goraya, G. S., & Ved, D. K. (2017). *Medicinal Plants in India: An Assessment of Their Demand and Supply*. National Medicinal Plants Board, Ministry of AYUSH, Government of India, New Delhi, and Indian Council of Forestry Research and Education, Dehradun.

Gowthami, R., Sharma, N., Pandey, R., & Agrawal, A. (2022). Status and consolidated list of threatened medicinal plants of India. *Genetic Resources and Crop Evolution, 68*, 2235–2263.

Hazarika, T. K., Marak, S., Mandal, D., & Shukla, A. C. (2016). Underutilized and unexploited fruits of Indo-Burma hotspot, Meghalaya, north-east India: Ethno-medicinal evaluation, socio-economic importance and conservation strategies. *Genetic Resources and Crop Evolution, 63*, 289–304.

ICAR-NBPGR. (2020). Memorandum of undertaking signed between National Medicinal Plants Board and ICAR-National Bureau of Plant Genetic Resources on 06–07–2020.

IUCN. (2020). *The IUCN Red List of Threatened Species* (Version 2020–1). https://www.iucnredlist.org (accessed on 12th August 2025).

Jackson, W. P. (1999). Plenary address. BGCI 4th International Congress on Education in Botanic Gardens, Thiruvananthapuram, Kerala, India.

Jaisankar, I., Velmurugan, A., Swarnam, T. P., & Singh, A. K. (2018). Hotspots: An introduction and role in conservation. In C. Sivaperuman & K. Venkataraman (Eds.), *Indian Hotspots* (pp. 1–21). Singapore: Springer.

Khandel, K. A., Ganguly, S., Bajaj, A., & Khan, S. (2012). New records, ethno-pharmacological applications and indigenous uses of *Gloriosa superba* L. (Glory Lily) practices by tribes of Pachmarhi Biosphere Reserve, Madhya Pradesh, Central India. *Nature and Science, 10*(5), 23–48.

Kirtikar, K. R., & Basu, B. D. (1935). *Indian Medicinal Plants* (Vol. II). Lalit Mohan Publication, Allahabad.

KSCSTE-JNTBGRI. (2019). Jawaharlal Nehru Tropical Botanic Garden and Research Institute–Herbal garden.

Kumari, G. P., Joshi, C., & Tewari, L. M. (2011). Diversity and status of ethno-medicinal plants of Almora district in Uttarakhand, India. *International Journal of Biodiversity and Conservation, 3*, 298–326.

Mao, A. A., & Hynniewta, T. M. (2000). Floristic diversity of North East India. *Journal of the Assam Science Society, 41*(4), 255–266.

Margules, C. R., & Pressey, R. L. (2000). Systematic conservation planning. *Nature, 405*, 243–253.

Mazid, M., Khan, T. A., & Mohammad, F. (2012). Medicinal plants of rural India: A review of use by Indian folks. *Indo Global Journal of Pharmaceutical Sciences, 2*, 286–304.

Mitter, M. R. A., Gill, P. R., & Hoffman, M. (2011). *Hotspots Revisited: Earth's Biologically Richest and Most Endangered Terrestrial Ecoregions*. Mexico: CEMEX.

Myers, N. (1988). Threatened biotas: Hotspots in tropical forests. *The Environmentalist, 8*, 187–208.

NMPB. (2020b). National Medicinal Plants Board (NMPB). https://www.nmpb.nic.in (accessed on 12th August 2025).

Pao, N. T., & Upadhaya, K. (2017). Effect of fragmentation and anthropogenic disturbances on floristic composition and structure of subtropical broad leaved humid forest in Meghalaya, northeast India. *Applied Ecology and Environmental Research, 15*(4), 385–407.

Pelletier, T. A., Carstens, B. C., Tank, D. C., Sullivan, J., & Espíndola, A. (2018). Predicting plant conservation priorities on a global scale. *Proceedings of the National Academy of Sciences USA, 115*, 13027–13032.

Pimm, S. L., & Brooks, T. M. (2000). The sixth extinction: How large, where, and when? In P. H. Raven & T. Williams (Eds.), *Nature and Human Society: The Quest for a Sustainable World* (pp. 46–62). Washington, DC: National Academy Press.

Sharma, N., & Pandey, R. (2013). Conservation of medicinal plants in tropics. In M. N. Normah, H. F. Chin, & B. M. Reed (Eds.), *Conservation of Tropical Plant Species* (pp. 437–487). New York: Springer.

Sharma, N., Pandey, R., & Gowthami, R. (2020). *In vitro* conservation and cryopreservation of threatened medicinal plants of India. In P. E. Rajasekharan & S. H. Wani (Eds.), *Conservation and Utilization of Threatened Medicinal Plants* (pp. 181–228). India: Springer.

Sharma, S. C., & Goel, A. K. (1994). Role of botanic gardens in modern times. *Indian Journal of Forestry, 17*, 230–238.

Singh, J. P., Kumar, S., Venkatesan, K., & Kulloli, R. N. (2016). Conservation status and utilization of *Caralluma edulis*: An important threatened medicinal plant species of the Thar Desert, India. *Genetic Resources and Crop Evolution, 63*, 721–732.

Towill, L. E. (1985). Low temperature and freeze/vacuum–drying preservation of pollen. In K. K. Kartha (Ed.), *Cryopreservation of Plant Cells and Organs* (pp. 171–198). Florida, USA: CRC Press.

Tynsong, H., Dkhar, M., & Tiwari, B. (2020). Traditional ecological knowledge of tribal communities of North East India. *Biodiversitas Journal of Biological Diversity, 21*, 3209–3224.

Ved, D. K., & Goraya, G. S. (2008). *Demand and Supply of Medicinal Plants in India*. Bishen Singh Mahendra Pal Singh, Dehradun, and FRLHT, Bangalore, India.

Wiienvis. (2019). Envis Centre on Wildlife and Protected Areas–Protected areas of India.

Young, K. R., Ulloa-Ulloa, C., Luteyn, J. L., & Knapp, S. (2002). Plant evolution and endemism in Andean South America: An introduction. *The Botanical Review, 68*, 4–27.

CHAPTER 9

Safe Seed Storage Including Ecofriendly Techniques Used in Medicinal and Aromatic Plants

RAJDEEP MOHANTA[1,*] and PAYAL MAITI[2]

[1]*Department of Agriculture, Brainware University, Barasat, Kolkata, West Bengal, India*

[2]*Department of Post-Harvest Management, Bidhan Chandra Krishi Viswavidyalaya, Mohanpur, Nadia, West Bengal, India*

*Corresponding author

ABSTRACT

In India, nearly 70% of the population depends on agriculture for their livelihood, making it the primary occupation for people worldwide. Thanks to advances in technology, seed production in India has been growing significantly; yet, post-harvest loss stays above 10%. Since adequate storage facilities are unavailable, losses incurred during storage account for around 6% of all losses. In addition to having a significant role in the sociocultural, medical, and spiritual spheres for rural Indians, medicinal plants are currently in high demand worldwide. Since ancient times, medicinal and aromatic plants (MAPs) have been used to treat a wide range of illnesses in conventional medical systems. They are also used as sources of natural dye, fat, essential oil, biopesticide, resin, protein, vitamin, condiment, spice, timber, fiber, and other useful materials, as well as for therapeutic, fragrant, and flavoring products in the pharmaceutical and cosmetic industries. The World Health Organization (WHO) estimates that 80% of individuals in underdeveloped nations receive their

Seed Technology for Medicinal and Aromatic Plants: Production, Testing, Storage, and Marketing.
Priyanka Sharma, Kishore Kumar Sharma, and Gopal Shukla (Eds.)

DOI: 10.1201/9781779641199-9

primary medical treatment mostly from traditional medicine. Historically, women have been the primary creators of medical items derived from plants. According to the Botanical Survey of India (March, 2022), there are currently 8000 species of medicinal plants in India (BSI). The BSI has been conducting a survey and documenting all of the nation's plant resources, including MAP and herbs, as well as traditional knowledge associated with them at the national, regional, state, and ecosystem levels. For millions of rural people worldwide, particularly women and tribal peoples, medicinal plants offer vital sources of income.

Therefore, it is crucial to educate farmers about the benefits and usefulness of cultivating medicinal plants. Additionally, maintaining a relatively high viability status, germination, and seedling vigor even after significant seed storage under ambient conditions suggests that the seeds are suitable for this species' regeneration and cultivation.

9.1 INTRODUCTION

The importance of growing MAP is continuously growing exponentially. The National Afforestation and Eco-development Board (NAEB), MoEF&CC, supports the cultivation of medicinal plants on degraded forest lands through various programs. Under the "Assistance to Botanic Gardens" Scheme of the MoEF&CC, numerous vulnerable species are conserved in botanical gardens affiliated with BSI and gardens affiliated with various universities/colleges. Both *in situ* (i.e., in their native habitat) and *ex situ* (i.e., using both traditional and cutting-edge methods, such as tissue culture) conservation are being used to preserve the vulnerable plant species. The following initiatives are funded by the Central Sector Scheme on "Conservation, Development and Sustainable Management of Medicinal Plants," which is being implemented (2022) by the Indian government. Those are:

- *In situ* conservation / *Ex situ* conservation.
- Livelihood linkages with Joint Forest Management Committees (JFMCs) / Panchayats / Van Panchayats / Biodiversity Management Committees (BMCs) / Self Help Groups (SHGs).
- IEC activities like Training / Workshops / Seminars/ Conferences and so forth.
- Research and Development.
- Promotion, marketing, and trade of medicinal plants produce.

Under the medicinal plants component of NAM (National AYUSH Mission) scheme, there is a provision to support semi-processing (small processing unit) / processing units, and the Ministry of Ayush had supported 25 semi-processing (small processing unit) / processing units throughout the country from the financial year 2015–16 to 2020–21. The significance of storing seeds has been acknowledged since the beginning of human domestication of plants. Farmers have required maintaining healthy seeds from one growing season to the next since the beginning of commercial agriculture (short-term seed storage, usually 3 to 9 months but occasionally up to 18 months). The goals and the type of plant being stored will influence how long the process takes. Different plant species' seeds react differently to their surroundings both before and after storage.

According to Roberts (1972), there are currently three primary categories of seed storage behavior: orthodox, recalcitrant, and a third category that falls in between, which is commonly known as intermediate seeds (Ellis et al., 1991). In suitable conditions, seeds of plant species with conventional seed storage behavior can be successfully preserved in *ex situ* conditions for an extended period of time. It is challenging to maintain the viability of seeds that exhibit insensitive and intermediate seed storage behavior. However, if the environment is well-defined, seeds can be stored for a medium amount of time with intermediate storage behavior. Since a seed is a living thing, its quality eventually gets affected by the numerous environmental obstacles it faces. The viability and vigor of seeds in storage are influenced by a variety of physico-chemical factors, including moisture content, temperature, atmospheric relative humidity, seed quality at initial planting, seed composition (both chemical and physical), gaseous exchange, storage structure, and packaging materials (Doijode, 1988). Temperature and relative humidity are the two main ones among them, and they cause the seed to decline substantially. Many crops have been the subject of numerous attempts to devise techniques for extending the time that seeds remain viable and vigorous while being stored (Ader, 1978; Agrawal, 1980; Arati, 2000; Caneppele et al., 1995; Doijode, 1997; Padma and Reddy, 2002). It is now commonly known that the viability and vigor of seeds are positively impacted by the materials chosen for seed coating or treatment, storage containers, and storage environment. According to Roberts (1972), damage to the seed membrane, enzyme, proteins, and nucleic acid causes the deterioration of seeds during storage. Over time, the accumulation of these degenerative alterations leads to the total disarray of membranes and cell organelles, which ultimately results in the seed's death and loss of germination. The rising interest in and acceptance of herbal medicine has led to a global increase in the demand for medicinal plants. The majority of the demand is satisfied by gathering significant quantities of usable parts or medicinal plants from natural environments. Consequently, the pace of exploitation of these ethno-medicinal plant species surpasses the rate of natural regeneration in the area. Natural habitats have been found to be diminishing in recent years (Pushpangadan and Nair, 2011; Rao and Rajasekharan, 2002) due to deforestation, urbanization, climate change, introduction of invasive species, illegal logging and wildlife trade, and especially conversion of natural habitats into agricultural land, pastureland, or industrial zones reduces the extent and quality of habitats, threatening the survival of many species. Consequently, the development and execution of conservation measures for exploited species of medicinal plants is imperative. Alternatively, some plant species' commercial cultivation could lessen their depletion. This is mostly dependent on the cultivator's access to viable seeds. The present research looked at how various storage settings affected the survival and germination of seeds from ethnobotanically significant medicinal plant species.

"Preservation of seed with minimum loss in viability and vigor until needed for sowing" is the definition of seed storage. In India, it is possible to produce any kind of crop at a low cost due to the availability of cheap labor and an environment that is suitable for the production of vigorous, bold, and high-quality seeds for both domestic and export markets. This will not only earn foreign exchange but also save it while providing the rural poor with skills that will enable them to create jobs and income. The best choice is to produce and export high-quality seeds. With hybrid varieties created domestically for domestic markets and commercial farmers, the seed industry is predicted to be vibrant and active in the future. Vegetable seed and diverse crops, such as aromatic and medicinal crops, will always have a great chance of success and be crucial to the economy. In the past, people would gather therapeutic plants from natural sources. Following several educational sessions, farmers are expressing interest in cultivating these plants which are presently in high demand on the global market for medicinal herbs. Commercial cultivation is used to grow medicinal herbs. It takes high-quality seed to grow therapeutic plants. Quality seed production of various medicinal plants and production technology are being worked on by the MAP Research Station in Rajendranagar, Hyderabad. Techniques to be followed for quality seed production in crops like Senna, Aswagandha, Kalmegh, Velvet bean, Basil, Palmarosa, and Isabgol are advised. Removing unwanted material from the seed lot and upgrading the quality of the seed by removing damaged and undersized seed with maximum efficiency in seed processing, which includes basic cleaning, grading, drying, and pre-conditioning, are the main steps in producing seed for medicinal and aromatic crops.

9.2 IMPORTANCE OF MEDICINAL PLANT

Medicinal crops provide better returns than traditional crops because:

- They can be grown as intercrops alongside traditional crops and on degraded lands.
- They can be stored for a long time and sold when the market is at its best. They are mostly resistant to pest attacks and diseases.
- They require fewer resources than traditional crops, so their cultivation costs are lower.
- They have a high demand both domestically and internationally.
- They can also be produced at better prices.

9.3 IMPORTANCE OF AROMATIC PLANTS

A sizable group of commercially significant plants includes aromatic plants. For the past 20 years, there has been a rise in demand for medications, essential oils, and fragrance compounds worldwide. Plants have aromatic substances in their roots, wood, bark, foliage, flowers, fruits, seeds, and other parts of the plant.

- Aromatic plants yield essential oils, fragrances, and flavors that have been used by human civilization for 1000s of years.
- Owing to its vast area and diverse agroclimatic conditions, it may be effectively grown commercially in many parts of India.
- They support a variety of human activities, including the use of soaps, cosmetics, and medications.
- They work great for preparation of agarbatti, ice cream, perfumed confections, aerated liquids, and disinfectants, among other things.

- There is a high demand in our nation for several significant fragrant plants, such as lemon grass, citronella, palmarose, vetiver, geranium, lavender, dawana, and so forth.

9.4 SEED QUALITY ENHANCEMENT IN AROMATIC CROPS

The goal of pre-sowing seed enhancement techniques is to maximize improvements in field stand establishment and yield by mobilizing the seeds' inherent resources and supplementing them with outside resources. Currently, a number of physical, physiological, and biochemical therapies are available to add value to seeds in order to accomplish this. The various methods that researchers have used to enhance seed quality are described below:

1. Dormancy-breaking treatments of *Carum carvi* seeds by soaking in 1% sodium hypochlorite (Hradilik and Cisarova, 1975).
2. *Saussurea lappa* seeds are soaked in GA_3 for keeping quality enrichment (Arora and Bhojwani, 1989).
3. Hydropriming of *Ocimum basilicum* (basil) for 12 h recorded the highest seedling vigor, germination percentage, and seedling dry weight (Hosseein and Kasra, 2011).
4. The highest germination percentage was observed when *Catharanthus roseus* (periwinkle) seeds were soaked in cold water (Shylaja et al., 1997).
5. Water soaking of *Tagetes erecta* (Mexican marigold) seeds for 18–24 h enhanced the seed quality of aged seed lots (Pramila et al., 2013).
6. Hydropriming for 12 h improved the germination and dry matter production in *Trigonella foenum-graecum* (fenugreek) seeds (Hosseein and Kasra, 2011).
7. Fresh seeds of *Rheum webbianum* were immersed in concentrated sulfuric acid (H_2SO_4) for dormancy breaking and germination improvement of seeds (Gautam et al., 2016).
8. Seeds of *Andrographis paniculata* (kalmegh) soaked in 100 ppm GA_3 for 6 h to improve germination (Fathima et al., 2003).
9. Soaking in 50 ppm GA_3 for 3 h recorded maximum vigor in *Asparagus sprengeri* (foxtail fern) seeds (Dhoran and Gudadhe, 2012).

10. *Cassia obtusifolia* (Nesavu) seeds were soaked in 200 ppm IAA for 6 h, which recorded higher germination (Arulmozhi, 2012).
11. Seeds of *Gymnema sylvestre* (Gymnema) soaked in 200 ppm Ethrel for 6 h and recorded maximum seedling growth (Harakumar, 1997).
12. Soaking of *Withania somnifera* (Ashwagandha) seeds in 1% KH_2PO_4 for quality improvement of seeds (Velmurugan et al., 2003).
13. *Aconitum balfourii* (Indian Atees) seeds were soaked in 50 to 100 mM KNO3, recording the maximum germination compared to the control (Pandey et al., 2000).
14. Prechilling (4°C) of *Descurainia sophia* (Flixweed) seeds for 10 days improved germination (Tavili et al., 2010).
15. *Bixa orellana* (Achiote) seeds were soaked in 25% coconut water for 6 h to enhance nursery emergence (Navamaniraj, 2005).
16. Seeds of *Descurainia sophia* soaked in prechilling at 4°C for 10 days stimulated germination and reduced the germination time (Tavili et al., 2010).
17. *Withania somnifera* (Ashwagandha) seeds treated with 1% sodium nitrate enhanced plant height, root length, and other yield parameters (Kattimani and Reddy, 2001).
18. Soaking the seeds of *Aconitum heterophyllum* in 130 mM thiourea recorded early germination (Pandey et al., 2000).

9.5 SEED QUALITY ENHANCEMENT IN MEDICINAL CROPS

Enhancement of seed quality in aromatic crops: One of the primary obstacles to the sustainable use of native medicinal plants of arid regions is their propensity to germinate easily in their natural habitat but to perform poorly in laboratory settings or when cultivation is tried (Gupta, 2003).

9.5.1 SEED PRIMING

A pre-sowing treatment for seeds is referred to particularly as "priming." Seeds are somewhat moistened during priming, and the following process involves drying them. Generally speaking, there are three stages to the best conditions for dry seed germination's water uptake. Phase I: Rapid imbibition is mostly caused by the matric pressures that the seed exerts. This stage can occur in seeds that are viable or not, dormant or not. The first phase involves the synthesis of proteins using messenger ribonucleic acid (mRNA) that has already been present, as well as the repair of DNA and mitochondria (McDonald, 2000). Phase II is the lag phase, during which there is a significant increase in metabolic activity but very little net water uptake, preparing viable, non-dormant seeds for radicle emergence. Proteins and mitochondria are synthesized by fresh mRNA during this phase. The activation phase is another name for phase II (Bradford, 1995). Phase III, or the last phase, is characterized by an increase in water intake and radicle elongation (Bewley and Black, 1994). Phases I and II of water uptake are completed by primed seeds, but the third phase is not reached by the seeds. Although not enough to allow for radicle protrusion, this hydration is enough to support pre-germinative metabolic events (Bradford, 1986).

9.5.1.1 TYPES OF PRIMING

There are different methods of seed priming:

- **Hydro-Priming:** soaking of seeds in water.
- **Halo-Priming:** hydration of seeds in inorganic salt solutions.
- **Osmo-Priming:** soaking of seeds in solutions such as polyethylene glycol (PEG).
- **Thermo-Priming:** treatment of seeds with low or high temperatures.
- **Matri-Priming:** treatment of seeds with solid matrices, such as hydrated sand, peat, and vermiculite (Taylor et al., 1998).
- **Bio-Priming:** coating seeds with bacteria, for example, *Trichoderma* spp. and *Pseudomonas aureofaciens* (Warren and Bennett, 1999).

In addition to all of these, sometimes seeds may be primed with plant growth regulators (Subedi and Ma, 2005).

9.5.2 SEED PRIMING AND GERMINATION

Medicinal plant seed germination is enhanced by seed priming. According to Hoseini et al. (2013), two landraces of lemon balm may benefit from seed priming procedures that boost antioxidant enzymes like cation ascorbate. Lipid peroxidation during germination time is decreased by these enzymes. Additionally, it was noted that *Heracleum candicans* seed germination is enhanced and improved by chemical treatments applied prior to sowing (Joshi and Dhar, 2003). Arjenaki et al. (2011) showed that osmo-priming with PEG-6000 improved the lengths of the radicle and shoots, as well as the percentage and pace of germination, in marigold (*Tagetes erecta*). Additionally, it was discovered that in *Echinacea purpurea*, seed priming increases the percentage and velocity of germination (Fariman et al., 2011). Increased activity of enzymes like lipase, amylase, and protease, which are important for breaking down macromolecules for embryonic growth and development, may be the cause of earlier germination of primed seeds (Dell-Aquila and Tritto, 1990). According to reports, the length of time in which seeds are soaked and the priming techniques used determine the final effect of priming (Ghassemi-Golezani et al., 2008). In Basil (*Ocimum basilicum* L), it was demonstrated that hydro-priming after 12 hours produced the best seedling vigor, germination percentage, and seedling dry weight (Aliabadi et al., 2011). It has been determined that pre-treating of *Salvia mirzayanii* seeds with sulfuric acid for 15 minutes has the greatest beneficial impact on the percentage and pace of germination. Some research indicates that soaking coriander (*Coriandrum sativum*) seeds in NaCl at a rate of 4 g L^{-1} for 12 hours produced the optimum germination percentage (Fredj et al., 2013). Hoseini et al. (2013) assessed how differently fennel (*Foeniculum vulgare*) responded to three distinct priming treatments: hydro-priming over 24 hours, gibberellic acid at 500 ppm, and potassium nitrate at 3%. They stated that seedlings that were soaked and treated with KNO_3 had the maximum germination percentage. When compared to the other treatments, the KNO_3 and GA_3 treatments had the greatest seedling weight and length.

In 2012, Dhorhan and Gudadhe reported that in *Asparagus sprengeri*, GA_3 seed priming at a concentration of 50 ppm was more effective than IAA, IBA, and NAA in promoting root growth and germination. In 2012, Takhti and Shekafandeh discovered that in thorn jujube (*Ziziphus spinachristi*), the germination rate of hydro-primed and osmo-primed (various concentrations of NaCl and ZnS) seeds was higher than that of control (seeds without treatment). The ZnS solution with an electrical conductivity (EC) of 8 dS/m had a higher germination rate than the ZnS solution with an EC of 16 dS/m.

It indicates that the germination rate was negatively impacted by a high ZnS content. According to reports, priming speeds up the germination process, although the efficiency of various priming agents differs depending on the crop species and priming solution content (Iqbal and Ashraf, 2005).

9.6 DIFFERENT SEED TREATMENT INFLUENCES THE SEED QUALITY

These technologies aim to combine the benefits of seed propagation and storage with clonal propagation and serve as a conduit for novel plant lines created via advancements in biotechnology.

9.6.1 VARIOUS TREATMENTS AND THEIR IMPACT ON SEED QUALITY

- **Seed Fortification:** Involves soaking seeds in water or diluted solutions of bioactive substances such as proteins, vitamins, growth regulators, and micronutrients.
- **Seed Infusion:** Using organic solvents rather than water, seeds are impregnated with bioactive organic compounds found in nature.
- **Seed Fumigations:** When volatile solvents or fungicides are burnt, fumes are discharged into the air, which kills the fungal infection by sublimating or evaporating, such as ethylene dibromide and methyl bromide.
- **Seed Pelleting:** The process of adding enough solid ingredients to seeds to create pelleted seed.
- **Irradiation:** The process of exposing hydrated seeds to light to induce germination or break seed dormancy.
- **Synthetic Seeds:** Somatic embryos, artificially encapsulated; shoot buds; cell aggregates; or any other tissue suitable for seed planting.

9.7 EFFECT OF DIFFERENT STORAGE MATERIALS ON SEED STORAGE OF MEDICINAL AND AROMATIC PLANTS

Traditional and eco-friendly practices for seed storage of MAP are enlisted below (Gowthami et al., 2016):

- Sun drying of grains
- Use of ash
- Sand application method
- Plastering of storage bins with clay and cow dung
- Storage of pulses with common
- Turmeric application method
- Use of garlic cloves
- Mixing of leaves
- Stepping method or Stamping method
- Use of salt and chili powder
- Use of Neem leaves
- Use of Neem leaves and dry chilies
- Use of Neem Oil
- Use of camphor
- Use of castor powder
- Sand mixture method
- Use of dried red chilies
- Use of lime powder
- Use of matchbox
- Fumigation of the go down
- Use of Neem seed powder
- Use of ginger rhizome
- Use of custard apple seed powder
- Use of Tulsi seeds

9.7.1 IMPORTANT TRADITIONAL AND ECO-FRIENDLY PRACTICES

- **Use of Garlic Cloves:** Garlic cloves have anti-feedant, bactericidal, fungicidal, insecticidal, nematicidal, and repellent qualities. They contain certain primary component kinds, such as diallyl di-, diallyl tri-, and diallyl sulphide (Prowse et al., 2006).

Layers of garlic cloves are stored in bins packed with seeds that have been proven to be efficient in protecting against a variety of pests during safe storage (Onu et al., 2015).

- **Use of Common Salt:** This approach involves physically mixing 1 kilogram of pulse seed with 150–200 g of regular salt, then storing the mixture in jute gunny bags for a duration of 6–8 months. By absorbing moisture, these methods help to keep the seeds dry and prevent spoiling by keeping insects away from the stored seeds. It is reasonably priced, reasonably effective, and ideal for short-term storage.
- **Turmeric Application Method:** Another effective substitute for treating the grains to keep insects and other pests away is to use powdered turmeric. Turmeric powder is combined with grains and seeds and then stored in jute bags or containers. This therapy is similarly safe to consume and offers protection for up to 6–8 months. This is the approach that many farmers use.
- **Use of Neem Products:**
  - **Neem Leaves:** Once the necessary number of leaves have been collected from the trees, it should be dried in the shade until it turns papery. Seeds are piled into three-quarters of the storage bins, and the surface is covered with a fresh cloth after a handful of neem leaves are scattered over it. Azadirachtin, the active component of neem leaves, has sterilizing, anti-fungal, non-toxic, and insect-repelling properties in addition to inhibiting insect eating. Farmers can safely store their seeds using this inexpensive and efficient method.
  - **Neem Oils:** This approach works better than neem leaf application. Apply 2–3 ml of neem oil per kg of seed to ensure that every seed is evenly coated. Neem seeds are protected from diseases and pests by their bitter flavor. Since it nearly eliminates the insect even in its egg stage, infestation is stopped early on (Rekha and Padmakar, 2014).
- **Using Tulsi Seeds:** Mix the dried Tulsi seeds and leaves with the seeds of aromatic and medicinal plants to prevent against weevil infestations during storage. The antifeedant and larvicidal

properties of Tulsi seeds are attributed to their palmitic, linolenic, linoleic, and stearic acids (Talukder, 2006).

- **Using Dry Red Chilies:** Red chilies have a strong scent that acts as an insect repellent, so adding 20–30 of them to a quintal seed bag of aromatic or medicinal plants helps keep stored grain pests at bay (Rekha and Padmakar, 2014). Dried chilies are stored in a container that is filled with seeds.
- **Use of Camphor:** Indigenous farmers occasionally use camphor to store medicinal and aromatic grains (Karthikeyan et al., 2009). In this procedure, the jute gunny sacks should contain 1–2 grams of camphor every 5 kilogram of seeds. The pungent smell of camphor kept the pests in storage at bay. This conventional storage technique allowed grains to be kept for up to three months. The grains are sun-dried once again after three months, and new bits of camphor are added for future preservation.

9.8 OTHER ECO-FRIENDLY METHODS

The development and introduction of eco-friendly alternatives for the postharvest handling of fruits and vegetables is a trend in postharvest studies due to the harm that petroleum-based packaging materials cause to the ecosystem and the adverse effects that agrochemicals have on human health and the environment. There are some eco-friendly packaging materials which include some agrochemicals and petroleum-based substitutes used for seed storage.

It is anticipated that these techniques will be improved and suitably adjusted in the future for the postharvest management of fruits and vegetables throughout the world.

9.8.1 EDIBLE COATINGS AND EDIBLE FILM PACKAGING

The term "edible film packaging" refers to a thin layer (less than 0.3 mm thick) made of a mixture of biopolymers and different additives that is prepared in advance, applied to the product surface, and then consumed as a component of the food product (Embuscado and Huber, 2009; Jeevahan and Chandrasekaran, 2019). Similar to edible films, edible coatings are created as a thin layer directly on the surface of the product (Guimaraes et al., 2018). In contrast to edible films, these are applied to the items in a liquid state by dipping them into the solutions. Due to their biodegradable qualities in comparison to plastic packaging materials, edible films and coatings are highly appealing to customers and environmentalists. It can be found in both plant and animal sources. Protein, lipid, polysaccharide, resin, or natural alone or in combination can be used as a coating material. When being processed, handled, and stored, it serves as a barrier against gases and moisture. To improve food stability, quality, functionality, and safety, edible coatings may contain functional ingredients including antioxidants, nutrients, and flavor. Typically, edible material is applied to fruits and vegetables by dipping, brushing, or spraying, creating a semi-permeable membrane that inhibits respiration, regulates moisture loss, and serves other purposes (Raghav et al., 2016). While oils and waxes are the most commonly employed lipid-based ingredients in edible films and coatings, wheat gluten, corn zein, and soy protein are also significant sources of protein. Conversely, plant-based polysaccharides (cellulose, gums, starch, pectin, etc.) and animal-based polysaccharides (chitosan and chitin) are widely utilized in the manufacturing of edible films and coatings (Wan et al., 2021). In food hydrocolloid applications, starch is a naturally occurring polysaccharide that is utilized due to its high functionality and affordability (Whistler, 1965). In general, starch film is colorless, tasteless, odorless, and has a low permeability to O_2. Pullulan is a biodegradable and edible extracellular polysaccharide of starch. Pullulan films have a high O_2 barrier and are tasteless, colorless, and odorless. Strawberries and kiwi fruits have been preserved using pullulan coating (Diab, 2001). Due to their rapid rate of respiration, fruits and vegetables respond well to starch coating.

9.8.2 ALOE VERA APPLICATION

The plant *Aloe vera* is both tropical and subtropical. *Aloe vera* has been used for ages and has therapeutic and medicinal qualities (Eshun, 2004). Fruit and vegetables can be coated with edible aloe vera gel. It can postpone food deterioration and lower transpiration and respiration. According to studies by Nabigol and Asghari (2013), the mycelium growth of *Aspergillus niger* and *Penicillium digitatum* can be effectively inhibited by aloe vera gel. In a similar vein, Kator et al. (2018) proposed that using aloe vera gel stops tomato fruit deterioration. Numerous other fruits, such as pineapples, nectarines, grapes, plums, strawberries, and others, can be included in the list of examples (Adetunji et al., 2012; Ahmed et al., 2009; Castillo et al., 2012; Martínez-Romero et al., 2017; Nasrin et al., 2017).

9.8.3 PROPOLIS APPLICATION

Western honey bees (*Apis mellifera*) gather and process propolis, a resinous material with variable color (green, red, yellow, and brown), from leaves, flower buds, stems, and cracks in the bark of many different tree species, including poplar, alder, birch, eucalyptus, acacia, and clusia. It is brought to the hive by bees, which combine it with beeswax to create a sticky compound (Bankova et al., 2014; Pellati et al., 2011). It has been used for many years in both conventional and scientific medicine, and since 300 BC, its capacity to cure wounds has been documented (Kahramanoglu et al., 2020). Pomegranate seed oil, at a 5% concentration, has been gelled with propolis as a structural agent to partially replace the fat phase in chocolate spread (Fayaz et al., 2017). Propolis extract-flavored orange soft drinks were produced. Its color changes, microbiological stability, and antioxidant activity were evaluated and contrasted with orangeade that contained potassium sorbate. When compared to soft drinks with potassium sorbate, those with propolis extract showed better antioxidant activity and phenolic content. The authors proposed that beverages containing propolis extract, a bioactive reagent, can be stored for over 16 weeks at a variety of temperatures without losing their quality or developing microbial deterioration (Vasilaki et al., 2019). Furthermore, it was said to offer significant advantages for managing fruits and vegetables after harvest. According to studies, the propolis extract's antimicrobial properties are derived from the high amounts of cinnamic acid, ferulic acid, and caffeic acid (Borges et al., 2013). As with the light and food materials, propolis's beneficial effects on pathogen control can function in two ways: either directly through control, or by strengthening the goods' resistance to infection (Kahramanoglu et al., 2020). Propolis extracts have been proposed as a means of reducing anthracnose on mango fruits (Mattiuz et al., 2015), gray mold (*Botrytis cinera*) on pomegranate fruits (Kahramanoglu et al., 2018), and *Penicillium digitatum* on orange fruits

(Badawy et al., 2016). Studies have indicated that propolis extracts, in addition to controlling postharvest pathogens, also mitigate weight loss and chilling injury, preserve soluble solids concentration, titratable acidity, ascorbic acid, total phenolic content, antioxidant activity, texture quality, and overall fruit acceptability, encompassing pomegranates, mangos, papayas, bananas, and oranges (da Cunha et al., 2018; Kahramanoglu et al., 2018; Mattiuz et al., 2015; Passos et al., 2016).

9.8.4 CHITOSAN APPLICATION TO STORED FRUITS

It is a deacetylated form of chitin, a linear polymer of 2-amino-2-deoxy-βD-glucan. It is a cationic biopolymer that occurs naturally (Bemiller, 1965; Davis, 1988). It is a part of the cell walls of fungi and insects, the skeleton of invertebrates, and the shell of crabs and shrimp (Raghav et al., 2016). It is widely used in many industries, including biotechnology, textile, food, medicine, and cosmetics. It has been suggested that adding chitosan to edible coatings and films has a significant chance of preserving product quality and inhibiting the growth of pathogens. According to Sharma et al. (2016), it is a biocompatible polysaccharide with inherent antibacterial properties. The European Union (Reg. EU 662014/563) authorized chitosan for use in plant protection. According to certain research, chitosan possesses three distinct qualities, making it a valuable postharvest substitute. It possesses strong antibacterial activity (Duan et al., 2019), the capacity to form biofilms on applied surfaces (Romanazzi et al., 2018), and the capacity to activate the products' defense mechanisms (Landi et al., 2017; Xoca-Orozco et al., 2019). An edible thin layer of coating is done directly on the surface mainly to protect the chitosan. Coatings and films that are edible can be eaten with the coated product and provide a protective layer around the fruit and vegetable (Kerch, 2015).

The development of a moisture and gas barrier in F&V preservation applications may result in weight loss and respiration rate decreases, delaying spoiling overall and extending the product's shelf life (Chien et al., 2007). Although edible chitosan films have very modest water vapor barrier properties, they are incredibly effective oxygen permeation barriers (Khan et al., 2012). Films and coatings acquire specific permeability properties, particularly to O_2, CO_2, and ethylene; they also permit some control over the respiration of fruits and inhibit the growth of microbes

(Dominguez-Martinez et al., 2017). For a very long time, coating techniques have been used to preserve citrus, apples (shellac and carnauba wax), tomatoes (mineral oil), and cucumbers (different waxes). Furthermore, according to Romanazzi et al. (2013), chitosan itself possesses the power to inhibit a few fungal diseases that reduce the quality of fruits while they are being stored. Research has also been done on papaya (*Carica papaya* L.), which loses 20% to 30% of its value after harvest (Romanazzi et al., 2017). Papaya deterioration postharvest is a microbiological process; the fruits lose their acceptability and shelf life as a result of being targeted by multiple market diseases. Likewise, it was discovered that chitosan coating worked well to improve the firmness, postpone ripening, and lessen browning of peach fruits (Ma et al., 2013).

9.8.5 ESSENTIAL OILS APPLICATION TO STORED FRUITS

Essential oils are concentrated, complex, hydrophobic liquids derived from plants that contain volatile aromatic compounds. Terpenes, phenols, esters, alcohols, nitrogen, and sulfur compounds make up the majority of these oils' composition (Johannessen, 2007). While steam distillation is the most used method for extracting essential oils, other methods include solvent extraction, cold pressing, and wax embedding. EOs are utilized in a variety of industries, such as food additives, air fresheners, cosmetics, and perfumes. According to numerous researches, various EOs are more effective than others at preventing postharvest illnesses (Boubaker et al., 2016). Fruits naturally have waxy coatings on their surfaces to provide protection. However, during postharvest treatment, this rapidly decreases or becomes thinner. Therefore, small layers of edible material are used to apply edible coatings to the fruit's surface, either in addition to or instead of the fruit's natural protective wax (Misir et al., 2014). It has been found that those edible coatings beautify the appearance, stop moisture and beneficial volatiles from escaping, and protect them from microbiological and mechanical harm (Arnon et al., 2015). This is due to the fact that antimicrobial essential oils can be contained in edible coatings, and the essential oils' vapor phases are gradually released. Because of this characteristic, edible coatings containing essential oils have been used to shield coated goods from microbiological deterioration. Essential oils can be applied directly to whole fruits and vegetables as well as fresh-cut produce. When added to an edible coating made of alginate, lemongrass (*Cymbopogon citratus* L.) essential oil (0.3% w/v) showed good effects on fresh-cut pineapple, reducing respiration rate, weight loss, and yeast and mold counts while maintaining firmness, color, and sensory qualities (Azarakhsh et al., 2014). Another fresh-cut study found that adding essential oils of eugenol and citral to coatings with sodium alginate or pectin as the base allowed the "Bravo de Esmolfe" apple to be stored for seven days at 4°C while maintaining its attractive appearance, sensory appeal, and nutritional value (Guerreiro et al., 2016). According to a recent study, the essential oils of black cumin (Kahramanoglu, 2021), lemongrass (Kahramanoglu, 2019), and myrtle (*Myrtus communis* L.) leaves (Bahadirli et al., 2020) have a strong potential to promote the storability of loquat, apricot, and strawberry fruits, respectively.

9.8.6 PLANT-DERIVED METHODS

In addition to essential oils, a variety of plant extracts and materials obtained from plants have been shown to have a major favorable impact on avoiding the deterioration of product quality as well as the proliferation and development of various postharvest diseases (Chen et al., 2019). For instance, it has been observed that ethanolic extracts of garlic can suppress *Penicillium* sp. in citrus fruits (Gong et al., 2018). In another study, it was proposed that banana fruit storability could be enhanced by guava leaf and lemon extracts (Tabassum et al., 2018). Another study found that neem, chinaberry, and marigold aqueous extracts (10% and 20%) enhanced the guava fruits' ability to be stored (Malik et al., 2015). Conversely, extracts of laurel, pimento, and cinnamon were investigated for their ability to inhibit postharvest grey mold on apple fruits. Fruit-related *in vivo* investigations were shown to be ineffectual; however, *in vitro* tests proved to be successful (Sernaite et al., 2020). The success of the plant extracts may be attributed to different types of secondary metabolites or the essential oils (as previously mentioned). The aforementioned examples demonstrate how biomaterials generated from plants significantly improve the quality of fruits and vegetables during preservation. Thus, the creation and application of materials produced from plants is a viable substitute for artificial chemicals that can enhance the product's storage quality.

9.9 SEED PACKAGING

Basically, the final step before marketing or storage is seed packaging or bagging. After being cleaned and processed, seeds are bundled into net-weight containers (Vanangmudi, 2014).

9.9.1 TYPES OF DIFFERENT PACKAGING MATERIAL USED FOR SEED STORAGE OF MEDICINAL AND AROMATIC PLANTS

- **Moisture Vapor Permeable Container:** Jute, cloth, paper, multi-wall, and non-woven materials are among the materials that can be used as bags. In the MAP Research Station, seed meant for counter sales is packaged in Rajendranagar fabric bags.
- **Moisture Vapor Resistant Container:** Polythene bags (200–300 gauge) or jute bags bonded with thin polythene film can be utilized.
- **Moisture Vapor Proof Containers:** Such as glass bottles, aluminum foil pouches, polythene bags (larger than 700 gauge), and tin cans. Aluminum containers are used to bundle seed batches at the Rajendranagar MAP Research Station.

9.10 CONCLUSION

The successful and safe seed storage of a variety of crops, such as aromatic and medicinal crops, will always be crucial to the economy. In the past, individuals would gather therapeutic plants from the woods. Following that, people began growing these plants and realized they needed to store them in an environmentally responsible manner so they could be used again without sacrificing quality. Reducing postharvest losses is essential to maintaining global food security, and managing postharvest losses sustainably is critical to maintaining food safety and sustainability. In order to lower postharvest losses across the value chain, it is important to employ environmentally friendly postharvest technology, such as edible films. In postharvest handling, choosing, and utilizing both conventional and novel methods is essential. A proper postharvest handling procedure should involve cleaning, selecting, grading, packaging, and finally storing.

Successful postharvest management requires curing and heat treatment prior to storage, pre-cooling, packaging with plastic or edible materials, and controlling the composition of the surrounding atmosphere. They are essential to maintaining the quality and shelf life of items along the supply chain. It is crucial to concentrate on environmentally favorable options for every application and to create cutting-edge, environmentally friendly technology to reduce postharvest losses.

KEYWORDS

- **aromatic plants**
- **ecofriendly treatments**
- **edible coatings**
- **medicinal plants**
- **seed storage**
- **traditional methods**

REFERENCES

Ader, F. (1978). Moisture content and storage conditions for pelleted vegetable seed and their influence on germination and plant growth. *Seed Science and Technology, 6*, 1033–1051.

Adetunji, C. O., Fawole, O. B., Arowora, K. A., Nwaubani, S. I., Ajayi, E. S., & Oloke, J. K. (2012). Effects of edible coatings from *Aloe vera* gel on quality and postharvest physiology of *Ananas comosus* (L.) fruit during ambient storage. *Global Journal of Science Frontier Research: Biotechnology and Genetics, 12*(5), 39–43.

Agrawal, P. K. (1980). Relative storability of seeds of ten species under ambient conditions. *Seed Research, 8*, 94–99.

Ahmed, M. J., Singh, Z., & Khan, A. S. (2009). Postharvest *Aloe vera* gel-coating modulates fruit ripening and quality of 'Arctic Snow' nectarine kept in ambient and cold storage. *International Journal of Food Science & Technology, 44*(5), 1024–1033. https://doi.org/10.1111/j.1365-2621.2008.01873.x (accessed on 12th August 2025).

Aliabadi, F. H., Moaveni, P., & Maroufi, K. (2011). Effect of hydropriming on seedling growth of basil (*Ocimum basilicum* L.). *Advances in Environmental Biology, 5*, 2258–2263.

Arati, P. (2000). Influence of containers and seed treatment on storability of chickpea (Master's thesis). University of Agricultural Sciences, Dharwad.

Arnon, H., Granit, R., Porat, R., & Poverenov, E. (2015). Development of polysaccharides-based edible coatings for citrus fruits: A layer-by-layer approach. *Food Chemistry, 166*, 465–472.

Arora, R., & Bhojwani, S. S. (1989). *In vitro* propagation and low temperature storage of *Saussurea lappa* C. B. Clarke–An endangered medicinal plant. *Plant Cell Reports, 8*, 44–47.

Arulmozhi, M. (2012). Seed technological studies in tagarai (*Cassia tora* L.) and nesavu (*Cassia obtusifolia* L.) (Master's thesis). Tamil Nadu Agricultural University, Coimbatore.

Azarakhsh, N., Osman, A., Ghazali, M. H., Tan, C. P., & Adzahan, M. N. (2014). Lemongrass essential oil incorporated into alginate-based edible coating for shelf-life extension and quality retention of fresh-cut pineapple. *Postharvest Biology and Technology, 88*, 1–7.

Badawy, I. F. (2016). Effect of ethanol extracted propolis on fruit quality and storability of balady oranges during cold storage. *Assiut Journal of Agricultural Sciences, 47*(4), 156–166.

Bahadirli, N. P., Kahramanoglu, I., & Wan, C. (2020). Exposure to volatile essential oils of myrtle (*Myrtus communis* L.) leaves for improving the postharvest storability of fresh loquat fruits. *Journal of Food Quality, 2020*, 8857669. https://doi.org/10.1155/2020/8857669 (accessed on 12th August 2025).

Bankova, V., Galabov, A. S., Antonova, D., Vilhelmova, N., & Di Perri, B. (2014). Chemical composition of propolis extract ACF® and activity against herpes simplex virus. *Phytomedicine, 21*, 1432–1438.

Bewley, J. D., & Black, M. (1994). Cellular events during germination and seedling growth. In J. D. Bewley & M. Black (Eds.), *Seeds: Physiology of Development and Germination* (pp. 147–197). New York, NY: Plenum Press.

Borges, A., Ferreira, C., Saavedra, M. J., & Simões, M. (2013). Antibacterial activity and mode of action of ferulic and gallic acids against pathogenic bacteria. *Microbial Drug Resistance, 19*(4), 256–265. https://doi.org/10.1089/mdr.2012.0244 (accessed on 12th August 2025).

Boubaker, H., Karim, H., El Hamdaoui, A., Msanda, F., Leach, D., & Bombarda, I. (2016). Chemical characterization and antifungal activities of four *Thymus* species essential oils against postharvest fungal pathogens of citrus. *Industrial Crops and Products, 86*, 95–101. https://doi.org/10.1016/j.indcrop.2016.03.036 (accessed on 12th August 2025).

Bradford, K. (1995). Water relations in seed germination. In J. Kigel & G. Galili (Eds.), *Seed Development and Germination* (pp. 351–396). New York, NY: Marcel Dekker.

Bradford, K. J. (1986). Manipulation of seed water relations via osmotic priming to improve germination under stress conditions. *Horticultural Science*, 21, 1105–1112.

Caneppele, M. A. B., RFDA, A. G. M., Junior, C. J. H., & Cardoso, A. A. (1995). Influence of packaging, environment and storage period on seed quality in onion (*Allium cepa* L.). *Revista Brasileira de Sementes, 17*, 249–257.

Castillo, S., Navarro, D., Zapata, P. J., Guillen, F., Valero, D., & Martínez-Romero, D. (2012). Using *Aloe vera* as a preharvest treatment to maintain postharvest organic table grape quality. *Acta Horticulturae, 933*, 621–625.

Chen, J., Shen, Y., Chen, C., & Wan, C. (2019). Inhibition of key citrus postharvest fungal strains by plant extracts *in vitro* and *in vivo*: A review. *Plants, 8*(2), 26. https://doi.org/10.3390/plants8020026 (accessed on 12th August 2025).

Chien, P. J., Sheu, F., & Yang, F. H. (2007). Effects of edible chitosan coating on quality and shelf life of sliced mango fruit. *Journal of Food Engineering, 78*(1), 225–229. https://doi.org/10.1016/j.jfoodeng.2005.09.022 (accessed on 12th August 2025).

da Cunha, M. C., Passos, F. R., Mendes, F. Q., Pigozzi, M. T., & de Carvalho, A. M. (2018). Propolis extract from different botanical sources in postharvest conservation of papaya. *Acta Scientiarum. Technology, 40*, e31074. https://doi.org/10.4025/actascitechnol.v40i1.31074 (accessed on 12th August 2025).

Dell'Aquila, T. V. (1990). Ageing and osmotic priming in wheat seeds: Effects upon certain components of seed quality. *Annals of Botany, 65*, 21–26.

Dhoran, V. S., & Gudadhe, S. P. (2012). Effect of plant growth regulators on seed germination and seedling vigour in *Asparagus sprengeri* Regelin. *International Research Journal of Biological Sciences, 1*(7), 6–10.

Diab, T., Biliaderis, C. G., Gerasopoulos, D., & Sfakiotakis, E. (2001). Physicochemical properties and application of pullulan edible films and coatings in fruit preservation. *Journal of the Science of Food and Agriculture, 81*, 988–1000.

Doijode, S. D. (1995). Effect of silica gel and storage containers on viability and vigor in onion. *Seed Research, 18*, 163–165.

Dominguez-Martinez, B. M., Martínez-Flores, H. E., & Berrios, J. D. J. (2017). Physical characterization of biodegradable films based on chitosan, polyvinyl alcohol and *Opuntia* mucilage. *Journal of Polymers and the Environment, 25*(3), 683–691. https://doi.org/10.1007/s10924-016-0851-y (accessed on 12th August 2025).

Duan, C., Meng, X., Meng, J., Khan, M. I., Dai, L., & Khan, A. (2019). Chitosan as a preservative for fruits and vegetables: A review on chemistry and antimicrobial properties. *Journal of Bioresources and Bioproducts, 4*(1), 11–21. https://doi.org/10.21967/jbb.v4i1.189 (accessed on 12th August 2025).

Ellis, R. H., Hong, T. D., Astely, D., & Kraak, H. L. (1991). Medium term storage of dry and ultra dry seeds of onion at ambient and subzero temperature. *Onion Newsletter for the Tropics, 6*, 56–58.

Embuscado, M. E., & Huber, K. C. (2009). *Edible Films and Coatings for Food Applications* (Vol. 9). New York, NY: Springer.

Eshun, K., & He, Q. (2004). *Aloe vera*: A valuable ingredient for the food, pharmaceutical and cosmetic industries. *Critical Reviews in Food Science and Nutrition, 44*, 91–96.

Fariman, Z. K., Azizi, M., & Noori, S. (2011). Seed germination and dormancy breaking techniques for *Echinacea purpurea* L. *Journal of Biology & Environmental Science, 5*, 7–10.

Fathima, G. A., Balasubramanian, M., Swaminathan, V., Balakrishnan, K., & Gururajan, B. (2003). Seed germination studies on kalmegh (*Andrographis pariculata*). In *National Seminar on New Perspective Species of Medicinal and Aromatic Plants* (pp. 27–29).

Fayaz, G., Goli, S. A. H., & Kadivar, M. (2017). Potential application of pomegranate seed oil oleogels based on monoglycerides, beeswax and propolis wax as partial substitutes of palm oil in functional chocolate spread. *LWT, 86*, 523–529. https://doi.org/10.1016/j.lwt.2017.08.036 (accessed on 12th August 2025).

Fredj, M. B., Zhani, K., Hannachi, C., & Mehwachi, T. (2013). Effect of NaCl priming on seed germination of four coriander cultivars (*Coriandrum sativum*). *Euro-Asian Journal of BioSciences, 7*, 21–29.

Ganji-Arjenaki, F., Amini-Dehghani, M., & Jabbari, R. (2011). Effects of priming on seed germination of marigold (*Calendula officinalis*). *Advances in Environmental Biology, 5*, 276–280.

Gautam, J., & Bhardwaj, S. D. (2001). Effect of seed size and presowing treatments on germination of ban oak (*Quercus leucotrichophora*). *Indian Journal of Forestry, 24*(3), 311–315.

Ghassemi-Golezani, K., Aliloo, A. A., Valizadeh, M., & Moghaddam, M. (2008). Effects of hydro and osmo-priming on seed germination and field emergence of lentil (*Lens culinaris* Medik.). *Notulae Botanicae Horti Agrobotanici Cluj-Napoca, 36*, 29–33.

Gong, M., Guan, Q., & Xu, S. (2018). Inhibitory effects of crude extracts from several plants on postharvest pathogens of citrus. In *AIP Conference Proceedings* (Vol. 1956, No. 1, p. 020043). Melville, NY: AIP Publishing LLC.

Gowthami, R., Sharma, N., Pandey, R., & Agrawal, A. (2021). Status and consolidated list of threatened medicinal plants of India. *Genetic Resources and Crop Evolution, 68*(6), 2235–2263.

Guerreiro, A. C., Gago, C. M. L., Faleiro, M. L., Miguel, M. G. C., & Antunes, M. D. C. (2016). Edible coatings enriched with essential oils for extending the shelf-life of 'Bravo de Esmolfe' fresh-cut apples. *International Journal of Food Science and Technology, 51*, 87–95.

Guimaraes, A., Abrunhosa, L., Pastrana, L. M., & Cerqueira, M. A. (2018). Edible films and coatings as carriers of living microorganisms: A new strategy towards bio preservation and healthier foods. *Comprehensive Reviews in Food Science and Food Safety, 17*(3), 594–614. https://doi.org/10.1111/1541-4337.12345 (accessed on 12th August 2025).

Gupta, V. (2003). Seed germination and dormancy breaking techniques for indigenous medicinal and aromatic plants. *Journal of Medicinal and Aromatic Plant Sciences, 25*, 402–407.

Harakumar, C. (1997). Seed technological studies in *Gymnema sylvistre* R. (Master's thesis). Tamil Nadu Agricultural University, Coimbatore.

Hoseini, M., Rahimzadeh-Khoei, F., & Mirshekari, B. (2013). Seed priming techniques improve germination and yield in two landraces of lemon balm in laboratory experiment and field study. *International Journal of Indigenous Medicinal Plants, 29*, 1144–1150.

Hosseein, A. F., & Kasra, M. (2011a). Effect of hydro priming on seedling vigor in basil (*Ocimum basilicum* L.) under salinity conditions. *Advances in Environmental Biology, 5*(5), 828–833.

Hosseein, A. F., & Kasra, M. (2011b). Hydropriming and NaCl influences on seedling growth in fenugreek (*Trigonella foenum-graecum*). *Advances in Environmental Biology, 5*(5), 821–827.

Hradilik, J., & Cisarova, H. (1975). Studies on the dormancy of caraway (*Carum carvi*) achenes. *Rostlinná Výroba, 21*, 351–364.

Iqbal, M., & Ashraf, M. (2005). Pre-sowing seed treatment with cytokinins and its effect on growth, photosynthetic rate, ionic levels and yield of two wheat cultivars differing in salt tolerance. *International Journal of Plant Biology, 47*, 1315–1325.

Jeevahan, J., & Chandrasekaran, M. (2019). Nanoedible films for food packaging: A review. *Journal of Materials Science, 54*(19), 12290–12318. https://doi.org/10.1007/s10853-019-03742-y (accessed on 12th August 2025).

Johannessen, G. S. (2007). Post-harvest strategies to reduce enteric bacteria contamination of vegetable, nut and fruit products. In *Handbook of Organic Food Safety and Quality* (pp. 433–453). Cambridge, England: Woodhead Publishing. https://doi.org/10.1533/9781845693411.4.433 (accessed on 12th August 2025).

Joshi, M., & Dhar, U. (2003). Effect of various presowing treatments on seed germination of *Heracleum candicans* Wall. ex DC.: A high value medicinal plant. *Seed Science & Technology, 31*, 737–743.

Kahramanoglu, I. (2019). Effects of lemongrass oil application and modified atmosphere packaging on the postharvest life and quality of strawberry fruits. *Scientia Horticulturae, 256*, 108527. https://doi.org/10.1016/j.scienta.2019.05.054 (accessed on 12th August 2025).

Kahramanoglu, I. (2021). Use of black cumin oil and liquorice syrup to maintain the postharvest quality of fully ripe apricot fruits var. 'Thyrinte.' *European Journal of Horticultural Science, 86*(3), 260–269. https://doi.org/10.17660/eJHS.2021/86.3.5 (accessed on 12th August 2025).

Kahramanoglu, I., Aktas, M., & Gunduz, S. (2018). Effects of fludioxonil, propolis and black seed oil application on the postharvest quality of "Wonderful" pomegranate. *PLoS One, 13*(5), e0198411. https://doi.org/10.1371/journal.pone.0198411 (accessed on 12th August 2025).

Kahramanoglu, I., Okatan, V., & Wan, C. (2020). Biochemical composition of propolis and its efficacy in maintaining postharvest storability of fresh fruits and vegetables. *Journal of Food Quality, 2020*, 8869624, 1–9. https://doi.org/10.1155/2020/8869624 (accessed on 12th August 2025).

Karthikeyan, C., Veeraragavathatham, D., Karpagam, D., & Ayisha, S. F. (2009). Traditional storage practices. *Indian Journal of Traditional Knowledge, 8*(4), 564–568.

Kator, L., Hosea, Z. Y., & Ene, O. P. (2018). The efficacy of *Aloe vera* coating on postharvest shelf life and quality of tomato fruits during storage. *Asian Research Journal of Agriculture, 8*(4), 1–9.

Kattimani, K. N., & Reddy, Y. N. (2001). A note on influence of presowing seed treatments on growth and root yield of aswagandha. *Karnataka Journal of Agricultural Science, 14*(3), 846–848.

Kerch, G. (2015). Chitosan films and coatings prevent losses of fresh fruit nutritional quality: A review. *Trends in Food Science & Technology, 46*(2), 159–166. https://doi.org/10.1016/j.tifs.2015.10.010 (accessed on 12th August 2025).

Khan, A., Khan, R. A., & Salmieri, S. (2012). Mechanical and barrier properties of nanocrystalline cellulose reinforced chitosan-based nanocomposite films. *Carbohydrate Polymers, 90*(4), 1601–1608. https://doi.org/10.1016/j.carbpol.2012.07.037 (accessed on 12th August 2025).

Landi, L., De Miccolis Angelini, R. M., Pollastro, S., Feliziani, E., Faretra, F., & Romanazzi, G. (2017). Global transcriptome analysis and identification of differentially expressed genes in strawberry after preharvest application of benzothiadiazole and chitosan. *Frontiers in Plant Science, 8*, 235. https://doi.org/10.3389/fpls.2017.00235 (accessed on 12th August 2025).

Ma, Z. X., Yang, L. Y., & Yan, H. X. (2013). Chitosan and oligochitosan enhance the resistance of peach fruit to brown rot. *Carbohydrate Polymers, 94*(1), 272–277. https://doi.org/10.1016/j.carbpol.2013.01.012 (accessed on 12th August 2025).

Malik, A. A., Bhat, A., Ahmed, N., & Kaul, R. (2015). Effect of postharvest application of plant extracts on physical parameters and shelf life of guava. *Asian Agri-History, 19*(3), 185–193.

Martínez-Romero, D., Zapata, P. J., Guillen, F., Paladines, D., Castillo, S., & Valero, D. (2017). The addition of rosehip oil to aloe gels improves their properties as postharvest coatings for maintaining quality in plum. *Food Chemistry, 217*, 585–592. https://doi.org/10.1016/j.foodchem.2016.09.035 (accessed on 12th August 2025).

Mattiuz, B. H., Ducamp-Collin, M. N., Mattiuz, C. F., Vigneault, C., Marques, K. M., & Sagoua, W. (2015). Effect of propolis on postharvest control of anthracnose and quality parameters of 'Kent' mango. *Scientia Horticulture, 184*, 160–168. https://doi.org/10.1016/j.scienta.2014.12.035 (accessed on 12th August 2025).

Misir, J., Brishti, F. H., & Hoque, M. M. (2014). *Aloe vera* gel as a novel edible coating for fresh fruits: A review. *American Journal of Food Science and Technology, 2*(3), 93–97.

Nabigol, A., & Asghari, A. (2013). Antifungal activity of *Aloe vera* gel on quality of minimally processed pomegranate arils. *International Journal of Agronomy and Plant Production, 4*(4), 833–838.

Nasrin, T. A., Rahman, M. A., Hossain, M. A., Islam, M. N., & Arfin, M. S. (2017). Postharvest quality response of strawberries with *Aloe vera* coating during refrigerated storage. *The Journal of Horticultural Science and Biotechnology, 92*(6), 598–605. https://doi.org/10.1080/14620316.2017.1324326 (accessed on 12th August 2025).

Navamaniraj, N. K. (2005). Research focuses on seed handling techniques in *Bixa orellana* L. (Ph.D. thesis). Tamil Nadu Agricultural University, Coimbatore.

Onu, F. M., Ogu, E., & Ikehi, M. E. (2015). Use of neem and garlic dried plant powders for controlling some stored grain pests. *Egyptian Journal of Biological Pest Control, 25*(2), 507–512.

Padma, V., & Reddy, M. B. (2002). Storage of brinjal seed under ambient conditions at two moisture levels. *Journal of Research ANGRAU, 30*(2), 6–10.

Pandey, H., Nandi, S. K., Nadeem, M., & Palni, L. M. S. (2000). Chemical stimulation of seed germination in *Aconitum heterophyllum* Wall. and *Aconitum balfourii*: Important Himalayan species of medicinal value. *Seed Science & Technology, 28*, 39–48.

Passos, F. A., Mendes, F. I., da Cunha, M. C., da Cunha, M. I., & de Carvalho, A. E. (2016). Propolis extract coated in Pera orange fruits: An alternative to cold storage. *African Journal of Agricultural Research, 11*(23), 2043–2049.

Passos, F. R., Mendes, F. Q., Cunha, M. C., Pigozzi, M. T., & Carvalho, A. M. (2016). Propolis extract in postharvest conservation banana 'Prata.' *Revista Brasileira de Fruticultura, 38*(2), 1–11. https://doi.org/10.1590/0100-29452016931 (accessed on 12th August 2025).

Pellati, F., Orlandini, G., Pinetti, D., & Benvenuti, S. (2011). HPLC-DAD and HPLC-ESIMS/MS methods for metabolite profiling of propolis extracts. *Journal of Pharmaceutical and Biomedical Analysis, 55*(5), 934–948.

Prakash, B. G., Raghavendra, K. V., & Gowthami, R. (2016). Indigenous practices for eco-friendly storage of food grains and seeds. *Advances in Plants & Agriculture Research, 3*(4), 101–107. https://doi.org/10.15406/apar.2016.03.00101 (accessed on 12th August 2025).

Pramila, C. K., Prasanna, K. P. R., Puttaraju, T. B., Ramegowda, S., Narayanaswamy, P., & Balakrishna, P. (2013). Influence of priming treatments on germination and vigor in aged seed lot of marigolds (*Tagetes erecta*). In *XIII National Seed Seminar on Innovations in Seed Research and Development* (p. 88). Bengaluru.

Prowse, G. M., Galloway, T. S., & Andrew, F. (2006). Insecticidal activity of garlic juice in two dipteran pests. *Agricultural and Forest Entomology, 8*(1), 1–6.

Pushpangadan, P., & Nair, K. N. (2001). Future of systematics and biodiversity research in India: Need for a national consortium and national agenda for systematic biology research. *Current Science, 80*, 631–638.

Raghav, P. K., Agarwal, N., & Saini, M. (2016). Edible coating of fruits and vegetables: A review. *International Journal of Scientific Research and Modern Education, 1*, 188–204.

Rao, V., & Rajasekharan, P. E. (2002). Threatened medicinal plant resources and conservation needs. *Botanica, 52*, 52–63.

Rekha, B. K., & Padmakar, C. K. (2014). Indigenous food grain storage practices followed by tribal farmers of Nandurbar district. *Golden Research Thoughts, 4*(4), 2–4.

Roberts, E. H. (1972). *Viability of seeds*. London: Chapman and Hall Co. Ltd.

Romanazzi, G., Feliziani, E., & Banos, S. B. (2017). Shelf-life extension of fresh fruit and vegetables by chitosan treatment. *Critical Reviews in Food Science and Nutrition, 57*(3), 579–601. https://doi.org/10.1080/10408398.2014.900474 (accessed on 12th August 2025).

Romanazzi, G., Feliziani, E., & Santini, M. (2013). Effectiveness of postharvest treatment with chitosan and other resistance inducers in the control of storage decay of strawberry. *Postharvest Biology and Technology, 75*, 24–27. https://doi.org/10.1016/j.postharvbio.2012.07.007 (accessed on 12th August 2025).

Romanazzi, G., Feliziani, E., & Sivakumar, D. (2018). Chitosan, a biopolymer with triple action on postharvest decay of fruit and vegetables: Eliciting, antimicrobial and film-forming properties. *Frontiers in Microbiology, 9*, 2745. https://doi.org/10.3389/fmicb.2018.02745 (accessed on 12th August 2025).

Sernaite, L., Rasiukeviciute, N., & Valiuskaite, A. (2020). Application of plant extracts to control postharvest gray mold and susceptibility of apple fruits to *Botrytis cinerea* from different plant hosts. *Foods, 9*(10), 1430. https://doi.org/10.3390/foods9101430 (accessed on 12th August 2025).

Sharma, S., Barman, K., & Siddiqui, M. W. (2016). Chitosan: Properties and roles in postharvest quality preservation of horticultural crops. In *Eco-Friendly Technology for Postharvest Produce Quality* (pp. 269–296). Cambridge, MA: Academic Press. https://doi.org/10.1016/B978-0-12-804313-4.00009-8 (accessed on 12th August 2025).

Shylaja, M. R., Sankar, M. A., Nair, G. S., & Mercy, K. A. (1997). Influence of seed treatments and storage on the germination of *Catharanthus roseus* L. (G. Don). *Indian Cocoa, Arecanut and Spices Journal, 21*(2), 45–47.

Subedi, K. D., & Ma, B. L. (2005). Seed priming does not improve corn yield in a humid temperate environment. *Agronomy Journal, 97*, 211–218.

Tabassum, P., Khan, S. A., Siddiqua, M., & Sultana, S. (2018). Effect of guava leaf and lemon extracts on postharvest quality and shelf life of banana cv. Sabri (*Musa sapientum* L.). *Journal of the Bangladesh Agricultural University, 16*(3), 337–342. https://doi.org/10.3329/jbau.v16i3.39489 (accessed on 12th August 2025).

Takhti, S., & Shekafandeh, A. (2012). Effect of different seed priming on germination rate and seedling growth of *Ziziphus spina-christi*. *Advances in Environmental Biology, 6*, 159–164.

Talukder, F. A. (2006). Plant products as potential stored–product insect management agents– A mini review. *Emirates Journal of Agricultural Science, 18*(1), 17–32.

Tavili, A., Hossein, P., Asgha, F., Salman, Z., Chahooki, Z., & Ali, M. (2010). The effect of different treatments on improving seed germination characteristics in medicinal species of *Descurainia sophia* and *Plantago ovata*. *African Journal of Biotechnology, 9*(39), 6588–6593.

Taylor, A. G., Allen, P. S., Bennet, M. A., Bradford, K. J., Burris, J. S., & Misra, M. K. (1998). Seed enhancements. *Seed Science & Research, 8*, 245–256.

Vanangamudi, K. (2014). *Seed Science and Technology*. New Delhi: New India Publication Agency (NIPA).

Vasilaki, A., Hatzikamari, M., & Stagkos-Georgiadis, A. (2019). A natural approach in food preservation: Propolis extract as sorbate alternative in non-carbonated beverage. *Food Chemistry, 298*, 125080. https://doi.org/10.1016/j.foodchem.2019.125080 (accessed on 12th August 2025).

Velmurugan, S., Vadivel, E., & Paramaguru, P. (2003). Studies on seed germination in ashwagandha (*Withania somnifera*). In *National seminar on new perspective species of medicinal and aromatic plants*.

Wan, C., Kahramanoglu, I., & Okatan, V. (2021). Application of plant natural products for the management of postharvest diseases in fruits. *Folia Horticulture, 33*(1), 203–215. https://doi.org/10.2478/fhort-2021-0016 (accessed on 12th August 2025).

Warren, J. M., & Bennett, M. A. (1999). Bio-osmopriming tomato (*Lycopersicon esculentum* Mill.) seeds for improved stand establishment. *Seed Science & Technology, 27*, 489–499.

Whistler, R. L., & Paschall, E. F. (1965). *Starch: Chemistry and Technology* (Vol. 1). New York: Academic Press.

Xoca-Orozco, L. A., Aguilera-Aguirre, S., Vega-Arreguin, J., Acevedo-Hernandez, G., Tovar-Perez, E., & Stoll, A. (2019). Activation of the phenylpropanoid biosynthesis pathway reveals a novel action mechanism of the elicitor effect of chitosan on avocado fruit epicarp. *Food Research International, 121*, 586–592. https://doi.org/10.1016/j.foodres.2018.12.023 (accessed on 12th August 2025).

CHAPTER 10

Nanopriming for Enhanced Seed Germination in Medicinal Plants

KHUSHBU KUMARI,[1,2] ANITA KUMARI,[1,2] SHUBHAM JOSHI,[1,2] SUMAN GUSAIN,[1,2] MEGHNA PATIAL,[1] MEENAKSHI RAWAT,[1] AKSHITA PATIAL,[1,3] RAKHI CHANDOLIYA,[1,3] VIKAS SHARMA,[3] and ROHIT JOSHI[1,2,*]

[1]*CSIR–Institute of Himalayan Bioresource Technology (CSIR–IHBT), Palampur, Himachal Pradesh, India*

[2]*Academy of Scientific and Innovative Research (AcSIR), Ghaziabad, Uttar Pradesh, India*

[3]*Department of Biotechnology, School of Bioengineering and Biosciences, Lovely Professional University, Phagwara, Punjab, India*

*Corresponding author

ABSTRACT

A range of abiotic and biotic stresses result in reduced plant productivity by inhibiting its growth and development. Seed priming is an effective method for breaking seed dormancy and promoting plant growth. Ultimately, this approach contributes to enhanced productivity across diverse environmental conditions and stressors. Nanotechnology has the capability to improve several aspects of agriculture, including reinforcing plant defense, monitoring plant growth, and improving seed quality. Because of the small dimensions and unique physicochemical properties of nanoparticles, nanopriming stands out among the other seed priming methods and is very effective. Nanopriming is an innovative method of seed priming that activates amylase-mediated starch breakdown and consequently

Seed Technology for Medicinal and Aromatic Plants: Production, Testing, Storage, and Marketing.
Priyanka Sharma, Kishore Kumar Sharma, and Gopal Shukla (Eds.)

DOI: 10.1201/9781779641199-10

accelerates the seed germination process. Mild accumulation of reactive oxygen species (ROS) driven by nanopriming acts as a prime signaling factor for downstream cascades involved in the production of secondary metabolites and the improvement of stress tolerance. Although the field of nanopriming and its impact on the germination of medicinal plant seeds remains largely unexplored, with only a few studies conducted so far. This chapter summarizes the potential processes through which nanopriming promotes seed germination, the impact of different nanoparticles on defense response, and their mechanism regulating the germination process.

10.1 INTRODUCTION

The germination of seeds is the first step in the beginning of plant life and is crucial for the survival and sustainability of plant species, mainly in the field of agriculture and rangeland ecosystems (Manjaiah et al., 2019). In the field of commercial agriculture, fast and dependable seed germination is crucial for obtaining optimal crop establishment, providing financial stability, and the effective use of production resources (Chen and Arora, 2013). Seed priming is a farmer-friendly, efficient, and cost-effective technique to enhance crop yield significantly (MS et al., 2021; Rhaman et al., 2020). Seed priming is a pre-treatment method that involves either exposing seeds to minor stressors (heat, cold, or UV) or partial hydration with synthetic (sodium hypochlorite, sodium nitroprusside, etc.) or natural chemicals (plant extracts, chitosan, polyamines, mannose, etc.). The purpose of this process is to increase the ability of plants to withstand abiotic stresses (Devika et al., 2021). There are several ways to reduce seed dormancy, including treatment of seeds with salt solutions (halo-priming), water (hydro-conditioning), osmotic agents (osmo-priming), plant hormone solutions (hormonal priming), beneficial microbe solutions (bio-priming), exposure to a magnetic field (magneto-priming) (De La Torre-Roche et al., 2020; Waqas et al., 2019). These advanced techniques show promising outcomes for crop improvement. Such advantages include improved germination rates, germination vigor, growth and development, higher resilience to environmental and biological stressors, increased crop yields, and increased micronutrient levels in cereals (Acharya et al., 2020). The use of nanoparticles (Titanium NPs, Silver NPs, Copper NPs, etc.) in priming, known as nano-priming, has shown greater promise compared to traditional priming methods in achieving viable agricultural yields (Khalaki et al., 2021). Nano-priming employs nanoparticles (NPs) smaller than 100 nm in size, and the term "priming" refers to activate defense response against environmental stress that can occur in future (Chandrasekaran et al., 2020).

Previous researchers extensively explored the impact of nano-priming on seed germination, with a focus on the utilization of different metal nanoparticles (NPs) such as zinc, zinc oxide, iron, titanium oxide, gold, and silver NPs, as demonstrated by various studies (Castiglione et al., 2011; El-Badri et al., 2021; Latef et al., 2017; Mahakham et al., 2017; Panyuta et al., 2016). These studies have repeatedly shown the beneficial impact of NPs on the germination of seeds and plant development. The small size and unique physio-chemical characteristics of nanoparticles, which make them an effective seed priming tool, are responsible for this efficiency (Dasgupta et al., 2017). Additionally, nano-priming activates particular metabolic processes that normally take place during the initial phase of seed germination, leading to more rapid seedling emergence, increased growth, higher crop yields, and improved crop quality (Acharya et al., 2020). Although nano-priming shows a promising effect on accelerating seed germination and improving stress tolerance in the face of minor and periodic difficulties, it has not yet undergone extensive study. Research on nano-priming is still regarded as an emerging discipline in the field of plant biology. This chapter provides a brief overview of nano-priming methods, the utilization of different nanoparticles in seed priming, their mechanism, and crosstalk with other cellular processes to improve seed germination and seedling development in medicinal plants.

10.2 VARIOUS APPROACHES OF SEED PRIMING

The definition of "germination" is surprisingly broad; it often refers to the process of dry seeds absorbing water, which is accomplished when the embryonic axis lengthens. The act of treating seeds through various methods to enhance pre-germinative metabolic and defense activity prior to radicle emergence is known as priming. These seeds can be dried, and upon subsequent ingestion, they usually show faster rates of radicle emergence and robust seedling development (Mirmazloum et al., 2020). For the purpose of enhancing the germination of numerous crops, various seed pretreatment procedures have been used, including hydropriming, osmopriming, halopriming, nanopriming, thermopriming, and biopriming (controlled hydration of seeds via the application of microbes or their biological components).

10.2.1 HYDROPRIMING

Hydropriming involves soaking seeds in pure water without any other additives. Water is used to condition hydroprimed seeds to a predetermined moisture level, and then the seeds are totally or partially dried to reduce the moisture content (Singh et al., 2017). Water accessibility inside the system may be strictly controlled or unrestricted. Hydropriming is a straightforward, low-cost, and safe way to improve germination. The majority of farmers (>80%) found that priming improved early germination, development, and yield, and the majority (>95%) planned to prime in subsequent years. Singh et al. (2017) observed how the Bael (*Aegle marmelos*) seeds will react when exposed to drought stress. As the level of osmotic stress increased, it significantly decreased the rate of seed germination, the ultimate germination percentage, and the amount of water that the seeds consumed, whereas seed hydropriming resulted in faster seed germination than seeds that were not primed. Twenty-five days after the final count, seeds that had been stimulated for 48 or 24 hours emerged sooner than control seeds (without any treatment). Total germination percentage revealed that primed seeds considerably outperformed unprimed seeds in terms of germination. This suggests that hydropriming is a crucial element to improve seed germination, uniform plant emergence, and tolerance to unfavorable environmental conditions (light, temperature, and water), due to hydration of seeds, a process in which hydrolytic enzymes converted nutrients stored in the seeds into chemicals that could be used for metabolism and resulted in embryo formation. Reduced lipid peroxidation and higher anti-oxidative activities during seed imbibition may be partly responsible for better seed development after priming. Hydroprimed seeds showed a notable improvement in germination performance.

The chosen approach and length of treatment determine the final effect of priming. Due to the use of distilled water as the priming medium, no specific technical equipment is needed. It is an easy and affordable method of priming seeds. In a similar vein, hydro priming was described by Venudevan et al. (2019) as likely the least expensive way of seed priming. Hydro priming treatment has been shown to accelerate seedling emergence by dissolving plant hormones that act as endogenous regulators (ethylene, auxin, and cytokinins) prior to sowing in the soil (Singh et al., 2017). As demonstrated in *Medicago truncatula* (Barrel medic) growing in contaminated soil, hydropriming alters the expression of genes involved in antioxidant defense and metal chelation in addition to boosting germination properties (Forti et al., 2020). However, hydropriming is defined as a control and as a reference treatment for correct assessment of the influence of priming chemicals suspended in water solutions in various research analyzing priming effects towards metallic stress tolerance. Hydropriming method affordability and potential to affect a vast population of primed plants are all unquestionably advantageous. It has long been used in agriculture to boost productivity and synchronize seed germination (Waqas et al., 2019).

10.2.2 OSMOPRIMING

Glycophytes are mostly dependent upon fertile soil and are moderately resistant to saline environments. However, seed germination is a significant issue in the cultivation of medicinal plants. Because there was no rain at the time of planting, the seeds were occasionally placed in seedbeds with unfavorable moisture, which led to poor and disorganized seedling emergence. Soil salinity is a prevalent issue in Iran's irrigated regions with little rainfall, and a significant barrier to seed germination. Seed germination may be impacted by soil salinity in one of two ways: either by establishing an external osmotic potential that prevents water intake, or by the toxic effects of Na^+ and Cl^- ions on the germination of the seed (Noori et al., 2022). Both suppression or delayed seed germination and seedling establishment are brought on by osmotic and salt stressors. The amount of water that is absorbed during imbibition decreases under stressed conditions, and salt stress may result in an increased ion uptake (Soleymani et al., 2018). Numerous crops, vegetables, and small-seeded grasses have shown improved germination and emergence when seeds were primed (Noori et al., 2022).

Osmopriming, also known as osmoconditioning, involves immersing seeds in osmotic solutions containing substances like mannitol, polyethylene glycol (PEG), sorbitol, glycerol, potassium nitrate, potassium phosphate, or potassium chloride. These osmotic solutions have a low water potential, causing seeds to slowly absorb water (Raj et al., 2019). This facilitates seed imbibition and activates the early stages of germination while inhibiting radicle protrusion. Typically, the osmotic potential of the priming solution falls within the range of –1 to –2 MPa, although this may vary depending on the plant species and the duration of the priming treatment. Polyethylene glycol, owing to its large molecular size, is a common chemical used in osmopriming treatments. In a study conducted by Arif et al. (2008), it was reported that osmopriming with PEG significantly improved seed emergence and grain yield in soybean when compared to untreated control seeds (Raj et al., 2019). In order to begin the pre-germination metabolism, the applied solutes are typically dissolved in water in different concentrations. Before the major root or radicle emerges, the primed seeds are subsequently expelled from the osmoticum (Paparella et al., 2015). Osmopriming is a commercially available method for enhancing seed vigor and germination. It entails carefully imbibing seeds to initiate the early stages of germination, followed by drying the seed to its original weight. Osmopriming has a number of benefits, such as quick and uniform plumule emergence, enhanced seedling growth, and better establishment in adverse environmental or soil conditions (Falahhosseini et al., 2017).

10.2.3 HALOPRIMING

Halopriming involves immersing seeds in non-naturally occurring salt solutions, such as NaCl or KCl. This technique promotes consistent seed germination and enhanced crop yields, even when faced with challenging environmental conditions, like extreme temperatures and limited oxygen availability. Besides boosting early vigor, seedling emergence rate, and seedling tolerance to environmental circumstances, seed halopriming also stimulates antioxidant enzymes production (Kumari et al., 2017). In this study, halopriming improved the morphological characteristics of seedlings while raising their salinity tolerance threshold. By boosting metabolism, halopriming can improve seed germination in a saline environment. Fresh and dry shoot biomass of any crop plant was affected by different priming agents such as $CaCl_2$, KCl, and NaCl (Kumari et al., 2017). It is determined that salinity stress has a negative impact on *Hibiscus sabdariffa* (Roselle) early developmental stages, but seed priming with NaCl mitigates this effect. According to Kumari et al. (2017), halopriming may improve tolerance towards salinity during germination and enable the establishment of these plants in saline lands. Halopriming is a cost-effective and straightforward agricultural method that ensures superior seedling emergence and uniform crop growth across diverse environmental vagaries.

10.2.4 BIOPRIMING

Due to its compatibility with, and support for, sustainable and organic approaches to plant production, biological priming, a technique that involves injecting plants with advantageous microbes to increase stress tolerance, is gaining attention on a global scale. This novel approach to seed treatment seamlessly mixes biological components, such as the introduction of helpful organisms, with physiological features, such as seed hydration (Rakshit et al., 2015). As an alternate strategy for managing soil- and seed-borne diseases, biopriming has become a recent and successful method for crop improvement. Typically, biopriming involves the use of advantageous microorganisms, with a focus on plant growth-promoting bacteria (PGPB) that are resilient in a variety of difficult environmental circumstances. These helpful microbes can be found in soils with high metal contamination, particularly in the rhizosphere of plants that can withstand metals, as well as in a variety of above-ground plant organs. Numerous bacterial genera, including *Alcaligenes*, *Curtobacterium*, *Bacillus*, *Microbacterium*, *Paenibacillus*, *Pseudomonas*, *Streptomyces*, and *Tetrathiobacter*, are known to have characteristics that promote plant growth (Benidire et al., 2016; Desoky et al., 2020; Roman-Ponce et al., 2017). There are also rhizosphere endophytes among the fungi, including *Piriformospora indica*, *Acrocalymma vagum*, and *Scytalidium lignicola* (Dabral et al., 2019; Hou et al., 2020). The ability of bacteria to produce and assimilate nitrogen or release chemicals that promote plant growth, including indole-3-acetic acid (IAA), siderophores, and a variety of volatile organic compounds, is thought to be the cause of enhanced plant growth in the presence of metal stress (Qadir et al., 2020; Wang et al., 2019; Zhou et al., 2017).

On the other side, several types of fungi (*Glomus intraradices, Funneliformis mosseae, Rhizophagus irregularis*, etc.) help plants to obtain less accessible nutrients by enhancing their antioxidant defenses, avoiding cell damage, and boosting water absorption (Jajoo and Mathur, 2021). Seed coating, pelleting, and hardening are examples of seed biopriming strategies, which include a variety of ways to improve seed germination by using microorganisms and chemicals that are produced by microbes (Prasad et al., 2016). The microbial communities that are present on seeds' exterior surfaces naturally have the ability to enter the seed itself, thereby facilitating seed disinfestation and seed disinfection. This can increase the effectiveness of seed hydration, shorten the imbibition period, and ultimately increase the germination rate (Mahmood et al., 2016). The use of microbial inoculants containing PGPB has already been shown to significantly boost crop productivity, by anywhere around 12% to 20%. Additionally, it has been demonstrated that biopriming seeds with several advantageous bacteria, particularly PGPB, increases plant resistance and growth efficiency when exposed to challenging environments (Pérez-Jaramillo et al., 2016).

According to Rakshit et al. (2015), the use of biopriming with advantageous microorganisms represents a cutting-edge method of crop protection. This technique promotes sustainable agricultural output by improving seed quality, seedling vigor, and the plant's capacity to thrive under adverse environments. A review by Mahmood et al. (2016) confirmed that biopriming speeds up germination, guarantees consistent seedling emergence, and subsequently improves crop yields and quality. Recent advancements in this area include the use of seed priming agents that are nutrient-solubilizing bacteria in conjunction with media that is nutrient-rich to increase crop nutrient uptake. For instance, Rehman et al. (2018a, b) showed that zinc seed priming with endophytic bacteria or *Pseudomonas*-aided zinc treatment increased the productivity and biofortification of *Triticum aestivum* (bread wheat).

10.2.5 NANOPRIMING

Nano-priming is an emerging technique for seed priming that increases crop yields by promoting seed germination, seedling growth, and plant resilience to a variety of stresses (Joshi et al., 2022). Nano-priming is noticeably more effective than other seed priming techniques. Nanoparticles (NPs) play a crucial role in seed priming by promoting electron exchange and augmenting surface reactions within different components of plant cells and tissues. Nano-priming fosters the creation of nanopores in the shoot, facilitating water absorption, triggering ROS and antioxidant mechanisms in seeds. It also generates hydroxyl radicals, contributing to cell wall loosening and serving as an inducer for the swift hydrolysis of starch. The broad use of nanotechnology in a variety of scientific fields is evidence from the in-depth research in this area. Nanotechnology has a significant impact on almost every scientific field, and its use in agriculture has the potential to significantly improve a number of agricultural and environmental issues (Joshi et al., 2022; Kumari et al., 2023). According to Fraceto et al. (2016), nanotechnology undoubtedly plays a key role in transforming the landscape of agriculture and food production and offers the possibility of reviving the agricultural industry. Materials at the nanoscale stand out due to their unique properties, making them ideal for the development of cutting-edge instruments to support sustainable agriculture.

According to Parisi et al. (2015), the use of nanomaterials in agriculture shows tremendous potential for tackling sustainability issues brought on by the looming problems of a continuously expanding global population and climate change. By outperforming conventional fertilizers in terms of efficacy and efficiency, nanoparticles have shown to be an appealing choice for creating nano-fertilizers. The excessive application of chemical fertilizers to crops could be controlled with the use of nanotechnology and nanomaterials in agriculture. According to Upadhyaya et al. (2017), this in turn supports the idea of precision farming by reducing fertilizer waste and reducing environmental degradation brought on by less-than-ideal agricultural practices. The ability of nanoparticles to improve crop nutrition and boost plants' ability to withstand stress is the factor that is driving their expanding use in industry. Nanoparticles smaller than 100 nm have shown promising results in previous studies (Joshi et al., 2022; Kumari et al., 2023). They display the ability to improve a variety of crops' germination rates, seedling dry weights, and general vigor. The higher penetration through the seed coat is thought to have permitted the increased water and nutrient absorption that was observed in nanopriming, which in turn improved germination and seedling growth rates.

10.3 UTILIZATION OF NANOPARTICLES IN SEED PRIMING OF MEDICINAL PLANTS

Nowadays, nanotechnology has become an important tool to enhance growth, resilience, and therapeutic potential of plants (Table 10.1). Nanoparticles (NPs) are small aggregates ranging between 1–100 nm. Various studies have shown the positive effect of nanoparticles as nano-pesticides and nano-fertilizers in agricultural practices. NPs are used as priming agents due to their unique physiochemical properties. Various nanoparticles such as silver (Mahakham et al., 2017), copper, and zinc oxide (Raja et al., 2019) are widely reported in agriculture for seed growth and emergence. Besides, agriculturally important crop nano-priming plays a significant role in the cultivation of medicinal plants. Various studies have shown the positive role of seed nano-priming in medicinal plants. Nanoparticles facilitate seed germination by increasing enzymatic activity and utilization of starch reserve in seeds. They also enhance the tolerance of plants against adverse environmental conditions. Seed nano-priming induces various physiological, biochemical, and molecular changes in plants such as regulating ROS concentration in the cell (Tripathi et al., 2017), antioxidant activity, seed germination, and vigor (Kandhol et al., 2022). However, it is necessary to optimize the concentration and timing of nanoparticles to ensure their safe and effective use to minimize any potential threat to plants and the environment.

10.3.1 COPPER NANOPARTICLES

Copper (Cu) plays a crucial role in regulating plant growth and metabolism. Copper regulates various processes in plants, such as photosynthetic reaction, hydrocarbon, and protein exchange. It is a cofactor of various enzymes, such as polyphenol oxidase (PPO), that carry out oxidation of phenolic compounds (Leng et al., 2015). Application of CuNPs in *Artemisia absinthium* (wormwood) resulted in increased biomass, antioxidant potential, and biosynthesis of secondary metabolites (Hayat et al., 2021). Similarly, copper nanoparticles along with cold atmospheric plasma improved the growth and physiological parameters of *Dracocephalum moldavica* (Moldavian dragonhead) (Haddadian et al., 2018). Applications of subtoxic concentrations of CuNPs in *Bacopa monnieri* (water hyssop) resulted in increased antioxidant capacity and level of phenol, saponins, alkaloids, and flavonoids (Lala et al., 2020). CuNPs also enhance the disease resistance in medicinal plants that protect them against soil borne pathogens and microbial threat. Therefore, use of copper nanoparticles can be useful for cultivation of medicinally important plants.

10.3.2 TITANIUM NANOPARTICLES

Titanium nanoparticles play a significant role in seed germination. TiO_2 facilitates gaseous exchange, and the absorption of water and nutrients in the cell to stimulate seed germination (Dehkourdi et al., 2013). TiO_2 also maintains ROS homeostasis in the cell. TiO_2 nano-priming in *Paeonia suffruticosa* (Tree peony) mitigated the adverse effects of salinity stress, such as germination rate, antioxidant enzyme activity, photosynthetic activity, and seed biomass (Liu et al., 2021). Similarly, TiO_2 mitigates the effect on *Origanum majorana* (Sweet marjoram) under salinity stress. It enhanced essential oil content and free radical scavenging activity in plants under salinity stress (Jafari et al., 2021). Besides this, TiO_2 NPs reduce cadmium toxicity in plants. A study on *Coriandrum sativum* L. showed that TiO_2 primed seedlings display a reduced concentration of cadmium and improved agronomic traits. Titanium dioxide lowered the MDA content and electrolyte leakage in the cell. Besides this, treated seedlings showed an increased concentration of osmolytes such as proline and reduced oxidative damage in the cell (Sardar et al., 2022). TiO_2 nanoparticles also showed a positive effect on *Crocus sativus* (saffron) cultivation. It resulted in an increase in stigma dry weight and promoted corm and flower yield (Nazari et al., 2021). Therefore, TiO_2 nano-seed priming can be a promising technique to increase the cultivation of medicinal plants. The optimum concentration of TiO_2 triggers the plant defense mechanism and biosynthesis of secondary metabolites such as geranial, z-citral, geranyl acetate, and geraniol (Gohari et al., 2020).

10.3.3 SILVER NANOPARTICLES

Recently, silver nanoparticles (AgNPs) have gained a lot of attention for their potential role in enhancing biosynthesis of secondary metabolites in medicinal plants. Silver nanoparticles promote seed germination via increased water absorption or by breaking dormancy. AgNPs also trigger plant defense and increase production of bioactive compounds. However, careful application of AgNPs is essential, as they cause toxicity in plants at higher concentrations. A study on *Juniperus procera* (African juniper) showed the positive effect of AgNPs on bioactive compounds such as gallic acid, hesperidin, ferruginol, and rutin (Salih et al., 2022). Similarly, AgNPs have increased artemisinin content in *Artemisia annua* (Sweet wormwood) (Zhang et al., 2013). A study on *Stevia rebaudiana* (Candyleaf) showed that AgNPs have stimulated glycoside synthesis (Majleshi et al., 2018). Apart from this, AgNPs are used to break seed dormancy in medicinal plants. A study on *Withania somnifera* (Ashwagandha) showed the role of deoxycholic acid capped AgNPs in breaking physical seed dormancy. Treated seeds showed uniform germination and enhanced germination percentage (Thangavelu et al., 2019). These studies suggest the potential role of silver nanoparticles in enhancing the bioactive compounds in medicinal plants that can be further exploited for therapeutic purposes.

10.3.4 ZINC OXIDE NANOPARTICLES

Zinc is a key element for the growth of plants because it helps plants produce chlorophyll and carbohydrates. The intake of harmful heavy metals is reduced, and their toxic effects on plants are lessened when zinc levels in plants are improved. Zinc oxide nanoparticles are employed in different applications, including biosensors, electrodes, household goods, and industries. Zinc nanoparticles, on the other hand, are primarily utilized in agriculture (Awasthi et al., 2017), and they are essential for the anatomical and physiological responses of plants. ZnO is crucial for numerous enzyme functions, including dehydrogenases and superoxide dismutase, and it controls how plant hormones are metabolized (Narendhran et al., 2016). Zinc oxide nanoparticles can be used as chemical absorbents, antibacterial agents, and catalysts due to their minimal toxicity, durability, significant specific surface area, high pore volume, and capacity for photodegradation. Recent research (Burman et al., 2013; Marslin et al., 2017) has concentrated on the impact of ZnO NPs on the germination of seeds from various plant species.

10.3.5 SILICA NANOPARTICLES

After oxygen, silicon (Si) is regarded as the second most prevalent element, making up about 31% of the Earth's crust (Dietz and Herth, 2011). Under stressful circumstances, silicon (Si) may promote plant growth and development (Ali et al., 2016; Alzahrani et al., 2018; Ullah et al., 2018; Zhang et al., 2018). According to reports (Gao et al., 2018; Pereira et al., 2018), silicon can reduce the toxicity of different heavy metals in crops, including Cd stress. According to certain researchers, Si may improve plants that are suffering from salt stress (Wang et al., 2011). Silicon contributes to increased photosynthetic capacity, improves light absorption, strengthens leaf structural integrity, lowers evapotranspiration, and increases plant resilience to both biotic and abiotic stimuli. Nano silica is a significant substance which is used in a variety of technological and scientific sectors, including electronics and biomedicine. Application of silicon fertilizers improves crop yield and quality, according to several field tests conducted with different plants under diverse climatic and soil conditions. Silicon nanoparticles (SiNPs) have unique physiological characteristics that enable them to influence plant metabolism (Rastogi et al., 2019). According to Azimi et al. (2014), nano-silica improves germination characteristics such as germination rate, radicle height, and plant dry weight. Numerous crops, including wheat (Rehman et al., 2018), rice (Gao et al., 2018), and peas (Rehman et al., 2017) have shown silicon-mediated reductions in Cd toxicity.

10.3.6 GOLD NANOPARTICLES

Since the 16th century, AuNPs, metallic nanoparticles, have been researched and employed for medical purposes (Kumar et al., 2013). Gold nanoparticles are the most stable and neutral metals because they are made of an inert core material; surface coatings can be added to functionalize them and keep the gold from interacting with the environment. Furthermore, several plant tissues have been found to contain gold nanoparticles with diameters between 0.5 and 100 nm (Judy et al., 2012). It was shown earlier that low concentrations of AuNPs improve the activity of antioxidant enzymes, the amount of photosynthetic pigment, the speed of electron transport, the metabolism of water, and plant development (Kumari et al., 2023). Although gold nanoparticles can build up in plants and have the ability to change secondary metabolism and plant growth, it is important to comprehend how they interact with plants (Joshi et al., 2022). It was shown earlier that seed priming with gold NPs increased the seed germination of maize (Mahakham et al., 2016) and rice (Mahakham et al., 2017).

TABLE 10.1 Nano-Priming and its Advantageous Impact on the Growth and Development of Different Medicinal Plant Species

S. No	Medicinal Plant Species	Nanoparticle used	Effect	References
1.	*Paeonia suffruticosa* (Tree peony)	Titanium dioxide	Enhance seed germination and early seedling growth under salt stress	Liu et al. (2021)
2.	*Solanum lycopersicum* (Tomato)	Titanium dioxide	Promote seed germination and reduce germination time	Singh et al. (2016)
3.	*Hordeum vulgare* (Barley)	Titanium dioxide	Improves growth parameters and vigor index	Hassan et al. (2016)
4.	*Spinacia oleracea* (Spinach) *Daucus carota* (Carrot)	Iron oxide	Rapid seed germination	Ndaba et al. (2022)
5.	*Stevia rebaudiana* Bertoni (Stevia)	Silicon dioxide	Improves seed germination, antioxidant system and starch metabolism	Hasanaklou et al. (2023)
6.	*Thymus kotschyanus* (Thyme)	Silicon dioxide/ Silver	Improves seed germination traits	Abbasi Khalaki et al. (2016)
7.	*Psophocarpus tetragonolobus* (Winged beans)	Silver	Enhance germinability, antioxidant enzymes and biochemical activities during seed germination	Kumar et al. (2020)
8.	*Withania somnifera* (Ashwagandha)	Silver	Improves total germination percentage, fresh weight, radicle length and cotyledon length	Thangavelu et al. (2019)
9.	*Brassica napus* (Rapeseed)	Selenium and Zinc oxide	Positive effect on germination percentage, germination rate, seed microstructure, and antioxidant enzyme activities	El-Badri et al. (2021)
10.	*Boswellia ovalifoliolata* (South Indian Olibanum)	Silver	Seed germination and seedling growth	Savithramma et al. (2012)
11.	*Phaseolus vulgaris* (Common bean)	Silica	Improves germination percentage and Vigor Index	Alsaeedi et al. (2017)
12.	*Thymus vulgaris* L. and *Thymus daenensis* Celak (Denaian thyme)	Sliver	Increased germination percent, germination rate and root length	Ghavam (2018)
13.	*Astragalus squarrosus* (Indian milkvetch)	Silicon dioxide	Improves germination	Azimi et al. (2016)
14.	*Guizotia abyssinica* (Niger)	Silicon dioxide/ Titanium dioxide	Increase water absorption of seeds and seedling growth	Eskandarinasab et al. (2019)

10.4 TRANSIT PATHWAYS AND MECHANISTIC OPERATION OF NANOPARTICLES

After stomata and cuticle, the primary barrier for the uptake of nanoparticles (NPs) in plants is their cell wall. The porosity of the cell wall varies from species to species, and the uptake of nanoparticles depends on physical properties like shape, size, and surface charge. However, nanoparticles can pass through the cell wall by modulating various cell wall components (Hu et al., 2020). After exposure to the plant root, nanoparticles penetrate into the root epidermis via the cell wall and cell membrane, which subsequently transfers to the vascular bundle, following the stele via symplastic movement, and finally translocate to the leaves. Xylem plays a vital role in the distribution and translocation of nanoparticles in plants (Aslani et al., 2014). The acquisition of water and other mineral nutrients occurs through the cell wall, which is generally a porous network of polysaccharides and fibers (Carpita and Gibeaut, 1993). Various studies show that the plant cell wall is a semi-permeable barrier and allows the facilitation of selected particles (Miralles et al., 2012).

After the cell wall, the secondary barrier for the uptake and translocation of nanoparticles inside the plant cell is the plasma membrane (PM), which generally acts as a barricade for the passage of molecules due to the presence of a phospholipid bilayer. The uptake of NPs is generally considered an active process, which involves various cellular processes like signaling, regulation, and recycling of the PM (Etxeberria et al., 2009). There are three different mechanisms reported thus far for the uptake of nanoparticles. The first mechanism states that nanoparticles can easily pass through the PM by diffusion; the uptake of NPs is concerned with their physical and chemical properties (Auría-Soro et al., 2019). According to the second mechanism, there is an active uptake of nanoparticles by a process known as endocytosis (Foroozandeh and Aziz, 2018). The third mechanism states that the uptake of nanoparticles is mediated through various transmembrane proteins (Li et al., 2020). Endocytosis involves two different pathways: clathrin-dependent and clathrin-independent. During endocytosis, there occurs the formation of a clathrin-coated vesicle from the PM, which helps in the internalization of the NPs (Miralles et al., 2012). It was reported that both clathrin-dependent and independent means are present in the plant endocytosis process.

Onelli et al. (2008) observed that internalization of gold nanoparticles in *Nicotiana tabacum* (tobacco) was through both clathrin-dependent as well as independent means. Various modes of uptake and transport of NPs from cell to tissue are through foliar application into seeds, leaves, as well as through roots. This transport of NPs is mediated via either the symplastic or apoplastic route. Nanoparticles, after foliar application, translocate through the phloem and get accumulated in different plant organs (Su et al., 2019). Apoplastic transport allows NPs to move through the PM to reach the central pith and then to the above-ground part of the plant. However, in the case of symplastic movement, NPs move through plasmodesmata with water and other nutrients (Lutts et al., 2016). During seed priming, a small amount of water gets imbibed into seeds to initiate the pre-requisite metabolites for early germination with radical projection. During nano-priming of seeds, NPs enter the cotyledons via intercellular spaces present in the parenchymatous tissue (Ma et al., 2010). Spatio-temporal localization of ROS is necessary for the communication between two cells, as well as it is also used to break the bonds between polysaccharides present in the endosperm (Oracz and Karpiński, 2016).

Another study shows that the accumulation of NPs on the seed coat causes the elevation of ROS, which in downstream is responsible for the activation of various mechanisms (Guha et al., 2018). In *Suaeda salsa* (seepweeds) seeds, phenolics present in the seed coat are endogenously regulated by gibberellic acid and abscisic acid (ABA) content, which further helps in the facilitation of nutrients across different seed compartments (Xu et al., 2016). In the case of *A. thaliana* (thale cress), ZnO NPs hinder seed germination; however, SiO_2 NPs and Al_2O_2 NPs have no effect on the growth and germination of seeds (Lee et al., 2010). Other than the PM, entry of NPs inside the plant cell is also facilitated by a carrier protein, aquaporin, present in the plasmodesmata (Madeira et al., 2016). Other than this, various ion channels, stomata, vasculature, and stomata can act as other possible means for the transportation of NPs inside the plant (Kolitsi and Yiantsios, 2020). In a study, gold nanoparticles (AuNPs) were found to get delivered inside the plant through plasmodesmata (Zhai et al., 2014). Eichert and Goldbach (2008) observed that in the case of large hydrophilic NPs ($\geq$40 nm), transfer of NPs occurs through stomata. Nanoparticles can also enter inside the plants through some other entry points, viz. pit membranes, cell wall junctions, Casparian strips, hydathodes, and extracellular spaces (Ruttkay-Nedecky et al., 2017). Most of the nanoparticles which are unstable in nature get aggregated after foliar application or in the soil due to their high ionic strength. However, organic macromolecules like fulvic acid, humic acid, extra cellular polymeric substances, and citric acid can help to elevate the stability of nanoparticles and subsequent reduction in deposition or sedimentation (Su et al., 2019). Toxicity of NPs can be observed due to blockage of pores and apoplastic barriers, which reduces the uptake of other nutrients, photosynthetic activity, generation of ROS, and DNA damage (Aken, 2015; Miralles et al., 2012).

10.5 SURPASSING SEED DORMANCY BY NANOPARTICLES-INDUCED ROS GENERATION AND PLANT GROWTH REGULATOR CROSSTALK

Seed dormancy is the state when a viable seed is unable to germinate under a favorable environment. It is an innate property to overcome the unfavorable abiotic conditions or premature germination. Seed dormancy can be of any type, according to morphological and physiological factors: physiological dormancy is caused when inhibitory compounds inside and outside the seeds prevent germination; when seeds contain underdeveloped embryos, it leads to morphological dormancy; morpho-physiological dormancy is caused when both physical and morphological factors inhibit seed germination; physical dormancy involves the suppression of water absorption by the seeds (Chai et al., 2016). Priming seeds with nanoparticles provides a promising mechanism to improve seed vigor, germination, and development. Several NPs such as ZnO NPs (Faizan et al., 2021), Cu NPs (Van Nguyen et al., 2021), Au NPs (Alshehddi et al., 2020), Ag NPs (Thiruvengadam et al., 2015), FeS_2 NPs (Srivastava et al., 2014), TiO_2 NPs (Gohari et al., 2020), and carbon-based NPs like carbon nanotube (Rahimi et al., 2016) and fullerene (Shafiq et al., 2019) have been used for seed priming for enhancing seed germination.

Nanoparticles have the ability to trigger certain metabolic pathways that affect germination and development of seedlings. Nanoparticles affect the physiology of seeds by the production of ROS. ROS is a wide group of reactive molecules such as H_2O_2, superoxide, singlet oxygen, and hydroxyl radicals. ROS can diffuse within and between the cells through membrane aquaporin channels (Tamma et al., 2018). Additionally, apoplastic ROS play a crucial role in physiological processes like cell growth.

ROS involvement is especially vital in enhancing the extensibility of the cell wall during root hair and pollen tube elongation, as well as in leaf growth (Novaković et al., 2018). In seeds, ROS is produced by metabolically active cells from embryogenesis to germination. The accumulation of NPs in the seeds enhances the oxidative respiration resulting in the formation of ROS, which is necessary to overcome seed dormancy. The elevated levels of ROS create oxidative stress which loosens the walls of the seed coat and induces germination (Mittler, 2017). ROS mainly interacts with GA_3 and ABA (Bailly, 2019). Signaling of ROS activates GA synthesis which stimulates the production of alpha amylase, thereby degrading the starch and stimulating seed germination. In nano-primed seeds, hormone regulation during germination includes different GA-induced processes like reserve mobilization, weakening of the endosperm, and elongation of the embryo cell. In primed seeds, increased activity of alpha amylase has been observed in germinating seeds. Silver nanoparticles act as a nano-catalyst and enhance the activity of alpha amylase, thereby increasing the enzymatic digestion of starch which leads to hastened seed germination (Mahakham et al., 2017).

When ROS increases, the aquaporin signaling pathway is also activated. This results in enhanced water and nutrient uptake and the transport of CO_2, NH_3, and H_2O_2 by modifying the phosphorylation sites of aquaporin proteins (Boursiac et al., 2008). Seeds primed with silver nanoparticles have enhanced seed germination because increased levels of ROS activate aquaporin genes (Mahakham et al., 2017). The aquaporin signaling pathway is regulated by ABA through *TIP3, TIP4, PIP2,* and *NIP1* during seed germination (Footitt et al., 2019). The balance between ABA and GA regulates the on/off of dormancy. This balance affects the absorption of water by the seeds. Water absorption is necessary for seeds to start the metabolism and growth of cells. Additionally, the ABA and GA balance also controls the phenols in the seed coat, which facilitates nutrient movement outside of the compartments in seeds (Xu et al., 2016). Nanoparticles also affect the reactivity of plant growth hormones, especially auxin, and enhance the phyto-stimulatory actions. Zinc oxide nanoparticles could elevate the level of IAA in germinating seeds, thereby increasing the rate of growth in seedlings. Quenching of free radicals by nanoparticles also enhances the physiological response in seeds (Korishettar et al., 2017). Nano-TiO_2 and Nano-SiO_2 enhance the nitrate reductase activity in seeds, thereby increasing the ability of seeds to absorb and utilize water, and also stimulate their antioxidant system, ultimately leading to enhanced seed germination and growth (Yang et al., 2006). Nanoparticles induce early embryo growth via ROS production. H_2O_2 causes changes in biochemistry during seed germination, which generates ROS. This ROS activates the enzymes to stimulate the process of germination. The elevated level of ROS leads to the breakdown of polymers by free radicals and subsequently helps in cell expansion by loosening of the cell wall. Silver nanoparticles induce the formation of OH^- radicals, which loosen cell walls of the seed coat and induce seed germination.

Metallothionein genes, which maintain the homeostasis of metal ions in plants, have been found associated with ROS signaling in nano-primed seeds (Anand et al., 2019). Elevated levels of ROS also elevate the levels of antioxidant enzymes like superoxide dismutase and catalase in seeds (Latef et al., 2020). This prevents the seeds from oxidative damage stimulated during seed germination. Increased levels of antioxidants enhance the germination rate, seedling growth, and post-germination changes. Complete seed germination takes place only when the ROS content is within an oxidative window that allows ROS signaling. The oxidative window accounts for the dual role of ROS in seed germination. Levels of ROS should be within the 'oxidative window' in order to progress towards germination (Feizi et al., 2013). H_2O_2, the key ROS molecule, acts as a signaling molecule for the regulation of seed germination. The levels of H_2O_2 content are regulated by antioxidants, which is necessary to maintain a balance between oxidative signaling that stimulates germination and oxidative damage that prevents or delays germination (Feizi et al., 2013). Thus, nano-primed seeds with elevated levels of H_2O_2 serve as a signal molecule and are in accordance with the concept of the oxidative window, resulting in better germination and faster seedling growth (Figure 10.1).

10.6 CONCLUSION AND FUTURE PERSPECTIVES

In today's world of advanced technology, it is crucial to make sure that seeds are healthy and germinate easily in order to grow strong seedlings that ultimately result in increased plant production. To accomplish this, both natural and synthetic substances are used as priming agents to improve seed sprouting and encourage the seedlings' development. Numerous studies have consistently shown that seed priming is a sensible and efficient

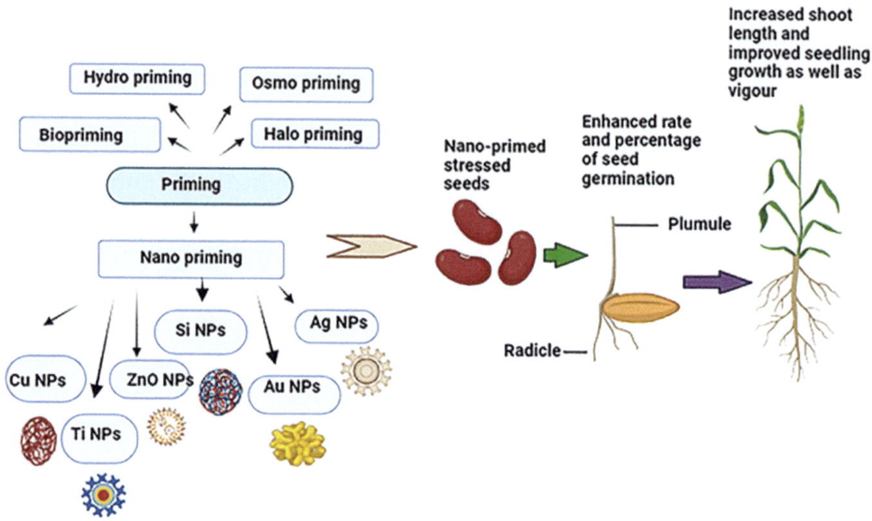

FIGURE 10.1 A schematic representation of recent developments in the field of nano-priming for enhancing seed germination by using various nanoparticles.

strategy, with the potential to increase germination rates, decrease seedling emergence times, improve seedling establishment, promote robust plant growth, increase survival rates, and strengthen plant resilience against a variety of biotic and abiotic stresses. In order to improve seed germination and strengthen resilience against various environmental stresses, nano-priming has made use of a variety of nanomaterials, including metal and metal oxide nanoparticles (NPs). Nanopriming synchronizes germination, encourages plant growth, and improves resistance to non-living stimuli, much like conventional priming techniques. The formation of ROS by NPs has been shown in numerous studies to be a key mechanism for triggering the synthesis of secondary metabolites. In plants' defense mechanisms, ROS play a crucial signaling role that eventually activates secondary metabolites that are particular to the stressor. Our knowledge of NP adhesion, the interaction of phytohormones in nano-priming-induced seed germination, and the complex signaling pathways involved in the production of secondary metabolites during nano-priming is still lacking. To understand the common regulatory networks that react to nanoparticles, extensive genome-wide transcriptome investigations under varied nano-priming circumstances are crucial. It is also important to remember that past studies were mainly concentrated on horticultural and agricultural crops. MAP, on the other hand, have only recently been the subject of research, despite the fact that their important pharmaceutical characteristics make them deserving of more attention.

ACKNOWLEDGMENTS

The authors thank the Director, CSIR-IHBT, Palampur, India, for infrastructure and support. Khushbu Kumari (NTA Ref. No.: 191620158661), Anita Kumari (NTA Ref. No.: 191620063807), and Suman Gusain (UGC Ref. No.: 574/CSIR-UGC NET JUNE 2019) are thankful to the UGC for providing JRF. Shubham Joshi thanks DBT for providing SRF (DBT/JRF/BET-19/1/2019/AL/212), and Meenakshi Rawat (DST/INSPIRE Fellowship/[IF200333]) thanks DST for providing INSIRE-JRF. The authors acknowledge the financial support from CSIR (MLP-201) and DBT (BT/PR45280/NER/95/1918/2022). This manuscript represents CSIR-IHBT communication no. 5407.

KEYWORDS

- **nanoparticles**
- **reactive oxygen species**
- **seed dormancy**
- **seed germination**
- **seed priming**
- **starch and lipid metabolism**

REFERENCES

Abbasi Khalaki, M., Ghorbani, A., & Moameri, M. (2016). Effects of silica and silver nanoparticles on seed germination traits of *Thymus kotschyanus* in laboratory conditions. *Journal of Rangeland Science, 6*(3), 221–231.

Abdel Latef, A. A. H., Abu Alhmad, M. F., & Abdelfattah, K. E. (2017). The possible roles of priming with ZnO nanoparticles in mitigation of salinity stress in lupine (*Lupinus termis*) plants. *Journal of Plant Growth Regulation, 36*, 60–70.

Acharya, P., Jayaprakasha, G. K., Crosby, K. M., Jifon, J. L., & Patil, B. S. (2020). Nanoparticle-mediated seed priming improves germination, growth, yield, and quality of watermelons (*Citrullus lanatus*) at multi-locations in Texas. *Scientific Reports, 10*(1), 5037.

Alsaeedi, A. H., El-Ramady, H., Alshaal, T., El-Garawani, M., Elhawat, N., & Almohsen, M. (2017). Engineered silica nanoparticles alleviate the detrimental effects of Na+ stress on germination and growth of common bean (*Phaseolus vulgaris*). *Environmental Science and Pollution Research, 24*, 21917–21928.

Alshehddi, L. A. A., & Bokhari, N. (2020). Influence of gold and silver nanoparticles on the germination and growth of *Mimusops laurifolia* seeds in the South-Western regions in Saudi Arabia. *Saudi Journal of Biological Sciences, 27*(1), 574–580.

Anand, A., Kumari, A., Thakur, M., & Koul, A. (2019). Hydrogen peroxide signaling integrates with phytohormones during the germination of magnetoprimed tomato seeds. *Scientific Reports, 9*(1), 8814.

Aslani, F., Bagheri, S., Muhd Julkapli, N., Juraimi, A. S., Hashemi, F. S. G., & Baghdadi, A. (2014). Effects of engineered nanomaterials on plants growth: An overview. *The Scientific World Journal, 2014*.

Auría-Soro, C., Nesma, T., Juanes-Velasco, P., Landeira-Viñuela, A., Fidalgo-Gomez, H., Acebes-Fernandez, V., Gongora, R., Almendral Parra, M. J., Manzano-Roman, R., & Fuentes, M. (2019). Interactions of nanoparticles and biosystems: Microenvironment of nanoparticles and biomolecules in nanomedicine. *Nanomaterials, 9*(10), 1365.

Awasthi, A., Bansal, S., Jangir, L. K., Awasthi, G., Awasthi, K. K., & Awasthi, K. (2017). Effect of ZnO nanoparticles on germination of *Triticum aestivum* seeds. *Macromolecular Symposia, 376*(1), 1700043.

Azimi, R., Borzelabad, M. J., Feizi, H., & Azimi, A. (2014). Interaction of SiO nanoparticles with seed prechilling on germination and early seedling growth of tall wheatgrass (*Agropyron elongatum* L.). *Polish Journal of Chemical Technology, 16*(3), 25–29.

Azimi, R., Heshmati, G. A., & Kavandi, H. R. (2016). Evaluation of SiO_2 nanoparticle effects on seed germination in *Astragalus squarrosus*. *Journal of Rangeland Science, 6*, 135–143.

Bailly, C. (2019). The signaling role of ROS in the regulation of seed germination and dormancy. *Biochemical Journal, 476*(20), 3019–3032.

Benidire, L., Pereira, S. I. A., Castro, P. M. L., & Boularbah, A. (2016). Assessment of plant growth promoting bacterial populations in the rhizosphere of metallophytes from the Kettara mine, Marrakech. *Environmental Science and Pollution Research, 23*, 21751–21765.

Boursiac, Y., Prak, S., Boudet, J., Postaire, O., Luu, D. T., Tournaire-Roux, C., Santoni, V., & Maurel, C. (2008). The response of *Arabidopsis* root water transport to a challenging environment implicates reactive oxygen species- & phosphorylation-dependent internalization of aquaporins. *Plant Signaling and Behavior, 3*(12), 1096–1098.

Burman, U., Saini, M., & Kumar, P. (2013). Effect of zinc oxide nanoparticles on growth and antioxidant system of chickpea seedlings. *Toxicological & Environmental Chemistry, 95*(4), 605–612.

Carpita, N. C., & Gibeaut, D. M. (1993). Structural models of primary cell walls in flowering plants: Consistency of molecular structure with the physical properties of the walls during growth. *The Plant Journal, 3*(1), 1–30.

Chai, M., Zhou, C., Molina, I., Fu, C., Nakashima, J., Li, G., Zhang, W., Park, J., Tang, Y., Jiang, Q., & Wang, Z. Y. (2016). A class II KNOX gene, KNOX4, controls seed physical dormancy. *Proceedings of the National Academy of Sciences, 113*(25), 6997–7002.

Chandrasekaran, U., Luo, X., Wang, Q., & Shu, K. (2020). Are there unidentified factors involved in the germination of nano primed seeds? *Frontiers in Plant Science, 11*, 832.

Chen, K., & Arora, R. (2013). Priming memory invokes seed stress-tolerance. *Environmental & Experimental Botany, 94*, 33–45.

Dabral, S., Varma, A., Choudhary, D. K., Bahuguna, R. N., & Nath, M. (2019). Biopriming with *Piriformospora indica* ameliorates cadmium stress in rice by lowering oxidative stress and cell death in root cells. *Ecotoxicology and Environmental Safety, 186*, 109741.

Dasgupta, N., Ranjan, S., & Ramalingam, C. (2017). Applications of nanotechnology in agriculture and water quality management. *Environmental Chemistry Letters, 15*, 591–605.

De La Torre-Roche, R., Cantu, J., Tamez, C., Zuverza-Mena, N., Hamdi, H., Adisa, I. O., Elmer, W., Gardea-Torresdey, J., & White, J. C. (2020). Seed biofortification by engineered nanomaterials: A pathway to alleviate malnutrition? *Journal of Agricultural & Food Chemistry, 68*(44), 12189–12202.

Dehkourdi, E. H., & Mosavi, M. (2013). Effect of anatase nanoparticles (TiO_2) on parsley seed germination (*Petroselinum crispum*) *in vitro. Biological Trace Element Research, 155*, 283–286.

Desoky, E. S. M., Merwad, A. R. M., Semida, W. M., Ibrahim, S. A., El-Saadony, M. T., & Rady, M. M. (2020). Heavy metals-resistant bacteria (HM-RB): Potential bioremediators of heavy metals-stressed *Spinacia oleracea* plant. *Ecotoxicology and Environmental Safety, 198*, 110685.

Devika, O. S., Singh, S., Sarkar, D., Barnwal, P., Suman, J., & Rakshit, A. (2021). Seed priming: A potential supplement in integrated resource management under fragile intensive ecosystems. *Frontiers in Sustainable Food Systems, 5*, 654001.

Dietz, K. J., & Herth, S. (2011). Plant nanotoxicology. *Trends in Plant Science, 16*(11), 582–589.

Eichert, T., & Goldbach, H. E. (2008). Equivalent pore radii of hydrophilic foliar uptake routes in stomatous and astomatous leaf surfaces–further evidence for a stomatal pathway. *Physiologia Plantarum, 132*(4), 491–502.

El-Badri, A. M., Batool, M., Mohamed, I. A., Khatab, A., Sherif, A., Wang, Z., Salah, A., Nishawy, E., Ayaad, M., Kuai, J., & Zhou, G. (2021). Modulation of salinity impact on early seedling stage via nano-priming application of zinc oxide on rapeseed (*Brassica napus* L.). *Plant Physiology and Biochemistry, 166*, 376–392.

El-Badri, A. M., Batool, M., Wang, C., Hashem, A. M., Tabl, K. M., Nishawy, E., Kuai, J., Zhou, G., & Wang, B. (2021). Selenium & zinc oxide nanoparticles modulate the molecular & morpho-physiological processes during seed germination of *Brassica napus* under salt stress. *Ecotoxicology and Environmental Safety, 225*, 112695.

Eskandarinasab, M., Rafieiolhossaini, M., Roshandel, P., & Tadayon, M. R. (2019). Investigation of seed germination indices and anthocyanin content of niger (*Guizotia abyssinica*) seedling under the effect of three nanoparticles. *Iranian Journal of Seed Research, 5*(2), 73–89.

Etxeberria, E., Gonzalez, P., & Pozueta, J. (2009). Evidence for two endocytic transport pathways in plant cells. *Plant Science, 177*(4), 341–348.

Faizan, M., Faraz, A., Mir, A. R., & Hayat, S. (2021). Role of zinc oxide nanoparticles in countering negative effects generated by cadmium in *Lycopersicon esculentum. Journal of Plant Growth Regulation, 40*, 101–115.

Falahhosseini, L., Alizadeh, M. A., & Vazan, S. (2017). Priming effect on the enhancement of germination traits in aged seeds of chamomile (*Matricaria chamomilla* L.) preserved in medium and long-term storage. *Journal of Medicinal Plants and By-product, 6*(1), 1–9.

Feizi, H., Moghaddam, P. R., Shahtahmassebi, N., & Fotovat, A. (2013). Assessment of concentrations of nano and bulk iron oxide particles on early growth of wheat (*Triticum aestivum* L.). *Annual Research and Review in Biology, 2013*, 752–761.

Footitt, S., Clewes, R., Feeney, M., Finch-Savage, W. E., & Frigerio, L. (2019). Aquaporins influence seed dormancy and germination in response to stress. *Plant, Cell & Environment, 42*(8), 2325–2339.

Foroozandeh, P., & Aziz, A. A. (2018). Insight into cellular uptake and intracellular trafficking of nanoparticles. *Nanoscale Research Letters, 13*, 1–12.

Forti, C., Shankar, A., Singh, A., Balestrazzi, A., Prasad, V., & Macovei, A. (2020). Hydropriming and biopriming improve *Medicago truncatula* seed germination and upregulate DNA repair and antioxidant genes. *Genes, 11*(3), 242.

Fraceto, L. F., Grillo, R., de Medeiros, G. A., Scognamiglio, V., Rea, G., & Bartolucci, C. (2016). Nanotechnology in agriculture: Which innovation potential does it have? *Frontiers in Environmental Science, 4*, 20.

Gao, M., Zhou, J., Liu, H., Zhang, W., Hu, Y., Liang, J., & Zhou, J. (2018). Foliar spraying with silicon and selenium reduces cadmium uptake and mitigates cadmium toxicity in rice. *Science of the Total Environment, 631*, 1100–1108.

Ghavam, M. (2018). Effect of silver nanoparticles on seed germination and seedling growth in *Thymus vulgaris* L. and *Thymus daenensis* Celak under salinity stress. *Journal of Rangeland Science, 8*(1), 93–100.

Gohari, G., Mohammadi, A., Akbari, A., Panahirad, S., Dadpour, M. R., Fotopoulos, V., & Kimura, S. (2020). Titanium dioxide nanoparticles (TiO_2 NPs) promote growth and ameliorate salinity stress effects on essential oil profile and biochemical attributes of *Dracocephalum moldavica*. *Scientific Reports, 10*(1), 912.

Guha, T., Ravikumar, K. V. G., Mukherjee, A., Mukherjee, A., & Kundu, R. (2018). Nanopriming with zero valent iron (nZVI) enhances germination and growth in aromatic rice cultivar (*Oryza sativa* cv. Gobindabhog L.). *Plant Physiology and Biochemistry, 127*, 403–413.

Haddadian, K., Iranbakhsh, A., Khavari-Nejad, R. A., & Ghoranneviss, M. (2018). Copper nanoparticles together with cold atmospheric plasma improve the growth and physiological indices of *Dracocephalum moldavica* in hydroponic system. *South Asian Journal of Experimental Biology, 7*(3), 130–139.

Hasanaklou, N., Mohagheghi, V., Hasanaklou, H. T., Ma'mani, L., Malekmohammadi, M., & Moradi, F. (2023). Seed nanopriming using silica nanoparticles: Effects in seed germination and physiological properties of *Stevia rebaudiana* Bertoni. *Chemical and Biological Technologies in Agriculture, 10*, 96.

Hassan, M., Ahangar, A. G., & Mir, N. (2016). Superior performance of dye-sensitized versus conventional TiO_2 nanoparticles for promoting germination and early growth of barley: From photovoltaic to biotechnological application. *Journal of Nano Research, 35*, 77–91.

Hayat, K., Ali, S., Ullah, S., Fu, Y., & Hussain, M. (2021). Green synthesized silver and copper nanoparticles induced changes in biomass parameters, secondary metabolites production, and antioxidant activity in callus cultures of *Artemisia absinthium* L. *Green Processing and Synthesis, 10*(1), 61–72.

Hou, L., Yu, J., Zhao, L., & He, X. (2020). Dark septate endophytes improve the growth and the tolerance of *Medicago sativa* and *Ammopiptanthus mongolicus* under cadmium stress. *Frontiers in Microbiology, 10*, 3061.

Hu, P., An, J., Faulkner, M. M., Wu, H., Li, Z., Tian, X., & Giraldo, J. P. (2020). Nanoparticle charge and size control foliar delivery efficiency to plant cells and organelles. *ACS Nano, 14*(7), 7970–7986.

Jafari, L., Abdollahi, F., Feizi, H., & Adl, S. (2022). Improved marjoram (*Origanum majorana* L.) tolerance to salinity with seed priming using titanium dioxide (TiO_2). *Iranian Journal of Science and Technology, Transactions A: Science, 46*, 361–371.

Jajoo, A., & Mathur, S. (2021). Role of arbuscular mycorrhizal fungi as an underground saviour for protecting plants from abiotic stresses. *Physiology and Molecular Biology of Plants*, 1–15.

Joshi, S., Dar, A. I., Acharya, A., & Joshi, R. (2022). Charged gold nanoparticles promote *in vitro* proliferation in *Nardostachys jatamansi* by differentially regulating chlorophyll content, hormone concentration, and antioxidant activity. *Antioxidants, 11*(10), 1962.

Judy, J. D., Unrine, J. M., Rao, W., Wirick, S., & Bertsch, P. M. (2012). Bioavailability of gold nanomaterials to plants: Importance of particle size and surface coating. *Environmental Science & Technology, 46*(15), 8467–8474.

Kandhol, N., Singh, V. P., Ramawat, N., Prasad, R., Chauhan, D. K., Sharma, S., Grillo, R., Sahi, S., Peralta-Videa, J., & Tripathi, D. K. (2022). Nano-priming: Impression on the beginner of plant life. *Plant Stress, 5*, 100091.

Kolitsi, L. I., & Yiantsios, S. G. (2020). Transport of nanoparticles in magnetic targeting: Comparison of magnetic, diffusive and convective forces and fluxes in the microvasculature, through vascular pores and across the interstitium. *Microvascular Research, 130*, 104007.

Korishettar, P., Vasudevan, S. N., Shakuntala, N. M., Doddagoudar, S. R., Hiregoudar, S., & Kisan, B. (2017). Influence of seed polymer coating with Zn & Fe nanoparticles on storage potential of pigeonpea seeds under ambient conditions. *Journal of Applied and Natural Science, 9*(1), 186–191.

Kumar, V. K., Muthukrishnan, S., & Rajalakshmi, R. (2020). Phytostimulatory effect of phytochemical fabricated nanosilver (AgNPs) on *Psophocarpus tetragonolobus* (L.) DC. seed germination: An insight from antioxidative enzyme activities and genetic similarity studies. *Current Plant Biology, 23*, 100158.

Kumar, V., Guleria, P., Kumar, V., & Yadav, S. K. (2013). Gold nanoparticle exposure induces growth and yield enhancement in *Arabidopsis thaliana*. *Science of the Total Environment, 461*, 462–468.

Kumari, A., Joshi, S., Dar, A. I., & Joshi, R. (2023). Physio-biochemical integrators and transcriptome analysis reveal nano-elicitation associated response during *Dendrocalamus asper* (Schult. and Schult. f.) Backer ex K. Heyne micropropagation. *Genes, 14*(9), 1725.

Kumari, N., Rai, P. K., Bara, B. M., & Singh, I. (2017). Effect of halo priming and hormonal priming on seed germination and seedling vigor in maize (*Zea mays* L.) seeds. *Journal of Pharmacognosy and Phytochemistry, 6*(4), 27–30.

Lala, S. (2020). Enhancement of secondary metabolites in *Bacopa monnieri* (L.) Pennell plants treated with copper-based nanoparticles *in vivo*. *IET Nanobiotechnology, 14*(1), 78–85.

Lee, C. W., Mahendra, S., Zodrow, K., Li, D., Tsai, Y. C., Braam, J., & Alvarez, P. J. (2010). Developmental phytotoxicity of metal oxide nanoparticles to *Arabidopsis thaliana*. *Environmental Toxicology and Chemistry: An International Journal, 29*(3), 669–675.

Leng, X., Jia, H., Sun, X., Shangguan, L., Mu, Q., Wang, B., & Fang, J. (2015). Comparative transcriptome analysis of grapevine in response to copper stress. *Scientific Reports, 5*(1), 17749.

Li, J. H., Au, B., Rentsch, J., Block, S., & Ewers, H. (2020). Directed manipulation of membrane proteins by fluorescent magnetic nanoparticles. *Biophysical Journal, 118*(3), 313a.

Liu, L. I., Cao, Y., Guo, Q., & Zhu, Z. (2021). Nanosized titanium dioxide seed priming enhances salt tolerance of an ornamental and medicinal plant *Paeonia suffruticosa*. *Pakistan Journal of Botany, 53*(4), 1167–1175.

Lutts, S., Benincasa, P., Wojtyla, L., Kubala, S., Pace, R., Lechowska, K., & Garnczarska, M. (2016). Seed priming: New comprehensive approaches for an old empirical technique. In S. Araujo & A. Balestrazzi (Eds.), *New Challenges in Seed Biology—Basic and Translational Research Driving Seed Technology* (Vol. 46). IntechOpen.

Ma, X., Geiser-Lee, J., Deng, Y., & Kolmakov, A. (2010). Interactions between engineered nanoparticles (ENPs) and plants: Phytotoxicity, uptake and accumulation. *Science of the Total Environment, 408*(16), 3053–3061.

Madeira, A., Moura, T. F., & Soveral, G. (2016). Detecting aquaporin function and regulation. *Frontiers in Chemistry, 4*, 3.

Maghsoudi, K., Emam, Y., Ashraf, M., & Arvin, M. J. (2019). Alleviation of field water stress in wheat cultivars by using silicon and salicylic acid applied separately or in combination. *Crop and Pasture Science, 70*(1), 36–43.

Mahakham, W., Sarmah, A. K., Maensiri, S., & Theerakulpisut, P. (2017). Nanopriming technology for enhancing germination and starch metabolism of aged rice seeds using photosynthesized silver nanoparticles. *Scientific Reports, 7*(1), 8263.

Mahmood, A., Turgay, O. C., Farooq, M., & Hayat, R. (2016). Seed biopriming with plant growth promoting rhizobacteria: A review. *FEMS Microbiology Ecology, 92*(8), fiw112.

Majlesi, Z., Ramezani, M., & Gerami, M. (2018). Investigation on some main glycosides content of *Stevia rebaudiana* B. under different concentrations of commercial and synthesized silver nanoparticles. *Pharmaceutical and Biomedical Research, 4*, 8.

Manjaiah, K. M., Mukhopadhyay, R., Paul, R., Datta, S. C., Kumararaja, P., & Sarkar, B. (2019). Clay minerals and zeolites for environmentally sustainable agriculture. In M. Mercurio, B. Sarkar, & A. Langella (Eds.), *Modified Clay and Zeolite Nanocomposite Materials* (pp. 309–329). Elsevier.

Marslin, G., Sheeba, C. J., & Franklin, G. (2017). Nanoparticles alter secondary metabolism in plants via ROS burst. *Frontiers in Plant Science, 8*, 832.

Miralles, P., Church, T. L., & Harris, A. T. (2012). Toxicity, uptake, and translocation of engineered nanomaterials in vascular plants. *Environmental Science and Technology, 46*(17), 9224–9239.

Mirmazloum, I., Kiss, A., Erdélyi, É., Ladányi, M., Németh, É. Z., & Radácsi, P. (2020). The effect of osmopriming on seed germination and early seedling characteristics of *Carum carvi* L. *Agriculture, 10*(4), 94.

Mittler, R. (2017). ROS is good. *Trends in Plant Science, 22*(1), 11–19.

MS, A., Sridharan, K., Puthur, J. T., & Dhankher, O. P. (2021). Priming with nanoscale materials for boosting abiotic stress tolerance in crop plants. *Journal of Agricultural and Food Chemistry, 69*(35), 10017–10035.

Narendhran, S., Rajiv, P., & Sivaraj, R. (2016). Toxicity of ZnO nanoparticles on germinating *Sesamum indicum* (Co-1) and their antibacterial activity. *Bulletin of Materials Science, 39*, 415–421.

Nazari, N., & Feizi, H. (2021). Magnetic fields and titanium dioxide nanoparticles promote saffron performance: A greenhouse experiment. *Journal of Horticulture and Postharvest Research, 4*(Special Issue—Recent Advances in Saffron), 33–42.

Ndaba, B., Roopnarain, A., Vatsha, B., Marx, S., & Maaza, M. (2022). Synthesis, characterization, and evaluation of *Artemisia afra*-mediated iron nanoparticles as a potential nano-priming agent for seed germination. *ACS Agricultural Science and Technology, 2*(6), 1218–1229.

Noori, H., Moosavi, S. G., Seghatoleslami, M., & Rostampour, M. F. (2022). Responses of cumin (*Cuminum cyminum* L.) to different seed priming methods under osmotic stress. *Notulae Botanicae Horti Agrobotanici Cluj-Napoca, 50*(1), 12600–12600.

Novaković, L., Guo, T., Bacic, A., Sampathkumar, A., & Johnson, K. L. (2018). Hitting the wall—Sensing and signaling pathways involved in plant cell wall remodeling in response to abiotic stress. *Plants, 7*(4), 89.

Onelli, E., Prescianotto-Baschong, C., Caccianiga, M., & Moscatelli, A. (2008). Clathrin-dependent and independent endocytic pathways in tobacco protoplasts revealed by labeling with charged nanogold. *Journal of Experimental Botany, 59*(11), 3051–3068.

Oracz, K., & Karpiński, S. (2016). Phytohormones signaling pathways and ROS involvement in seed germination. *Frontiers in Plant Science, 7*, 864.

Panyuta, O., Belava, V., Fomaidi, S., Kalinichenko, O., Volkogon, M., & Taran, N. (2016). The effect of pre-sowing seed treatment with metal nanoparticles on the formation of the defensive reaction of wheat seedlings infected with the eyespot causal agent. *Nanoscale Research Letters, 11*, 1–5.

Paparella, S., Araújo, S. S., Rossi, G., Wijayasinghe, M. A. L. A. K. A., Carbonera, D., & Balestrazzi, A. (2015). Seed priming: State of the art and new perspectives. *Plant Cell Reports, 34*, 1281–1293.

Parisi, C., Vigani, M., & Rodríguez-Cerezo, E. (2015). Agricultural nanotechnologies: What are the current possibilities? *Nano Today, 10*(2), 124–127.

Pérez-Jaramillo, J. E., Mendes, R., & Raaijmakers, J. M. (2016). Impact of plant domestication on rhizosphere microbiome assembly and functions. *Plant Molecular Biology, 90*, 635–644.

Prasad, R. S., Kamble, U. R., Sripathy, K. V., Udaya Bhaskar, K., & Singh, D. P. (2016). Seed bio-priming for biotic and abiotic stress management. In D. P. Singh, H. B. Singh, & R. Prabha (Eds.), *Microbial Inoculants in Sustainable Agricultural Productivity* (pp. 211–228). Springer.

Qadir, M., Hussain, A., Hamayun, M., Shah, M., Iqbal, A., & Murad, W. (2020). Phytohormones producing rhizobacterium alleviates chromium toxicity in *Helianthus annuus* L. by reducing chromate uptake and strengthening antioxidant system. *Chemosphere, 258*, 127386.

Rahimi, D., Kartoolinejad, D., Nourmohammadi, K., & Naghdi, R. (2016). Increasing drought resistance of *Alnus subcordata* CA Mey. seeds using a nano priming technique with multi-walled carbon nanotubes. *Journal of Forest Science, 62*(6), 269–278.

Raj, A. B., & Raj, S. K. (2019). Seed priming: An approach towards agricultural sustainability. *Journal of Applied and Natural Science, 11*(1), 227–234.

Raja, K., Sowmya, R., Sudhagar, R., Moorthy, P. S., Govindaraju, K., & Subramanian, K. S. (2019). Biogenic ZnO and Cu nanoparticles to improve seed germination quality in blackgram (*Vigna mungo*). *Materials Letters, 235*, 164–167.

Rakshit, A., Sunita, K., Pal, S., Singh, A., & Singh, H. B. (2015). Bio-priming mediated nutrient use efficiency of crop species. In A. Rakshit, H. B. Singh, & A. Sen (Eds.), *Nutrient use Efficiency: From Basics to Advances* (pp. 181–191). Springer.

Rastogi, A., Tripathi, D. K., Yadav, S., Chauhan, D. K., Živčák, M., Ghorbanpour, M., El-Sheery, N. I., & Brestic, M. (2019). Application of silicon nanoparticles in agriculture. *3 Biotech, 9*, 90.

Rehman, A., Farooq, M., Naveed, M., Nawaz, A., & Shahzad, B. (2018b). Seed priming of Zn with endophytic bacteria improves the productivity and grain biofortification of bread wheat. *European Journal of Agronomy, 94*, 98–107.

Rehman, A., Farooq, M., Naveed, M., Ozturk, L., & Nawaz, A. (2018a). *Pseudomonas*-aided zinc application improves the productivity and biofortification of bread wheat. *Crop and Pasture Science, 69*(7), 659–672.

Rehman, M. Z., Rizwan, M., Hussain, A., Saqib, M., Ali, S., Sohail, M. I., Shafiq, M., & Hafeez, F. (2018). Alleviation of cadmium (Cd) toxicity and minimizing its uptake in wheat (*Triticum aestivum*) by using organic carbon sources in Cd-spiked soil. *Environmental Pollution, 241*, 557–565.

Rhaman, M. S., Rauf, F., Tania, S. S., & Khatun, M. (2020). Seed priming methods: Application in field crops and future perspectives. *Asian Journal of Research in Crop Science, 5*(2), 8–19.

Roman-Ponce, B., Reza-Vázquez, D. M., Gutierrez-Paredes, S., María de Jesús, D. E., Maldonado-Hernandez, J., Bahena-Osorio, Y., Estrada-De los Santos, P., Wang, E. T., & Vásquez-Murrieta, M. S. (2017). Plant growth-promoting traits in rhizobacteria of heavy metal-resistant plants and their effects on *Brassica nigra* seed germination. *Pedosphere, 27*(3), 511–526.

Ruffini Castiglione, M., Giorgetti, L., Geri, C., & Cremonini, R. (2011). The effects of nano-TiO_2 on seed germination, development and mitosis of root tip cells of *Vicia narbonensis* L. and *Zea mays* L. *Journal of Nanoparticle Research, 13*, 2443–2449.

Ruttkay-Nedecky, B., Krystofova, O., Nejdl, L., & Adam, V. (2017). Nanoparticles based on essential metals and their phytotoxicity. *Journal of Nanobiotechnology, 15*(1), 1–19.

Salih, A. M., Al-Qurainy, F., Khan, S., Nadeem, M., Tarroum, M., & Shaikhaldein, H. O. (2022). Biogenic silver nanoparticles improve bioactive compounds in medicinal plant *Juniperus procera in vitro*. *Frontiers in Plant Science, 13*, 962112.

Sardar, R., Ahmed, S., & Yasin, N. A. (2022). Titanium dioxide nanoparticles mitigate cadmium toxicity in *Coriandrum sativum* L. through modulating antioxidant system, stress markers and reducing cadmium uptake. *Environmental Pollution, 292*, 118373.

Savithramma, N., Ankanna, S., & Bhumi, G. (2012). Effect of nanoparticles on seed germination and seedling growth of *Boswellia ovalifoliolata* an endemic and endangered medicinal tree taxon. *Nano Vision, 2*(1), 61–68.

Shafiq, F., Iqbal, M., Ali, M., & Ashraf, M. A. (2019). Seed pre-treatment with polyhydroxy fullerene nanoparticles confer salt tolerance in wheat through upregulation of H_2O_2 neutralizing enzymes and phosphorus uptake. *Journal of Soil Science and Plant Nutrition, 19*, 734–742.

Singh, P., Singh, R., Borthakur, A., Srivastava, P., Srivastava, N., Tiwary, D., & Mishra, P. K. (2016). Effect of nanoscale TiO_2-activated carbon composite on *Solanum lycopersicum* (L.) and *Vigna radiata* (L.) seeds germination. *Energy, Ecology and Environment, 1*, 131–140.

Singh, R. (2017). Effects of hydro priming on seed germination and vigor of *Aegle marmelos*. *Journal of Pharmacognosy and Phytochemistry, 6*(5), 446–449.

Soleymani, A., & Shahrajabian, M. H. (2018). Changes in germination and seedling growth of different cultivars of cumin to drought stress. *Journal of Applied Life Sciences and Environment*, 91–100.

Srivastava, G., Das, C. K., Das, A., Singh, S. K., Roy, M., Kim, H., Sethy, N., Kumar, A., Sharma, R. K., Singh, S. K., Philip, D., & Das, M. (2014). Seed treatment with iron pyrite (FeS_2) nanoparticles increases the production of spinach. *RSC Advances, 4*(102), 58495–58504.

Su, Y., Ashworth, V., Kim, C., Adeleye, A. S., Rolshausen, P., Roper, C., & Jassby, D. (2019). Delivery, uptake, fate, and transport of engineered nanoparticles in plants: A critical review and data analysis. *Environmental Science: Nano, 6*(8), 2311–2331.

Su, Y., Wu, D., Xia, H., Zhang, C., Shi, J., Wilkinson, K. J., & Xie, B. (2019). Metallic nanoparticles induced antibiotic resistance genes attenuation of leachate culturable microbiota: The combined roles of growth inhibition, ion dissolution and oxidative stress. *Environment International, 128*, 407–416.

Tamma, G., Valenti, G., Grossini, E., Donnini, S., Marino, A., Marinelli, R. A., & Calamita, G. (2018). Aquaporin membrane channels in oxidative stress, cell signaling, and aging: Recent advances and research trends. *Oxidative Medicine and Cellular Longevity, 2018*.

Thangavelu, R. M., Munisamy, B., & Krishnan, K. (2019). Effect of deoxycholate capped silver nanoparticles in seed dormancy breaking of *Withania somnifera*. *Current Science, 116*(6), 952–958.

Thiruvengadam, M., Gurunathan, S., & Chung, I. M. (2015). Physiological, metabolic, and transcriptional effects of biologically synthesized silver nanoparticles in turnip (*Brassica rapa* ssp. *rapa* L.). *Protoplasma, 252*, 1031–1046.

Torney, F., Trewyn, B. G., Lin, V. S. Y., & Wang, K. (2007). Mesoporous silica nanoparticles deliver DNA and chemicals into plants. *Nature Nanotechnology, 2*(5), 295–300.

Tripathi, D. K., Singh, S., Singh, V. P., Prasad, S. M., Dubey, N. K., & Chauhan, D. K. (2017). Silicon nanoparticles more effectively alleviated UV-B stress than silicon in wheat (*Triticum aestivum*) seedlings. *Plant Physiology and Biochemistry, 110*, 70–81.

Upadhyaya, H., Roy, H., Shome, S., Tewari, S., Bhattacharya, M. K., & Panda, S. K. (2017). Physiological impact of zinc nanoparticle on germination of rice (*Oryza sativa* L.) seed. *Journal of Plant Science and Phytopathology, 1*, 62–70.

Van Aken, B. (2015). Gene expression changes in plants and microorganisms exposed to nanomaterials. *Current Opinion in Biotechnology, 33*, 206–219.

Van Nguyen, D., Nguyen, H. M., Le, N. T., Nguyen, K. H., Nguyen, H. T., Le, H. M., Nguyen, A. T., Dinh, N. T. T., Hoang, S. A., & Van Ha, C. (2021). Copper nanoparticle application enhances plant growth and grain yield in maize under drought stress conditions. *Journal of Plant Growth Regulation, 41*, 364–375.

Venudevan, B., & Geetha, R. (2019). Influence of seed polymorphism on physical and physiological seed quality in Senna KKM (Se) 1 (*Cassia angustifolia* Vahl). *Journal of Pharmacognosy and Phytochemistry, 8*(2S), 622–625.

Wang, C., Liu, Z., Huang, Y., Zhang, Y., Wang, X., & Hu, Z. (2019). Cadmium-resistant rhizobacterium *Bacillus cereus* M4 promotes the growth and reduces cadmium accumulation in rice (*Oryza sativa* L.). *Environmental Toxicology and Pharmacology, 72*, 103265.

Wang, H., Kou, X., Pei, Z., Xiao, J. Q., Shan, X., & Xing, B. (2011). Physiological effects of magnetite (Fe_3O_4) nanoparticles on perennial ryegrass (*Lolium perenne* L.) and pumpkin (*Cucurbita mixta*) plants. *Nanotoxicology, 5*(1), 30–42.

Wang, X., Yang, X., Chen, S., Li, Q., Wang, W., Hou, C., Gao, X., Wang, L., & Wang, S. (2016). Zinc oxide nanoparticles affect biomass accumulation and photosynthesis in *Arabidopsis*. *Frontiers in Plant Science, 6*, 1243.

Waqas, M. A., Kaya, C., Riaz, A., Farooq, M., Nawaz, I., Wilkes, A., & Li, Y. (2019). Potential mechanisms of abiotic stress tolerance in crop plants induced by thiourea. *Frontiers in Plant Science, 10*, 1336.

Waqas, M., Korres, N. E., Khan, M. D., Nizami, A. S., Deeba, F., Ali, I., & Hussain, H. (2019). Advances in the concept and methods of seed priming. In M. Hasanuzzaman & V. Fotopoulos (Eds.), *Priming and Pretreatment of Seeds and Seedlings: Implication in Plant Stress Tolerance and Enhancing Productivity in Crop Plants* (pp. 11–41). Springer.

Xiang, L., Zhao, H. M., Li, Y. W., Huang, X. P., Wu, X. L., Zhai, T., Yuan, Y., Cai, Q. Y., & Mo, C. H. (2015). Effects of the size and morphology of zinc oxide nanoparticles on the germination of Chinese cabbage seeds. *Environmental Science and Pollution Research, 22*, 10452–10462.

Xu, Y. G., Liu, R., Sui, N., Shi, W., Wang, L., Tian, C., & Song, J. (2016). Changes in endogenous hormones and seed-coat phenolics during seed storage of two *Suaeda salsa* populations. *Australian Journal of Botany, 64*(4), 325–332.

Yang, F., Hong, F., You, W., Liu, C., Gao, F., Wu, C., & Yang, P. (2006). Influence of nano-anatase TiO_2 on the nitrogen metabolism of growing spinach. *Biological Trace Element Research, 110*, 179–190.

Zhai, G., Walters, K. S., Peate, D. W., Alvarez, P. J., & Schnoor, J. L. (2014). Transport of gold nanoparticles through plasmodesmata and precipitation of gold ions in woody poplar. *Environmental Science and Technology Letters, 1*(2), 146–151.

Zhang, B., Zheng, L. P., Yi, W., & Wang, J. W. (2013). Stimulation of artemisinin production in *Artemisia annua* hairy roots by Ag-SiO_2 core-shell nanoparticles. *Current Nanoscience, 9*(3), 363–370.

Zhou, C., Zhu, L., Ma, Z., & Wang, J. (2017). *Bacillus amyloliquefaciens* SAY09 increases cadmium resistance in plants by activation of auxin-mediated signaling pathways. *Genes, 8*(7), 173.

CHAPTER 11

Principles of Seed Collection, Seed Processing, Drying Temperature, Storability, and Seed Germination of Some Seeds of Medicinal Plants

SHREYA DAS,[1] SUDDHASUCHI DAS,[2,*] and SUDHIR CH. DAS[3]

[1]*ICAR–Agricultural Technology Application Research Institute, Kolkata, West Bengal, India*

[2]*Malda Krishi Vigyan Kendra, Uttar Banga Krishi Viswavidyalaya, Malda, West Bengal, India*

[3]*Indian Forest Service (Retd.), West Bengal, India*

*Corresponding author

ABSTRACT

Medicinal plants have been utilized as a source of pharmaceuticals since ancient times. The Vedas, the Quran, and the Bible all make reference to the widespread usage of herbal medicines and medical preparations. Assurance of the preservation, efficacy, and quality of medicinal plants and herbal products has emerged as a critical concern in developing nations these days. Medicinal plants have been used to preserve and taste food, treat illnesses, and even halt the spread of sickness and epidemics for many years. Over the ages, knowledge of their medicinal properties has been transmitted. Plants have always been valued as a source of natural products to maintain human health, but in recent years, more detailed study has been done on natural medicines. Herbal medicine offers a very anodyne and effective technique, especially in the treatment of cancer,

Seed Technology for Medicinal and Aromatic Plants: Production, Testing, Storage, and Marketing.
Priyanka Sharma, Kishore Kumar Sharma, and Gopal Shukla (Eds.)

DOI: 10.1201/9781779641199-11

when compared to current western medicine. The compounds created during secondary metabolism are typically responsible for the biological characteristics of plant species that are utilized for a variety of objectives across the domain, such as the management of infectious diseases. The local resources that fall within their purview are what sustain the tribal communities that reside within and around the forests. When needed, they make use of the fruits, seeds, barks, roots, and leaves of the medicinal plants that grow right outside their door—in the forests. These plants interact with both endogenous and pharmacological chemicals in genuine ways and have a poly-chemical, polyvalent effect on human physiology.

11.1 INTRODUCTION

India has a geographical area of 302.9 million hectares, of which 75.81 million hectares are covered in forests that are home to roughly 45,000 different types of flowering species. About 2,563 species of trees are among the 18,664 vascular plant species that make up this group. In the pretense of progress, the devastation of the natural world has recently become the norm in our civilized society. Phyto-resources are the most utilized of all the natural resources, and for this reason, many species are facing extinction. Taking stock of phyto-resources and recording their traditional use patterns are thought to be crucial (Mukherjee and Bouri, 2011). Documenting the long-standing traditional knowledge of indigenous populations living in and around forest areas of the country about the medicinal use of plants has drawn particular attention from scientists. In the Indian subcontinent, the majority of the indigenous population is made up of tribal people, who are distributed throughout the country, mostly in central and north-east India, and who heavily rely on medicinal herbs. This prehistoric civilization, which lives near to the natural world, had special information of the traditional uses of the animals and plants in their surroundings. In view of the recent global recognition of the value of traditional knowledge, it is essential to emphasize the ethno-botanical significance of medicinal plants.

The demand for biological resources obtained from the wild has been rising steadily on a global scale. In 2022, 325,500 MT of botanical raw medicines was anticipated to be needed annually in India. The herbal medication industry currently uses around 6000 plants in one way or another.

About 1289 botanical raw medicines are derived from about 960 medicinal plants, of which 178 species are eaten in amounts more than 100 MT annually (Sarkar et al., 2012). The majority of this need is supplied by wild populations, with no consideration given to how long-term their true availability will be. This reliance on wild populations of plants to provide life-saving medications will not last for very long for at least two reasons: first, there are currently no technologies available to domesticate a wide range of species, particularly endangered ones, and growing competition and developmental activities (Swarup et al., 2009). The variety of significant medicinal plant species is seriously threatened by the indiscriminate harvesting of medicinal plants for pharmaceuticals and medications. The three endangered species of Asia and Africa—the glory lily (*Gloriosa superba*), sarpagandha (*Rauwolfia serpentina*), and is warmul (*Aristolochia indica*)—may become more widely available if appropriate propagation and conservation techniques are used (Siva and Krishnamurthy, 2000). The *Gloriosa* species possesses anticancer, anti-carcinogenic, and nematocidal effects (Pandey and Hassep, 1988) in addition to being utilized as a substitute plant for autumn crocus (*Colchicum autumnale*) for alkaloid colchicine's (Bharthi et al., 2006). Scientists must pay close attention to this nearly unknown field of standardization protocol for conservation and easy, rapid propagation following identification as separate genetic material if they are to save these endangered medicinal plants (Sarkar et al., 2011).

One of the most important requirements to preserve our natural resources is the conservation of biological diversity. With their indigenous traditional knowledge, the locals take care of and preserve the diversity of wild flora, crops, and medicinal plants. Their existence is threatened by the effects of modern society, which have altered environmental conditions and destroyed habitat. India ranks among the top 12 countries with the highest levels of mega-biodiversity, accounting for 8% of global biodiversity and 12.5% of global phyto-diversity (Anonymous, 2000). Unfortunately, overuse of resources and habitat destruction have put this rich biodiversity in grave danger. An increasing number of plant species are being listed in the Red Data Book as threatened. This calls for the preservation of wild flora in both botanical gardens and their native habitats.

A significant impact on biodiversity is caused by the wild harvesting of almost 95% of the raw materials. Approximately 1500 plant species are currently considered to be threatened; of these, 33 have become extinct,

157 are endangered, 114 are vulnerable, and 246 are unusual (Anonymous, 2000). According to Mudappa and Oommen (1998), the trade in medicinal plants is active for over 70% of the vulnerable species. About 235 species (or around 38%) of the 625 therapeutic plant species that are now traded are herbs, and collecting them is typically damaging (Mudappa and Oommen, 1998). Since the amount of biodiversity loss is so great, urgent action must be taken to stop this damage.

The traditional systems of medicine in India, including Ayurveda, Homoeopathy, Unani, and Siddha, are based on medicinal plant-based drugs. Tribal people living in and around the forests are dependent on the local resources available within their reach. They use roots, leaves, barks, fruits, and seeds of medicinal plants available in their doorstep (i.e., in forests) as and when required, like Kalmegh (*Andrographis paniculata*), Amla (*Emblica officinalis*), Haritoki (*Terminalia chebula*), Bahera (*Terminalia bellerica*), Arjuna (*Terminalia arjuna*), Neem (*Azadirachta indica*), Sarpagandha (*Rauwolfia serpentina*), Chirata (*Swertia chirayita*), Jamun (*Syzygium cumini*), Nayantara (*Catharanthus rosea*), Basak (*Justicia adhatoda*), Lajjaboti (*Mimosa pudica*), Sajne (*Moringa oleifera*), Seuli (*Nyctanthes arbortristis*), Bel (*Aegle marmelos*), Kurchi (*Holarrhena antidysenterica*), Nishinda (*Vitex negundo*), Asoke (*Saraca asoka*), Tulsi (*Ocimum sanctum*), Ghritakumari (*Aloe barbadensis*), Satamuli (*Asparagus racemosus*), Harjora (*Cissus quadrangularis*), Dhutra (*Datura metal*), Kesud (*Eclipta alba*), Glory Lily (*Gloriosa superba*), Anantamul (*Hemidesmus indicus*), Bherenda (*Jatropha curcas*), Amrul (*Oxalis corniculata*), Castor (*Ricinus communis*), Thuja (*Thuja orientalis*), Bhringaraj (*Wedelia calendulacea*), Ulatkambal (*Abroma augustum*), and so forth. In this chapter, we will discuss the principles of seed collection, seed processing, drying temperature, storability, and seed germination of six important medicinal plants, viz. sandalwood, neem, jamun, arjun, bahera, and amla.

11.2 SANDALWOOD (*SANTALUM ALBUM* L.)

White Sandal (Sandalwood) is the fragrant heartwood of *Santalum album*. About 400 species make up the 30 genera, including the commercially significant and extensively dispersed *Santalum* genus, which is part of the Santalaceae family. Many of these species are either fully or partially parasitic (John, 1947). Sandalwood is an evergreen big shrub that can grow to a height of 12–15 m for tall trees (*S. album* in India) and a girth of

FIGURE 11.1 Showing sandalwood tree, seeds and seedling.

1.0–2.4 m for moderately sized or small trees (*S. spicatum*) (Sen Sarma, 1982). The primary distribution of *S. album* is in the Deccan plateau of India. Its distribution spans about 9600 km^2, of which 8200 km^2 are in the states of Tamil Nadu and Karnataka (Srinivasan et al., 1992).

11.2.1 IMPORTANT USES

In the commercial domain, sandalwood is called "East Indian Sandalwood," and its essential oil, also called "East Indian Sandalwood Oil," is among the earliest recognized components in fragrances. Both wood and oil are used to make incense and medications, and they both have a large market value. Numerous serious illnesses, including piles, fever, dropsy, diabetes, mental and skin disorders, hemorrhages, toxin control, and a variety of other common uses for sandalwood, are treated with it.

Not only is sandalwood oil valuable as a fragrance, but it is also used in medicine to treat gonorrhea, dysuria, bronchitis, urinary tract infections, and as an antipyretic, antiscabietic, antiseptic, diuretic, expectorant, and stimulant (Jain, 1968; Okasaki and Oshima, 1953). However, its use as a fragrance foundation has much outweighed its application in the making of medications. According to Srinivasan et al. (1992), Indian sandalwood is highly prized for its aromatic heartwood that contains up to 6% oil and 90% α and β-Santalol. According to Srinivasan et al. (1992), it is mostly employed to create wonderfully beautiful idols, boxes, and other curios.

The moisturizing, hydrating, anti-wrinkling, and protecting properties of sandalwood oil are excellent when making soap. It has always been a blend of economic and traditional traditions. It has a three 1000 year history and is still utilized today in the Ayurvedic medical system by Susrutha and Charaka Samhita. The cells of the sandalwood plant and its reservoir of solar healing power produce the essential oil. Pharmaceuticals, as well as artifacts and elaborate sculptures, are made using sandalwood (Anonymous, 2000).

11.2.2 MEDICINAL USES

Jahan and Rahman (2014) studied that sandalwood dissolves tumors and stabilizes palpitation and inflammation. It is applied as a disinfectant for infections of the bronchi and urinary tract. For burns, headaches, and fever, wood paste is used. Additionally, it is used to treat biliousness, blood impurities, depression, acne, bronchitis, cough, menorrhagia, diarrhea, and sore throat. It is one of the possible anti-ulcer medications in the Unani medical system. Sandalwood is used to treat a number of different conditions, including vomiting-associated diarrhea, poisoning, hemorrhaging, hemorrhaging piles, intrinsic hemorrhage, the first stage of the pox, hiccups, urticaria, eye infections, and umbilicus irritation (Benencia and Courreges, 1999). α-santalol has been researched for its ability to prevent skin cancer in mouse models of skin carcinogenesis using human epidermoid carcinoma cells and the active ingredient in sandalwood oil. Ayurveda has used sandalwood powder and oil to treat a variety of conditions, including fever, respiratory disorders, diarrhea, gastric irritability, genito-urinary disorders, high blood pressure, heart palpitations, inducing sleep, wound cleaning, and various skin diseases Dwivedi et al. (2014).

11.2.3 DOMESTIC USES

Around 90% of sandalwood's constituent is β-santalol, which possesses antimicrobial qualities. It is mostly used to make soaps in aromatherapy. It can be used to remove blackheads and spots from the skin due to its anti-microbial properties, but it must always be diluted correctly with carrier oil. Sandalwood paste is used as an embalming paste on idols in temples. Sometimes, sandalwood paste—which is applied to the forehead during worship—is used to produce the bindi dot.

11.2.4 SEED COLLECTION AND PROCESSING

- Typically, sandalwood blooms twice a year, from September to November and from March to May. At first, flowers have a straw yellow color and no perfume, but as they mature, they become a deep purplish brown color. Cymose panicles with axillary or terminal flowers are common. The tree begins to flower at the age of 4 to 5 years, although seeds from plants older than 10 years should be harvested for multiplication purposes.
- When fully grown, the fruit is a single-seeded, purplish-black, succulent drupe. Fruits can be globose, ovate, or elongate in shape. When the fruit reaches maturity, its color shifts from green to a purple black.
- Fresh fruits with a purple hue should ideally be picked from trees or dropped to the ground. The fruit-gathering season is April through May and November through December.

11.2.5 SEED DRYING AND STORING

- The matured purple-colored fruits are collected during April-May and November-December. By rubbing, fruits are depulped. The depulped seeds are treated with fungicide, dried in shade, and stored in airtight containers.
- Seed weight is 6000 seeds/kg. When kept in an airtight container at a low temperature (50°C), the viability of the seed can be preserved for an extended length of time.
- Freshly collected seeds are dormant for a period of 2 months and remain viable up to 9 months.

11.2.6 DORMANCY AND SEED GERMINATION

Santalum album germination is irregular and takes 4–12 weeks to finish (Srimathi et al., 1995; Srinivasan et al., 1992). Sandalwood seeds have been seen to germinate quickly when the seed coat is fully removed, or they can be immersed in 0.05% (500 ppm) gibberellic acid for 16 hours (Nagaveni and Srimathi, 1981). For seed germination, Srinivasan et al. (1992) advised a nursery bed consisting of soil, sand, and rotten cow dung (potting mixture) in a 1:1:1 ratio. The suggested seed density per bed is 500 g/m^2. Under field circumstances, pre-treating seeds with GA_3 (500 ppm) for 16 hours produced 50–60% germination. Between the fourth and sixth leaf stage, typically, the seedlings are moved to 1500 cc (9" x 5") poly-bags that are filled with a 1:1:1 potting mixture of sand, soil, and FYM, using *Cajnus cajan* (arhar) as the pot host. In six to eight months, plantation seedlings with a dark brown stem and a height of 30–40 cm can be produced (Rai, 1990). Gunaga et al. (2014) discovered that seeds soaked in GA_3 solution at 300 ppm (68%) and 500 ppm (63%) for a whole night produced superior germination compared to the control (38%). They found that, in comparison to depulped seeds, the germination of bird-dispersed seeds produced the highest levels of seed germination in the shortest amount of time.

Under field conditions, germination is roughly 60% (ICFRE, 1992). According to Nagaveni and Srimathi (1980, 1981), consistent germination of up to 70% can be achieved in 60 days by soaking seeds in 0.05% gibberellic acid overnight before sowing. A kilogram of seeds weighs about 6000. According to Rai and Kulkarni (1986), the most often used techniques in the past for seeding mounds was the trench-mound method, which involved planting seeds in bushes with a success rate of typically between 30 and 40 percent. In sandalwood seeds, germination takes place over an extended length of time following the dormant phase. It begins 25 days after seeding and, with 500 ppm GA_3 soaking for 16–24 hours, only reaches 50–60% in 60 days (Das and Tah, 2013).

11.2.6.1 DETAILS OF PRE-TREATMENT BY GIBBERELLIC ACID

Matured seeds were collected from the sandalwood trees in November-December, depulped, dried in shade, and stored in polybags. The seeds were collected and tied with marking cloth. Five containers (1 liter each) were taken for four different concentrations (100 ppm, 300 ppm,

500 ppm, 1000 ppm) of gibberellic acid (GA_3) and one for control. In each container, two seed samples were dipped for 16 hours and 24 hours soaking. In the sand bed, the treated sandalwood seeds were then sown. Watering of sand beds will be done twice daily in the morning and afternoon. Germination will start after 26 days of seed sowing. The number of seeds germinated in each treatment was recorded, and the germination was continued up to 90 days after sowing (Das and Tah, 2013). At 3 to 4 leaf stage, the germinated seedlings were transplanted in polypod (9″x 5″) and 300 cc hycopots in nursery and kept in nursery without and with host (*Cajanus cajan*). Full grown plantable seedlings of 30–40 cm height will be available after 1 year.

11.3 NEEM (*AZADIRACHTA INDICA* L.)

Neem (*Azadirachta indica*), of the family Meliaceae, also called margosa, is a fast-growing deciduous tree of importance as a medicinal plant, as a good source of preparation of organic pesticides, and for its timber. Neem is indigenous to arid regions of South Asia and the Indian subcontinent. It has been introduced to many South and Central

FIGURE 11.2 Showing neem tree, flowering twig, seeds and seedling.

American countries, the Caribbean, and portions of Africa. Neem trees can reach 15–30 meters (49–98 feet) in height and have thick furrowed bark and attractive rounded crowns. The leaves are compound with toothed leaflets, and the tree becomes leafless immediately after winter (deciduous tree). The little fragrant white flowers are either bisexual or staminate, and they are borne in clusters in the leaf axils. The pulp of the silky, yellow-green drupe fruit smells delicious. While cuttings or root suckers can be used to propagate neem, seeds are often used for growth. Plants used in parks and gardens, this resistant and tenacious plant thrives in rocky, low-fertility soils. Neem can withstand a wide range of climatic conditions, although it cannot withstand freezing temperatures or waterlogging.

11.3.1 IMPORTANT USES

The plant has a long history of use in organic farming, cosmetics, folk and ayurveda medicine, and organic farming. All components of the neem tree are beneficial, and its antifungal and antibacterial qualities are the basis for many of its therapeutic and cosmetic uses. Neem is frequently found in shampoos for the treatment of dandruff and athlete's foot, as well as in soaps and lotions for skin ailments like psoriasis and acne. Young twigs are used as inexpensive toothbrushes in rural regions and, especially in the Indian subcontinent, as an ingredient in a variety of toothpastes and mouthwashes. Neem leaves have long been used in traditional diabetes treatments, and some clinical research suggests that neem may help regulate blood sugar levels. Neem leaves, bark, or oil should not be consumed by pregnant women since it can cause miscarriage.

It is possible to extract oil from the seeds, which can then be used directly as a fungicide, organic insecticide, mite, and insect repellant, and as the basis for many commercial pesticide products such as granules, dusts, and concentrates. The primary insecticidal component, azadirachtin, inhibits eating and interferes with hormones that prevent larvae from maturing into adults by causing mottling in the insects. By preventing mating and reproduction, neem oil minimizes the number of pests by inadvertently killing soft-bodied insects upon contact. To control blight, anthracnose, rust, black spot, mildew, and scab, neem oil is sprayed as a fungicide. Because neem-based insecticides are not hazardous to mammals, they are frequently utilized in organic farming applications.

11.3.2 SEED COLLECTION AND PROCESSING

Neem trees grow in all regions. It is an evergreen tree. In the dry locations, it sheds its leaves immediately after winter (during February); new leaves appear during February–March. Climate affects flowering differently. In March and April, flowers begin to bloom. Fruits mature between May and July. A fully grown tree can produce 5–10 kg of seeds. Seed weight: 3500–4000/kg. Ripe yellow-colored fruits are collected from the trees or from the ground when they have fallen. Neem fruits are collected and depulped using clean water and sun-dried for 2 days. To protect young saplings from fungal attack, the dried neem seeds are treated with fungicide or Trichoderma.

11.3.3 SEED DRYING, VIABILITY AND STORING

The seeds are depulped, allowed to dry in the sun for a day or two, and then placed in a shaded area with good ventilation. Neem seeds are only good for a week or 10 days. After 15 days, viability starts to drastically decrease, and after two or three months, it almost completely disappears. Neem seeds nearly lost their capacity to germinate at 75% relative humidity (RH) at 20°C in just six months, but viability only slightly diminished at 32% RH (Van der Plas and Hoekstra, 2000). According to early research, seeds that were equilibrated to 50% relative humidity at 20°C had the best chance of surviving. Later, it was discovered that neem seeds are susceptible to imbibitional stress, which may be somewhat mitigated by temperature, if they are dried any more than an equilibrium relative humidity of 50%. As seeds aged, their susceptibility to imbibitional stress rose. The fact that seeds can tolerate significant dehydration is shown by the fact that controlled rehydration at high temperatures (up to 40°C) might significantly increase the percentage of germination. This implies that neem seeds are conventional, even while their ability to tolerate rehydration without difficulty contributes to their desiccation tolerance.

Neem seeds are sensitive to temperatures of less than 10°C, according to results of cold storage. According to Plas and Hoekstra (2000), seeds exhibited a 50% average viability after two years of storage at temperatures ranging from –20°C to 20°C, indicating a significant increase in tolerance to low temperatures when moisture contents were 4–8%. An investigation of the lifespan of neem seeds under varying temperature and water content conditions revealed that, independent of the origin of the seeds (African or Asian), there were relatively few differences in the storage behavior of seed lots and batches. However, compared to the seeds from fully grown yellow fruits, the older or greener brown fruit seeds did not last as long.

11.3.4 SEED GERMINATION

Cuttings, root suckers, and seed are the three ways that neem is propagated; however, seed is the most practical. Neem seeds are viable for a very short time, so using yellow, ripe neem fruits is advised in order to collect viable neem seeds. It produces more than 95% germination in 2–4 months; however, depending on the weather and storage circumstances, that percentage can drop to 10%–15%. Neem fruits are collected, cleaned, and then depulped with clean water before being dried for two days under the shade. To prevent fungal invasion, the dried neem seeds are treated with trichoderma or fungicide. The treated seeds are then planted in polybags filled with the appropriate potting mixture of soil, sand, and organic matter (1:1:1) or on the nursery raised bed at a proper spacing of 2–5 cm (plant to plant) x 10–15 cm (row to row) at a depth of 2–2.5 cm. It is advised to use nine 300 cc hycopots or five-by-five black polypots with drain holes three inches up from the bottom for optimal root development. Germination begins in 7–10 days.

The multicultural practices that keep the soil well-aerated include weeding, watering, sorting, and shifting. In order to prevent rain and excessive water from dampening down illness, the nursery area will be covered with white plastic shade. Neem seedlings produced in the nursery bed must be removed after two to three months and transported to a planting location or into polybags. Plantings in poly-bags can be stored for a year, depending on the needs of the plantation. The ideal plantation height is 45–60 cm.

11.4 JAMUN (*SYZYGIUM CUMINI* L. *EUGENIA JAMBOLANA* L.)

Jam/Jamun, known as black plum, belongs to the family Myrtaceae. Jamun is indigenous to India. The tropical and subtropical regions are home to the highest concentration of jamun trees. It is also found up to 1600 meters in the Kumaon Mountains and up to 1300 meters in the lower Himalayan region. It is a tall and evergreen tropical tree favored for its timber, fruit,

FIGURE 11.3 Showing Jamun tree, fruits, seeds and seedlings.

and ornamental value. The bark is smooth, light grey with dark patches. The leaves are simple, opposite, variable in shape, with nerves joining in a distinct intra-marginal vein, gland-dotted, smooth, and shiny. The flowers are greenish-white in trichotomous panicles. The fruits are oblong, dark purple with pinkish juicy pulp, a single-seeded drupe. Even though the fruits are highly valued and enjoyed by a lot of people, the plant is still not cultivated as an orchard tree. When the scented leaves are young, they are pinkish, but as they become older, they turn a glossy, leathery dark green color with a yellow midrib.

It is a species that grows quickly, reaching heights of up to 30 meters (100 feet) and a lifespan of more than 100 years. It is grown for its beauty appeal, and its dense foliage provides shade. The bark is hard and dark grey towards the base of the tree and lighter grey and smoother further up. Kiln-dried wood has a water-resistant quality. For this reason, it is used in well motor installations and railway sleepers. The most popular kind produced in North India is called "Ram Jamun." When fully ripe, it yields large, rectangular fruits that are deep purple or bluish-black in color. The ripe fruit has a purple-pink pulp and is delicious and tasty.

11.4.1 IMPORTANT USES

It has been demonstrated that patients with diabetes considerably benefit from it. However, organized orcharding is still weak in India due to a lack of sufficient information on growing techniques and the unavailability of dwarf and high-producing types. Jambolana fruits can be eaten raw or turned into jam or sauces. They have a mildly acidic or sweet flavor. Fruits can be used to make jelly, juice, sorbet, or fruit salad. In the commercial sector, jamun is used to make wines, vinegar, jams, jellies, and other drinks. It's also used to make squash when combined with sugar, water, citric acid, and sodium benzoate. The final one serves as a preservative. Ripe jamuns are utilized to make exquisite wines, while unripe grapes are used to make acidic vinegar.

Jamun fruit has many nutritive values. Apart from sugars, minerals, and proteins, it is also a very important source of iron. Raw fruit is 16% carbohydrates, 1% protein, 83% water, and negligible fat. The Jamun fruit is a delicious treat for diabetics because it is low in calories and high in calcium, iron, phosphorus, vitamin C, magnesium, and folic acid. Additionally, this fruit's polyphenolic compounds are useful in the treatment of diabetes.

Jamun has a long medicinal history and is highly regarded as a successful treatment for a variety of illnesses in both ayurvedic and Unani medicines. Important for its medicinal properties, timber, and beauty, the Jamun tree is important among the shady trees in India. Sweet and acrid, the bark of the jamun tree has digestive properties and works as an astringent by destroying worms. An important source of blood purification, Jamun tree bark cures asthma, thirst, bronchitis, biliousness, dysentery, sore throat, and ulcers. Jamun phal can control constipation and other digestive disorders because of its diuretic properties. Jamun fruit has the potential to treat anemia since it increases the amount of red blood cells and hemoglobin in the blood.

An alkaloid, Jambosine, and glycoside jambolin or antimellin are present in seeds, which halts the diastatic conversion of starch into sugar. Seed extract has been reported to lower blood pressure by 34.6% due to the presence of ellagic acid. Seeds are also rich in flavonoids and are well-known antioxidants. Seeds are fairly rich in protein, calcium, and other minerals as well.

11.4.2 SEED COLLECTION AND PROCESSING

In Jamun, new vegetative shoots arise in two flushes, from February to May and from August to October. The February flush produces the greatest number of inflorescences per shoot, while the October flush produces the least amount (Mishra and Bajpai, 1971). In Jamun, a 20% sucrose solution may yield the maximum amount of pollen germination (Singh, 1978). When pollination occurs one day following anthesis, the maximum fruit set (32.6–36.0%) can be achieved; after that, a dramatic fall in fruit set may be seen. Berries are poly-embryonic, ovoid, tasty, and have a reddish color.

From March to April, *Syzygium cumini* (jamun) trees start flowering. The tiny, fragrant blossoms have a diameter of roughly 5 mm (0.2 in). The Syzygium species' fruit is referred to as "drupaceous;" it resembles huge berries and begins maturing by May or June. The fruit is ovoid, oblong. Unripe fruit looks green in color. Mature fruit first turns pink, then turns a brilliant scarlet red, and ultimately turns black. The fruit tends to turn the tongue purple and has a flavor that is sweet, slightly acidic, and astringent.

Planting after 8 to 10 years, seedling jamun plants begin to bear fruit, while grafted plants begin to produce fruits after 6 to 7 years. On the other hand, commercial bearing occurs 8 to 10 years after planting and lasts until the tree reaches 50 or 60 years of age. Ripe fruit at its largest size is primarily distinguished by its rich purple or black color. When the fruit reaches ripeness, it should be plucked right away because it cannot ripen further on the tree. Picking the ripe fruits by hand involves climbing the tree while carrying hung sacks over one shoulder. Fruits should be protected from any potential harm with extreme caution. A fully grown seedling tree typically yields between 80 and 100 kg of fruits, and 60 to 70 kg per year from a grafted one.

A common method of harvesting is by shaking the tree and collecting the fruits by holding a big piece of cloth or canvas under the tree. The fruits cannot remain on the tree in the ripe stage and start dropping to the ground immediately after ripening. As a result, a considerable amount of fruit is damaged and become unfit for fresh consumption. The ripe fruits are available during summer and disappear with onset of the monsoon.

11.4.3 SEED DRYING, VIABILITY AND STORING

"Ram-Jamun" is a common cultivar grown in North India. It yields large, oblong fruits that are 2.5–3.5 cm in length and 1.5–2.0 cm in diameter when fully ripe. The fruits are juicy and sweet tasting. These fruits with tiny stones ripen in June and July. Another late-maturing type has small, somewhat rounded fruits that are 1.5–2.0 cm long and 1.0–1.5 cm in diameter when fully mature. The fruits are deep purple or blackish in color. These cultivars possess rather large stones. The month of August is when fruits ripen. Spread the seeds out on a clean cloth and set them in the sun to dry. After two or three days in the sun, the outer shell becomes loose and easily peels off.

Since jamun is highly perishable, it is very challenging to store and sell in remote areas. The vitality of seeds can be increased by storing them in the refrigerator. In comparison to other treatments, the seeds stored in polybags and the seeds treated with polybag + Trichoderma harzianum (10 g/kg) showed better rates of seed germination. Research showed that the seeds' low viability was caused by their resistance. This could be avoided by storing jamun seeds in a container and at a temperature that is appropriate. Significant variations between the various storage and treatment methods were observed. The percentage of seeds that germinated when stored at 5°C in a refrigerator was higher than that of seeds stored under other circumstances. Following seeds treated with poly bag + charcoal powder and poly bag, higher germination was seen at 0, 15, 30, 45, 60, 75, and 90 days after extraction (DAE) (Devi et al., 2016). Seeds held at room temperature are not suggested for jamun seed storage, as the rate of seed deterioration is increased under such conditions within 15 days following extraction. Therefore, it is advised to employ moisture-holding media and bio-control agents as a means of increasing the viability and potential to boost the germination of jamun seeds.

11.4.4 SEED GERMINATION

The jamun is propagated by both seed and vegetative means. It can come true to the parent through seed thanks to polyembryony. The most common method for growing this fruit tree is from seed. Jamun seeds do not hibernate. Thus, new seeds can be planted in the ground in 7–10 days, 3–4 cm deep, 25 cm apart, and with the pointy end pointing down. Water the soil carefully and gently without disturbing the seeds. After seeding, germination happens in roughly 10 to 15 days. The seedlings are prepared for transplanting in the upcoming spring (February to March) or monsoon (July to August) for use as rootstock. If seeds are buried too deeply, they might not get enough air, which could lead to weak roots and a delayed seedling emergence. Polypots can also be used to sow seeds since they make handling rootstock and grafted plants easier. Nucellar seedlings can be utilized to create true-to-type plants because jamun has a 20–25% chance of polyembryony (Singh and Thakur, 1977). Under Indian circumstances, jamun seeds require 24 to 60 days to fully germinate (Singh and Thakur, 1977). According to Sastri et al. (2001), germination rates for large-sized seeds were higher (98–99%) than for small-sized seeds (79–89%).

It is nearly impossible to raise rootstock in nursery beds or lift budded plants with an earth ball in extremely sandy soils. Long-distance plant transportation can also result in substantial plant mortality, especially in semi-arid and arid environments. In order to expedite the rootstock raising process and prevent damage during handling and transit, commercial rootstocks may be raised in polybags (10" x 6") and polypots (9" x 5"). For drainage and aeration, small holes are cut in the bottom and sides of the polybags or polypots, which are then filled with porous rooting media (soil, sand, and organic manure in a ratio of 1:1:1).

11.5 ARJUN (TERMINALIA ARJUNA)

The huge deciduous tree Arjun, *Terminalia arjuna*, is a member of the Combretaceae family. It has a broad canopy, a buttressed trunk, a spreading crown, and drooping limbs. The country's tropical and subtropical moist regions are home to the tree in its natural habitat. The tree prefers soils with good drainage, high water retention, and loose, moist, and fertile characteristics, such as alluvial loamy or black cotton soils. Its natural habitat is found in streams, ravines, and the soils that border rivers. The plant can also grow in areas that are open, sunny, and have minimal rain-fall. Its height can reach up to 35 meters. Its smooth, thin, thick bark sheds in uneven sheets and has a color range of grey to pinkish green. Its smooth, grey bark and oblong, conical leaves are brown on the underside and green on top. Its pale yellow flowers bloom from March to June, and its 2.5–5 cm fibrous, woody fruit is glabrous, divided into five wings, and appears from September to November (Biswas et al., 2011). The tree is susceptible to *Phyllactinia terminale* and *Polystictus affinis* rot; however, it is not impacted by any serious diseases or pests. The tree can be found growing

FIGURE 11.4 Showing arjuna tree, flowering twig, fruits and seedlings.

on riverbanks or next to dry river beds in Bihar, West Bengal, Odisha, Uttar Pradesh, Maharashtra, Madhya Pradesh, South and Central India, Bangladesh, and Sri Lanka. The tree is mostly distributed throughout the Indian subcontinent. The *Antheraea paphia* feeds on the leaves to generate tassar silk, a wild silk with significant commercial value.

11.5.1 IMPORTANT USES

Bark from *Terminalia arjuna* is decoction. Based on the centuries-old observations of ancient physicians, it has been utilized in the Indian subcontinent for anginal discomfort, hypertension, congestive heart failure, and dyslipidemia. Medicinal plants have been a significant source of compounds used in medicine since ancient times. One of the most popular and effective medicinal plants in the traditional medical system for treating a wide range of severe ailments is *Terminalia arjuna*. The ethnomedical, phytochemical, pharmacogenetic, pharmacological, and therapeutic aspects of it are significant for a variety of illnesses, particularly cardiovascular issues. This plant has a strong safety record when used in conjunction with other conventional drugs. Through a variety of studies, this chapter demonstrates the many medical benefits of *T. arjuna*, including its hypotensive, anti-inflammatory, anti-carcinogenic, anti-mutagenic, antioxidant, and gastro-productive effects (Amalraj and Gopi, 2017).

Arjun trees begin to flower during the sixth year. The tree can live for 50 years or more in total. Bark is regularly scraped over the winter to serve as a raw material for medicines. Bark on healthy trees should be removed in spiral or vertical strips, ideally beginning in the tenth year, with widths of no more than 5 cm and lengths of no more than 25 cm. The collected bark is dried in well-ventilated shade and then stored in dry, airy places in boxes or polybags. It is reported that the tannin cardiac glycoside comes from both fruit and bark. In fruit, the typical range is 7%–20%, while in stem bark, it is 20%–24%.

From the tenth year onward, approximately 500 kg of dried bark can be collected from a 1 hectare planting annually. *T. arjuna* is a common agroforestry plant cultivated next to coconut and citrus trees. It also makes an excellent shade tree, especially on coffee fields. It yields good fuel and charcoal, with the sapwood having a calorific value of 5030 Kcal/kg and the heartwood having a calorific value of 5128 Kcal/kg. The heartwood is brown to dark-brown, strong, glossy, robust, and heavy (specific gravity 0.74; weight 816–865), whereas the sapwood is pinkish-white. In the area, wood is used to create carts, water troughs, houses, boats, electric poles, and traps. It is commonly grown in India, where the leaves are closely cropped, to produce tassar silkworms and feed animals. There is 9–11% crude protein and 14–20% crude fiber in the leaves. It is planted in parks and avenues for visual interest or shade. Arjun is typically employed in sand dune afforestation programs with *Casuarina* species, as well as in agro- and social forestry for the reclamation of alkaline soils, saline, and steep ravines.

11.5.2 SEED COLLECTION AND PROCESSING

The tree bears panicked spikes of pale yellow flowers, which bloom from March to June. The glabrous, 2.5–5 cm fibrous woody fruit, which is divided into five wings, ripens from September to November. The fruit is drupe-shaped, 2.5 cm long, oblong, thick, with five inflexible, longitudinal wings, and 0.6 cm broad. It has a notched area near the top and is dotted with striations that curve upward at an oblique angle. Fruits mature between February and April. Ripe fruits are picked in March, either by lopping the branches or by sweeping the previously cleaned ground. The seeds will keep for at least a year if stored in sealed tins.

11.5.3 SEED DRYING, VIABILITY AND STORING

Arjun fruits ripen in March. Collected ripe fruits are dried in sunlight for 2 days. Seed viability is about one year. We can store the seeds for 6–12 months when stored in sealed tins. The harvested seeds can be sown immediately after drying. Number of fruits /kg of seeds vary from 300 to 1200 depending on the fruit size. Plant percent is 40–50. Therefore, roughly 23 kg of seeds spaced 5 m by 5 m may be needed to raise 1 hectare of plantation.

11.5.4 SEED GERMINATION

The seeds can be pre-treated by soaking them in cool water for 48 hours. An alternative method is to immerse the seeds in boiling water, allow it to cool, and let it stay soaking for a whole day. The germination rate of untreated seeds is between 50% and 60%, but up to 90% of pre-treated seeds germinate. In nursery beds, seeds are sown in the early summer, typically right after collecting. Within 8 to 12 days, germination begins, and it takes 6–7 weeks to finish. The seeds that have germinated can be planted in polybags containing an equal amount of sand, soil, and manure. If not, the pre-treated fruits are immediately placed in polythene bags in April, with half of the fruit above ground. It is advised to use pre-germinated seeds in order to save time and obtain uniform germination. Both throughout the germination and growth stages, the seedlings are extremely vulnerable to drought. Shade may be supplied above the germination beds during the initial phase.

To produce pre-germination of seeds, a trench of 5’ (length) x 2’ (width) x 2’ (depth) size may be dug. At the bottom of the trench, the 1st 6” layer will be sand to facilitate drainage of excess water, the 2nd 4” layer will be seeds, the 3rd 4” layer will be dry leaves or straw, the 4th 4” layer will be seeds, and the 5th 4" layer will be dry leaves or straw. Then, watering will be done as and when required so that seeds will not be rotten due to excess water, or seeds will not be dry due to lack of moisture. Within 5–6 days, pre-germinated seeds will be produced which can be put in polypots/ hycopots.

In July and August, around 10-month-old saplings are placed in the field in trenches spaced five meters by five meters apart. When using an intercropping strategy, a hectare requires roughly 400 seedlings total. The crop can be planted as a single plantation or as a border crop during its early growth stage. *Celastrus paniculata* (Malkangni), *Operculina turpethum* (Trivrit), *Operculina petaloidea* (Nishoth), *Marsdenia tenacissima* (Moorva), and *Tinospora cordifolia* (Giloy) are among the climbers that can be utilized as medicinal plants. These climbers can be planted beneath the mature tree that serves as their host.

11.6 BAHERA (TERMINALIA BELLERICA L.)

The species *Terminalia bellerica* is commonly found in tropical and subtropical zones up to an altitude of 1500 m in the following countries: Bangladesh, Bhutan, Cambodia, China, India, Indonesia, Laos, Malaysia,

FIGURE 11.5 Showing Bahera tree, flowering twig, dry fruits and seedlings

Myanmar, Nepal, Sri Lanka, Thailand, and Vietnam. It grows on mountain slopes, sunny upper-layer trees of stream valleys, and lower seasonal rain forests, scattered forests. It is a member of the Combretaceae family. It is a 20–30 m tall perennial deciduous tree with thick, brownish-grey bark and a narrow longitudinal fissure. Its natural habitat is tropical Asia. It is extremely prevalent in mixed deciduous forests. Its crown is expanding and formed like a dome. It is found in clay soils as well as shaded ones. The tree has a significant need for light. For the young plants, however, some shade is helpful. Seeds are multiplied by fermentation. It can withstand some shade when it is young. Young seedlings are reported to be able to tolerate significant shade during their first and second years; however, with time, the plants finally perish from the shade. Frost can kill this plant very easily.

Leaves are glabrous, alternate, large, broadly elliptic to obovate-elliptical, 4–24 cm x 2–11 cm, rufous-sericeous but soon glabrescent, base rounded to cuneate, with 6–9 pairs of secondary veins. The leaf has a long petiole. Young leaves are copper red, then soon becoming parrot green, then dark green. The leaf venation is reticulate with a prominent midrib. Solitary flowers, small and 3–15 cm long, greenish white, axillary spikes, simple, flowers with new leaves, and have a strong smell like honey. Powdered drugs show vessels with simple pits and groups of sclerosis, and few fibers. Powdered drug is free-flowing with a granular appearance and brownish in color.

11.6.1 IMPORTANT USES

The bark, leaves, and fruit are the biological sources of *Terminalia bellerica*. Medicinal properties of the plant contain different chemical constituents and have different biological activities, like analgesic activity, anti-diarrheal activity, antioxidant activity, antihypertensive effect, anti-salmonella activity, anti-microbial activity, antispasmodic and bronchodilatory characteristics, antimicrobial and toxicity investigations, and wound-healing capabilities. Fruits have astringent, anthelmintic, laxative, and antipyretic properties; they are also used as a hair tonic and to treat conditions including asthma, dyspepsia, hepatitis, diarrhea, coughs, bronchitis, piles, hoarseness of voice, and eye disorders. Cough is treated with the green fruit decoction. Fruit pulp can be used to treat leprosy, dropsy, piles, and dysenteric diarrhea. Half-ripe fruit has purgative properties. The fruit kernel is narcotic. Fruits are used for menstrual disorders. Seed oil is used in rheumatism. The bark's gum is purgative and soothing. The fruits' triterpenoid content has strong antibacterial properties. With its purgative properties, kernel oil was well tolerated over an extended period of time in mice. Fruit pulp is used in chronic diarrhea, dysentery, flatulence, vomiting, and urinary disorder. It nourishes the brain and nerves and improves vision. The respiratory system is utilized to treat whooping cough, asthma, and dyspnea. The digestive system is used to treat flatulence, weakness, slow heartbeat, foul breath, constipation, and abdominal pain and distention. Skin: Burns and scalds can be treated with a fine powder paste. Ellagic acid, phyllemblin, β-sitosterol, mannitol, glucose, fructose, rhamnose, galloyl glucose, ethyl gallate, and chebulagic acid are among the tannins.

11.6.2 SEED COLLECTION AND PROCESSING

Along with the new leaves, pale greenish-white flowers with a rich honey scent develop in April and June. From November to February, fruits ripen. Fruit is a dry, fleshy drupe with a diameter of 1.3 to 1.9 cm, that is globose or ovoid, coated in woolly hairs that are grey and velvety, and it has a hard, one-seeded, pale yellow putamen that is surrounded by green tissue. *Terminalia bellerica* seed is ovoid, 5-ridged, and extremely hard. Every fruit has a single stone. The fruit is subsessile, silky, and grey in color, 2.5 cm in length and 2 cm in width. The fruit gets slightly ridged when it is dried.

Fruits are harvested as soon as they start to turn yellow. The best time to harvest fruits for their maximum content is in January; otherwise, the tannin content will be lower. If the collection is completed too soon, the tannin content will also be of similar quality. They shrivel significantly when they dry in the sun rather than being wet from rain. The crushed are infused with hot water in a series of vats for 8 to 10 hours in order to extract tannin and prepare extracts. While the wood is still green, it should be sawed. The seed is sun-dried before being stored after the pulp is removed right away. Following 7–8 years of growth, harvesting is completed. The period between mid-December and mid-January is the ideal time to collect.

11.6.3 SEED DRYING, VIABILITY AND STORING

November to February is when the fruits ripen. Ripe fruits are harvested in March by sweeping the previously cleaned area or by pruning the branches. Fruits are picked from strong trees, depulped, and the seeds are sun-dried before being stored. It can be stored for nearly a year, even if its viability and germination percentage drop from 65–70% to 40–50%. Fruits per kilogram of seed: 100–125. Table 11.1 shows that the drying temperature should be lower with higher seed moisture contents and vice versa.

Bahera must be disposed of within a year of the processing day; else, its valuable attributes would be lost. Bahera needs to be packed in dry, clean gunny bags and kept under storage in a dry location. It must not be stored in a damp area, or else fungi will infect it and cause discoloration.

TABLE 11.1 Drying Temperature of Seeds

Moisture Content	Drying Temperature
10% and below	110°F (43.3°C)
10–18%	100°F (42.2°C)
18–30%	90°F (32.2°C)

Damaged and insect-affected fruits should be avoided during collection. Before being stored, the fruits are dried and depulped. Two days of sun drying in a dry, well-ventilated area are required after deseeding. It can be kept for a year in this state of storage.

11.6.4 GERMINATION

Pre-treatment of seeds is done to soften the hard seed coat; seed is soaked in mild hot water for 24 hrs. Pre-treated fresh seed is sown in March-April in polythene bags. Germination commences in 2 weeks and is completed in one month. Pre-germinated seeds sown in polythene bag show good results. The development of the root is very fast. The seedlings are ready for planting out in July-August when they are five months old.

Trees are regenerated by coppicing and seeds. Trees can coppice quite successfully, yielding cuttings in May and June. Favorable characteristics include healthy seeds having a high terminal capacity for natural regeneration, good seed yields, and seedlings that can survive for the first year or two in shade. Direct sowing or transplanting one-year-old nursery-raised seedlings are two methods for raising the plant in the field. In June and July, when the monsoon rains arrive, seeds are sown. To soften the seed coat, the seed is steeped in water for a whole day. Sowing is done in lines 3 meters apart. The lines where the soil will be sown have been thoroughly worked. There is a light covering over the seeds.

11.7 AMLA (*EMBLICA OFFICINALIS* GAERTN. = *PHYLLANTHUS EMBLICA* L.).

Amloki/Amla is a small to medium-sized tree with spreading branches, a crooked trunk, and greyish-green bark that peels off in flakes; it is usually a deciduous tree. *Emblica officinalis*, commonly known as Amla, Amloki, or Indian Gooseberry, belongs to the family Euphorbiaceae. The tree is deciduous, with slender branches and an average height of 8 to 18 meters.

FIGURE 11.6 Showing Amla tree, Fruits, seeds and seedlings.

The bark is light grey and exfoliates in small, irregular flakes, revealing a fresh layer of a different color beneath the older bark. The main stem measures 70 cm in girth on average. Very close to the base, the main trunk is typically divided into 2 to 7 scaffolds. On thin branchlets, leaves are densely pinnate, distichous, and alternating. Its pinnate arrangement, measuring 10–13 mm in length and 3 mm in width, gives the branches a generally feathery appearance. Leaves emerge following fruit setting. Small, axillary, monoecious flowers with simple perianths and sepals that are either free or briefly connate. Disc glandular, ovary with three cells where carpels connate. They are 4 to 5 mm long, pale green in color, and unisexual. They are carried in leaf axils in clusters of six to ten. Big, meaty fruits with three bony, two-valved cocci are present. Fruits are round, fleshy, and nearly flattened. They weigh between 5.3 and 5.7 g, with a diameter of 2.1 and 2.4 cm, and a volume of 4.5 and 5.0 ml.

Fruit stone with six ribs that splits into three pieces, each of which typically contains two seeds. The seeds themselves weigh between 572 and 590 mg and are 4–5 mm long and 2–3 mm wide. The trees are planted in gardens, forests, and even avenue plantations. The fruit, known as Indian gooseberry or Amla, is native to the Indian subcontinent. Due to the rising demand for amla extract, *Emblica officinalis* is now grown commercially in many regions of the world. Amla has high concentrations of iron, calcium, phosphorus, vitamins, carbohydrates, and amino acids. Amla extract is mostly offered in pulp and powder form. Tamil Nadu, Gujarat, Madhya Pradesh, and Uttar Pradesh are the states that produce the most amla.

11.7.1 IMPORTANT USES

Amla fruits are a fantastic option for long-distance travel because they are durable. Because amla is bitter and extremely acidic, people do not like it when it is fresh, which is why it is processed into a range of goods. Amla is consumed in huge amounts. The market for healthy food products has grown significantly. Consumers love amla goods, and there are many different brands available, including candies, jams, pickles, juices, powders, capsules, and Chawanprash. The amla fruit is eaten raw or cooked in a number of cuisines, such as dal, which is a lentil preparation, and amle ka murabbah, a sweet dish made by soaking the berries in sugar syrup until they become candied.

It is traditionally eaten as Indian gooseberry pickle after meals. Ayurveda frequently uses both fresh and dried plant fruits as a component. The chemical components of these fruits, such as emblicanin A (37%), emblicanin B (33%), pedunculagin (14%), and punigluconin (12%), have a bitter flavor that may be due to a high density of ellagitannins and high levels of ascorbic acid or vitamin C. Phyllanemblinin A and punicafolin, as well as other polyphenols such ellagic acid, kaempferol, flavonoids, and gallic acid, are also present in amla. Numerous research demonstrate the anti-diabetic, anti-emetic, anti-microbial, anti-inflammatory, antioxidant, and hepatoprotective properties of amla. The medication is used both alone and as an ingredient in a number of compound formulations, including *Dhatryarishta, Amalaki Rasayana, Dhatriloha,* and *Chyavanaprasa.*

Fruits have a variety of medicinal uses, including those for pain relief, antiviral, hypoglycemic, antihepatotoxic, anti-nephrotoxic, antioxidant, antipyretic, and anti-inflammatory conditions. It is used for scurvy, anemia, leucorrhoea, biliary colic, hiccough, ophthalmia, cataract, constipation, erysipelas, piles, vomiting, biliousness, and high blood pressure. For jaundice, fresh fruits and root paste are employed. Asthma, bronchitis, leucorrhea, and vomiting are treated with seeds. In cases of ophthalmia and developing blindness, leaves are employed. When semen discharge is abnormal, root pastes are employed. Triphala is made out of a combination of amla, harra, and bahera and contains dried fruits as an ingredient.

11.7.2 SEED COLLECTION AND PROCESSING

After 3–4 years of planting, amla trees begin to give fruit. Because amla is an extremely perishable fruit, it is crucial to carefully harvest the fruit because damaged fruit cannot be processed. India typically produces amla between mid-September and mid-December. It can be grown in regions impacted by salinity and desolation. Amla is harvested from September to December. Fruits are harvested in February when they turn from light green to dull greenish yellow in order to extract seeds. Mature fruits need to be shaken hard because they are tough and will not fall when lightly handled. Using a knife to break up the mesocarp of the drupes, the stones were entirely removed (de-pulped) and let to air dry for a full day.

The stone was split open by lightly pressing it longitudinally. Four seeds were extracted from every fruit. A mature tree will produce 50–70 kg of fruit after 10 years. Fruits typically weigh between 60 and 70 g, and depending on their size, 1 kg can include 40 to 50 fruits. A tree with proper care can provide fruit for up to 70 years.

Because of its uneven shape, pulp's tight adherence, and processing requirements, manually removing the stone from amla fruit can be challenging. It is also laborious, causing 15–20% of the pulp to stay on to the stone and the fruit to split into segments. The processing industry needs a mechanical device to de-stone amla fruits in order to increase the effectiveness of this crucial segment separation stage. A device for manually destoning amla fruits was created by CISH. The machine's primary functional components are a plunger, a core, and a die. It operates on the principles of compression and coring. The machine can process between 10 and 12 kilograms of amla fruits in an hour. After this procedure, only roughly 6–8% of the pulp is still attached to the stones, leaving the amla fruit whole. The machine costs approximately Rs. 3000.00.

11.7.3 SEED DRYING, VIABILITY AND STORING

After splitting the drupes' mesocarp with a knife and removing all of the pulp, the stones were allowed to air dry for an entire day. Light pressure was applied lengthwise to split the stone open. Up to four seeds were taken out of each drupe. For a full day, the removed seeds were left to air dry. In order to assess the viability of the extracted seeds, they were placed in a jar with water and put through a flotation test. Pre-treatments were applied to the seeds that settled at the bottom. Seeds have very short viability. About 80–85 kgs of fruits gives 1 kg of seeds. The seeds lose viability after drying, and hence cannot be stored for a longer duration as seeds lose viability very rapidly. The semi-orthodox seeds of *P. emblica* exhibit a long dormancy period, which hinders natural sexual propagation.

11.7.4 DORMANCY AND SEED GERMINATION

In an experiment, Mawalagedera et al. (2014) used four different pre-treatments to induce germination in *Emblica* seeds in order to break their dormancy. Pre-treatments included leaving the seeds untreated (control), scarifying the seeds, partially removing the seed coat, and treating the seeds with 1% gibberellin. Among these four seed pre-treatments, not a single seed from the seeds that were scarified, had their seed coat removed, or were treated with 1% gibberellin exhibited germination even after three months. On the other hand, seeds that were pre-treated with 1% gibberellin and seed scarification sprouted precisely 85 days later, with a germination percentage of 43% (40–50%). Their experiment revealed that, within a week of emerging from dormancy, entire seedlings with cotyledons and radicles measuring about 5 cm in length emerged on the paper towel (within the zip-lock bag). This indicates that *P. emblica* seeds have a very poor natural germination potential, which can be overcome by pre-treating the seeds with gibberellin and scarifying them. For healthy seedlings, the pre-treated seeds are then planted in polypots.

11.8 CONCLUSION

One vital natural resource for the treatment of more chronic illnesses is medicinal plants. Worldwide, the use of phytomedicines is becoming just as common as that of conventional medicine. It is interesting to note that numerous plants have been found to contain cardio-protective principles; yet, these substances will not go to clinical studies or eventually be available for purchase. It is sometimes said that the synergistic bioactivity of phytochemicals in plant extracts is a benefit that is challenging for individual synthetic medications to match. Economists claim that individuals will not be motivated to conserve resources unless they reap the rewards of both their sustainable use and conservation. Only when property rights are safe and well established is it feasible.

The people who profit from exploitation, the people who have rights that are guaranteed, and the people who benefit from conservation must all be the same. As a result, we can conclude that, generally speaking, there are two main threats to medicinal plants: first, habitat loss due to land use conversion, agricultural expansion, and other factors, which leads to the extinction of both known and unknown species; and second, overexploitation of known species due to rising demand.

KEYWORDS

- **diseases, herbal remedies**
- **indigenous**
- **medicinal plants**
- **medicinal properties**
- **phyto-resources**
- **seed germination**

REFERENCES

Amalraj, A., & Gopi, S. (2017). Medicinal properties of *Terminalia arjuna* (Roxb.) Wight & Arn.: A review. *Journal of Traditional and Complementary Medicine, 7*(1), 65–78. https://doi.org/10.1016/j.jtcme.2016.02.003 (accessed on 12th August 2025).

Anonymous. (2000). The Biom News. Department of Biotechnology, GoI, Delhi, *1*(1), 1–6.

Benencia, F., & Courreges, M. C. (1999). Antiviral activity of sandalwood oil against herpes simplex viruses 1 and 2. *Phytomedicine, 6*(2), 119–123.

Bharthi, P., Philomina, D., & Chakkaravarthi, S. (2006). Estimation of colchicines in six different species of *Gloriosa* grown *in vivo*. *Indian Journal of Farm Sciences, 68*, 806–809.

Biswas, M., Biswas, K., Karan, T. K., Bhattacharya, S., Ghosh, A. K., & Haldar, P. K. (2011). Evaluation of analgesic and anti-inflammatory activities of *Terminalia arjuna* leaf. *Journal of Phytology, 3*(1), 33–38.

Das, S. C., & Tah, J. (2013). Effect of GA_3 on seed germination of sandal (*Santalum album* L.). *International Journal of Current Science, 8*, E79–E84.

Devi, C. A., Swamy, G. S. K., & Naik, N. (2016). Studies on storage and viability of jamun seeds (*Syzygium cuminii* Skeels). *Biosciences Biotechnology Research Asia, 13*(4). https://www.biotech-asia.org/?p=17333 (accessed on 12th August 2025).

Dwivedi, N. K., Indiradevi, A., Nair, R. A., & Suma, A. (2014). Ethnomedicinal uses, collection and conservation of sandalwood (*Santalum album* L.) germplasm. In *Abstracts of International Seminar on Sandalwood: Current Trends and Future Prospects*, IWST, Bangalore, February 26–28, 2014 (p. 50).

Gunaga, R. P., Hanumantha, M., & Narkhede, S. S. (2014). Seed germination studies in sandalwood (*Santalum album* L.): Problems and solution. In *Abstracts of International Seminar on Sandalwood: Current Trends and Future Prospects*, IWST, Bangalore, February 26–28, 2014 (p. 27).

ICFRE. (1992). *Sandal (Santalum album Linn.)*. Institute of Wood Science and Technology, Indian Council of Forestry Research & Education, p. 10.

Jahan, N., & Rahman, K. (2014). Medicinal value of sandalwood and its oil in Unani system of medicine. In *Abstracts of International Seminar on Sandalwood: Current Trends and Future Prospects*, IWST, Bangalore, February 26–28, 2014 (p. 65).

Jain, S. K. (1968). *Medicinal Plants*. National Book Trust, New Delhi, 123–125.

John, H. (1947). The history, present distribution and abundance of sandalwood on Oahu, Hawaiian Islands. Hawaiian plant studies. *Pacific Science, 1*, 5–20.

Mawalagedera, S. M. U. P., Perera, G. A. D., & Sooriyapathirana, S. D. S. S. (2014). Prolonged seed dormancy in *Phyllanthus emblica* L. can be overturned by seed scarification and gibberellin pre-treatment. *Open Journal of Forestry, 4*(1), 38–41.

Mishra, R. S., & Bajpai, P. N. (1971). Vegetative growth studies in the jamun. *Indian Journal of Horticulture, 28*, 273–277.

Mudappa, A., & Oommen, S. (1998). Problems and prospects of India's wild medicinal plants in trade: A current understanding. *Amruth, 2*(3), 10–13.

Mukherjee, A., & Bouri, T. (2011). Indigenous knowledge about the use of non-timber phyto-resources as documented from forest areas under Durgapur Forest Range, Burdwan District. *The Ecoscan, 1*, 59–62.

Nagaveni, H. C., & Srimathi, R. A. (1980). Studies on germination of the sandal seeds, *Santalum album* L. II. Chemical stimulant for germination. *Indian Forester, 106*(11), 792–799.

Nagaveni, H. C., & Srimathi, R. A. (1981). Studies on germination of the sandal (*Santalum album* L.): Pretreatment of sandal seeds. *Indian Forester, 107*(6), 348–354.

Okasaki, K., & Oshima, S. (1953). Antimicrobial effect of essential oils. *Journal of the Pharmaceutical Society of Japan, 73*, 344.

Rai, S. N., & Kulkarni, H. D. (1986). Sandalwood plantation. In Srivastava et al. (Eds.), *Plantation Crops. Vol. I: Opportunities and Constraints* (pp. 295–300). Oxford and IBH Publishing Co.

Sarkar, H. K., Lahri, T., Maity, T. K., & Bhattacharya, S. (2011). Evaluation of glory lily (*Gloriosa superba*) using morphological, anatomical and molecular parameters for genetic resource conservation. *Environment and Ecology, 29*(1A), 389–392.

Sarkar, H. K., Singh, P. K., & Bhattacharya, S. (2012). Conservation of some endangered medicinal plants of West Bengal. In *Medicinal Plants: Various Perspectives* (pp. 264–276). Department of Botany, University of Burdwan. ISBN 81-87259-85-X.

Sastri, G., Srimathi, P., & Malarkadi, K. (2001). Effect of seed size on seed quality in jamun. *Madras Agricultural Journal, 88*, 524–526.

Sen Sarma, P. K. (1982). Sandalwood–Its cultivation and utilization. In C. K. Atal & B. M. Kapur (Eds.), *Cultivation and Utilization of Aromatic Plants* (pp. 395–495). Regional Research Laboratory, CSIR, Jammu Tawi.

Singh, H. L. (1978). Studies on pollen germination in jamun (*Syzygium cuminii* Skeels). *Plant Science, 10*, 151–152.

Singh, R. K., & Thakur, S. (1977). Seed germination and seedling growth of jamun. *Proceedings of Bihar Academy of Agricultural Sciences, 25*, 139–142.

Siva Kumar, G., & Krishnamurthy, K. V. (2000). Micropropagation of *Gloriosa superba* L., an endangered species of Asia and Africa. *Current Science, 78*, 30–32.

Srimathi, R. A., Kulkarni, H. D., & Venkatesan, K. R. (1995). *Recent Advances in Research and Management of Sandal (S. album L.) in India*. Associated Publishing Co., New Delhi.

Srinivasan, V. V., Sivaramakrishnan, V. R., Rangaswamy, C. R., Ananthapadmanabha, H. S., & Shankaranarayana, K. H. (1992). *Sandal–Santalum album Linn.* Indian Council of Forestry Research and Education. ICFRE, Bangalore.

Swarup, R., Sarin, S., Nitish, S., & Gawshaiah, K. N. (2009). Mapping life. *Biotech News, 4*(1), 18–22.

Van der Plas, L. H. W., & Hoekstra, F. A. (2000). Stree, storage and survival of neem seeds. In M. Sacandé (Ed.), *Laboratory of Plant Physiology, EPS. Agricultural University* (p. 124). ISBN 9789058082107.

CHAPTER 12

Seed Conservation of Medicinal and Aromatic Plants

S. K. VERMA* and AMIT MISHRA

Department of Fruit Science, Acharya Narendra Deva University of Agriculture and Technology, Kumarganj, Ayodhya, Uttar Pradesh, India

*Corresponding author

ABSTRACT

Medicinal and aromatic plants (MAPs) are natural resources used in traditional healing, modern pharmaceuticals, and perfumes. India has a rich variety of these species, but many are now in danger due to habitat loss, overharvesting, and environmental decline. In this chapter, we will look at the urgent need to conserve the seeds of these plants. This is important for protecting their unique genetic traits and ensuring their survival for future generations. It will discuss challenges such as limited knowledge, poor seed viability, and a lack of infrastructure, and provide practical strategies for conservation. These include seed banks, on-site conservation, off-site conservation, cryopreservation, and farmer-led initiatives. India's efforts have support from institutions like CIMAP, TBGRI, and the National Medicinal Plants Board. These organizations are working to preserve and promote MAPs through research, documentation, and community involvement. This chapter also shows global collaborations like the Millennium Seed Bank Partnership. It emphasizes the shared responsibility of protecting these plants. Then, the chapter concludes with an invitation to incorporate not only science, policy, and grassroots involvement, but also to include making the healing capabilities of nature available not only for us, but for generations to come.

Seed Technology for Medicinal and Aromatic Plants: Production, Testing, Storage, and Marketing.
Priyanka Sharma, Kishore Kumar Sharma, and Gopal Shukla (Eds.)

DOI: 10.1201/9781779641199-12

12.1 INTRODUCTION

Aromatic and medicinal plants are members of the group of species that have various uses. Many species that are grown for food, ornamental purposes, timber, and so forth, are utilized for aromatic and medicinal uses. This fact frequently leads to disputes among workers regarding the proper way to classify these species within a clear, use-oriented crop category. Pharmaceuticals, the fragrance industry, and traditional medicine all heavily rely on medicinal and aromatic plants (MAPs).

Over 25% of the 30,000 known species of medicinal plants worldwide are found in India, with 8000 of those species discovered in forest environments. The diversity and variability of living things, the biological communities in which they coexist, together with the processes of ecology and evolution that support them, are all components of biodiversity. It is frequently a species-level indicator of biological system health that shows how likely it is that the totality of past species will survive as opposed to going extinct. Approximately 2142 species, or 11.53% of India's 18,532 vascular plants, are red listed. Eight are extinct species, 432 are threatened species, and about 54 species fall under the category of near threatened (IUCN, 2020).

Contrarily, conservation entails the meticulous preservation and defense of something, particularly the methodical administration of a natural resource with the goal of preventing abuse, misuse, or even destruction. The preservation of these plants' genetic diversity becomes essential as demand for them rises. The creation of *in situ* reserves is one strategy for maintaining crops' genetic diversity. These are environments that have been conserved and support wild populations of crop species. These types of activities are also the regions where custodian farmers continue native landraces in return for subsidies. A viable strategy for preserving the priceless genetic resources of aromatic and medicinal plants is seed conservation. Effective conservation methods for MAPs should be implemented in two domains: that is, *in situ* and *ex situ* conservation. In accordance with these initiatives, integrated conservation attempts are necessary.

12.1.1 IMPORTANCE OF SEED CONSERVATION

It is necessary to conserve the seeds of aromatic and medicinal plants in order to maintain their genetic diversity. These plants frequently contain distinctive chemical compounds that support their aromatic or medicinal qualities. Preserving this variety guarantees raw materials for the essential oil, pharmaceutical, and conventional medicine sectors. Since many MAPs are crucial to ecological processes, conservation also aids in preserving ecosystem stability.

Inaction has led to the current outcry against ecological genocide, genetic erosion, environmental degradation, fragmentation and destruction, and the loss of our biological legacy (Krikorian, 1998). There will always be a threat of overexploitation to medicinal plants and their natural habitats as long as forests are destroyed (Walter and Gillett, 1998). Preservation of vulnerable medicinal plant species, that is, *Saraca asoca, Picrorrhiza kurroa, Costus speciosus, Berberis aristata, Gloriosa superba,* and so forth, having natural habitats and sustainable exploitation of those species in less vulnerable areas are the ultimate goals of conservation (Cunningham, 1993).

Before considering how to conserve the plants, plant parts, or even the entire plant are carelessly taken from wild sources. Because of overexploitation and habitat destruction, many significant and useful species are in danger of going extinct. Over 95% of medicinal plants are harvested from the wild, and many of these have become endangered in their native environments. Encouragement of these plants' propagation and cultivation is necessary.

12.2 CHALLENGES IN SEED CONSERVATION

- **Limited Knowledge:** Gaining a thorough understanding of the biology, ecology, and seed characteristics of the many aromatic and medicinal plants that can be found in distant and varied ecosystems is difficult.
- **Threats to Habitats:** Rapid urbanization, deforestation, and climate change pose serious threats to the natural habitats of MAPs, leading to habitat loss and fragmentation.
- **Poor Seed Viability:** Some medicinal plants have short-lived seeds (*Hypercium scruglii*), making it difficult to store them for extended periods without loss of viability.
- **Inadequate Conservation Infrastructure:** Insufficient facilities and resources for seed banks and conservation programs hinder comprehensive efforts to preserve MAPs.

12.3 STRATEGIES FOR SEED CONSERVATION

12.3.1 SEED BANKS

A place where appropriate conditions are kept to preserve seed specimens of various plant species (wild or cultivated) is known as a seed bank or germplasm bank. The obvious objective is to ensure that as many plants as possible are preserved for future generations. The Norwegian government created the world's largest secure seed storage facility in February 2008 and is located 1300 kilometers beyond the Arctic Circle in the permafrost situated in a high-altitude mountain pass. The Defense Institute of High-Altitude Research and the NBPGR created the safe seed bank, Indian Seed Vault, which is the second-largest seed bank in the world. It is situated on Chang La in (Ladakh), India. About 54 wild and cultivated species of aromatic and medicinal plants as well as 401 crop species have been gathered at various NBPGR seed banks. It is essential to establish seed banks specifically for aromatic and medicinal plants. Providing a controlled environment for long-term storage, these banks act as repositories for seeds.

12.3.2 IN SITU CONSERVATION

Whether in the wild or in conventional farming systems, *in situ* techniques involve preserving genetic variation where it is found. Less glamorous species conservation, let alone the genetic conservation of species, is still the primary objective of only a small number of nature reserves and natural parks that were created to preserve large fauna or to preserve aesthetically pleasing landscapes (Hoyt, 1988). Identification, management, monitoring, and conservation of genetic diversity in a specific natural area are all part of the process of conserving wild species in a genetic reserve. This method can be relatively inexpensive when the management regime is minimal, making it the most suitable for the majority of wild species. Both orthodox and recalcitrant seeded plant species can benefit from genetic reserves since multiple taxa can be safeguarded in a single reserve, allowing for the continuation of species evolution. The conserved material is not immediately available for human exploitation, which is a drawback. Additionally, if the management regime is lax, there may not be much information available for characterization or evaluation.

Farmer-based conservation refers to the practice of farmers within traditional agricultural systems maintaining traditional crop or cultivation systems. Plants commonly referred to as "landraces" are planted and harvested on traditional farms. The farmers save a portion of the harvested seed for subsequent plantings each season, known as carry-over seeds. On a smaller scale, home garden plant conservation is a closely related variation of landrace conservation on farms. It entails growing a wider variety of species in backyard, dooryard, kitchen, and home gardens (e.g., tomatoes, peppers, digitalis, mint, oregano, and parsley). Vegetable, flavoring, and medicinal plants are the main emphasis of these home gardens. The main benefit of using the on-farm technique is that it guarantees the survival of wild species and highly adapted landraces and breeds, which are often essential to traditional agriculture.

Protecting natural habitats through *in situ* conservation is vital to ensure the survival of MAPs in their native environments. This involves sustainable management practices and the establishment of protected areas.

Biosphere reserves of MAP in India are located in several states like Meghalaya, Karnataka, Kerala, Tamil Nadu, Arunachal Pradesh, Uttar Pradesh, West Bengal, Andaman and Nicobar Islands, Assam, and Uttarakhand.

12.3.2.1 OUTLINE FOR IN SITU CONSERVATION

- A national policy pertaining to the protection and use of medicinal plants in protected areas should be drafted by the Parks Department.
- Determining the protected areas that are most crucial for medicinal plants.
- Goals and techniques for recording and monitoring medicinal plants in protected areas.
- Methods and protocols for gathering medicinal plants inside of protected zones.
- The extent to which the nation's medicinal plants are covered by the protected areas system ought to be evaluated by the Parks Department. Next, in order to guarantee the preservation of all the nation's medicinal plants, it should expand already-existing protected areas and establish new ones. Economic and social incentives for preserving wild species and their natural habitats

should be developed by the Parks Department. D. The administrators of the parks should make sure that site management plans include provisions for the use and protection of medicinal plants.

- It is best to re-introduce species that have been severely reduced by over exploitation into places where they were formerly wild.

12.3.3 EX SITU CONSERVATION

Ex situ, or outside of their natural habitat, methods for conserving medicinal plants include growing and caring for plants in parks, botanical gardens, and other appropriate locations. Additionally, plant propagules are kept in gene banks for a long time (seed, pollen, and DNA libraries, among others), plant tissue culture repositories, and cryopreservation.

12.3.3.1 FIELD GENE BANK

- Gene Bank: Seeds are stored at –20°C for base collection and +4°C to 0°C for active collection. In India, three national gene banks (NGB) have been set up to preserve aromatic and medicinal plants *ex situ*.
- CIMAP is located in Uttar Pradesh and is overseen by CSIR, Ministry of Science and Technology, Government of India.
- Palode, Thiruvananthapuram, Kerala, is home to the Tropical Botanical Gardens Research Institute (TBGRI).
- The National Bureau of Plant Genetic Resources, which is solely in charge of conserving the genetic variability of domesticated plants and their wild relatives, works under DARE of ICAR.

12.3.3.2 SEED GENE BANK

Conserving germplasm is more economical at the Seed Gene Bank. In order to preserve seeds for future generations, the NGB, situated at the NBPGR in New Delhi, is principally responsible for conserving the germplasm of wild species of agri-horticultural crops. These are placed in modules with a temperature of –20°C and are referred to as "Base Collection." The seeds are hermetically sealed in moisture-resistant aluminum foil packets once they have been dried to a moisture content of 4–6 percent. These seeds remain viable for up to 50–100 years after being stored. Only seed samples with a seed viability of more than 85% for the majority of crops are processed. When storing seeds in gene banks, it is ideal to adhere to the FAO/IPGRI-recommended standards.

12.3.4 NATIONAL ACTIVE GERMPLASM SITES

The essential part of the network is the national active germplasm sites (NAGS). Currently, there are 50 NAGS, which are situated at SAUs and ICAR institutes (crop-based institutes for a particular crop or group of crops). The responsibility for multiplying, evaluating, preserving, and distributing active collections to authorized users both domestically and internationally has been placed on the NAGS. These operational and dynamic collections are placed in storage modules that maintain a consistent +4°C temperature and 35–40% relative humidity (RH). Under these conditions, seeds ought to last 8–50 years. To store seeds for a medium amount of time, the moisture content is reduced to 8–10%. In order to support the various regions' ongoing efforts to conserve their germplasm, the NBPGR manages a network of 10 regional stations located throughout the country's agro-climatic zones.

12.3.5 CRYOPRESERVATION (IN LIQUID NITROGEN AT –165°C TO –196°C)

Cryopreservation or freeze preservation under liquid nitrogen.

12.3.5.1 SEED PRESERVATION

Seeds are among the most often used and favored forms of germplasm for the *ex situ* preservation of plant genetic resources (PGR). Typically, they are kept dry in seed banks at –20°C in accordance with global guidelines. Certain seeds, on the other hand, have short lifespans at –20°C and cannot withstand drying or storage at these conditions. For them, long-term preservation through cryopreservation is recommended. Seeds that can withstand drying at 32 ± 3% relative humidity at 18°C, often known as orthodox seeds, can be kept in the liquid nitrogen vapor phase. Adjusting drying and cooling techniques is necessary when seeds exhibit resistance to desiccation (also known as recalcitrant seeds) or have limited tolerance to desiccation and/or −20°C storage (also known as intermediate seeds). Cryoprotection is frequently also necessary in these situations.

12.3.5.2 POLLEN PRESERVATION

The primary purpose of pollen storage development was to facilitate regulated pollination of plants that flower at the same time. Furthermore, pollen storage has been regarded as an emerging technology with the purpose of maintaining genetic variety by preserving. In comparatively small areas, pollen is easily collected and cryopreserved in large quantities.

Plant quarantine issues are reduced when pollen exchanges germplasm. Recent years have seen the development of cryopreservation methods for pollen in numerous species, and several nations have established cryobanks of pollen for fruit tree species.

12.4 MEDICINAL PLANT CONSERVATION CENTERS IN INDIA

In India, as in many other nations, numerous initiatives pertaining to medicinal plants have been implemented, and various tactics are used to ensure their preservation. Many organizations are engaged in national-level activities, and major institutions supported by the Indian government play a significant role in these organizations.

The following CSIR institutions are involved in these programs:

National Botanical Research Institute (NBRI), Central Drug Research Institute (CDRI), Central Institute of Medicinal and Aromatic Plants (CIMAP), and Regional Research Laboratories (RRLs). In collaboration with the Indian Institute of Horticultural Research (IIHR), NRCMAP, National Bureau of Plant Genetic Research (NBPGR), and Indian Council of Agricultural Research (ICAR), an AICRP on MAP is carried out.

Established in various geographic regions of India, including Dehradun (Northern Circle), Allahabad, Shillong, Pune (Western Circle), Coimbatore, and Port Blair, the Botanical Survey of India (BSI) is supported by the Department of Environment and Forest, Government of India. Jodhpur, Gangtok, and Itanagar are the locations of the other three stations. Ten experimental botanical gardens are attempting to restore medicinal plants that are seriously endangered, according to a very recent report (Sanjappa, 2004). A program for creating "Vanaspati Van," or Reserve Forests, and cultivating medicinal plants has been started by the Indian Council of Forest Research (ICFRE). Tropical Botanical Garden and Research Institute (TBGRI) in Kerala has started a major program on the preservation and responsible use of the abundant medicinal plant resources in the Indian peninsula. G.B. Pant Institute of Himalayan Environment and Development, an institute of Government of India's Ministry of Environment and Forests, conserves the biological diversity of the North-Western Himalayan region. The creation of a display garden, field gene bank, *in vitro* gene bank, and seed gene bank are some of TBGRI's main initiatives. This is a component of the Department of Biotechnology (DBT), Government of India, G-15 GBMAP program. DBT has established three gene banks for medicinal plants through cryopreservation, namely NGB for Medicinal and Aromatic Plant (NGBMAP) Thiruvananthapuram, NGBMAP (New Delhi), and NGBMAP (Lucknow). The Jabalpur-based Tropical Forest Research Institute places a strong emphasis on the cultivation of medicinal plants that could be grown alongside trees. In addition to these public and non-public organizations, a number of businesses and private groups encourage farmers to cultivate medicinal plants for profit (Bhatt, 2002; Nair, 2002; Purohit and Vyas, 2004; Shah and Kalakoti, 1996).

12.5 RESEARCH AND DOCUMENTATION

The World Wide Web now provides access to a multitude of information sources that offer detailed knowledge on pharmacology, agriculture, natural resource management, economic properties of herbs, traditional herbal drugs, active ingredients, utilized plant parts, biological processes, medicinal applications, chemical components, and so forth. Although information on the use of medicinal plants has been computerized and covers a large area, there is little and inconsistent information available on conservation through electronic media (Table 12.1). As a result, several Indian groups have started working in this new field. Centralized online databases have been created by TBGRI's Bioinformatics Canter:

'Seed Pack' is a database on TBGRI's seed bank; 'Garden Info' contains information on plants conserved in the institute; and additionally, "Plant Info" provides information on Kerala's native medicinal plants and trees. An initiative network of several Indian organizations with knowledge of various aspects of medicinal plants, INMEDPLAN (Indian Medicinal Plants National Network of Distributed Databases) aims to create a multidisciplinary information pool by pooling resources (botanical, horticultural, pharmacological, and other). The data is given to the network secretariat at FRLHT, Bangalore, upon request. The "Encyclopedia of Indian Medicinal Plants" online database from FRLHT includes information on approximately 7361 plant species, while the "Medicinal Plant Conservation Concern" database concentrates on 880 species of traded medicinal plants in India.

12.6 COMMUNITY INVOLVEMENT

The sustainable maintenance of MAP species thus depends on the involvement of primary stakeholders, namely farmers, rural and forest communities. This involvement also serves as a means of facilitating bioprospecting of existing resources for better use and protection.

It is important to consider participation in light of the MAP species' marketing and promotion. The "filière" would be the most strategic component of such an approach. The term "filière" originates from French and refers to the connection between all parties involved and the processes that begin with data collection, use optimization, policy formulation, and end with marketing and commercialization. Programs to conserve seeds are more successful when local communities are involved. Indigenous people's knowledge of the plants and their environments can be extremely valuable.

12.7 CASE STUDIES

12.7.1 THE MILLENNIUM SEED BANK PARTNERSHIP

An international effort is spearheaded by the Royal Botanic Gardens (RBG), Kew, to preserve seeds from a variety of plant species, including medicinal plants.

TABLE 12.1 Existing Electronic Database Systems on Medicinal Plants in India

Sr. No.	Database	Full Data Base Name	Special Features of Database	Organizations and Institutions Developing Databases
1.	CITES Plants	Convention on International Trade in Endangered Species	Appendices of 3 categories of plants – threatened, vulnerable and that require close vigil with pictorial identification	BSI, ENVIS Howrah 711103, WB www.envisbsi.org/ citesplant.htm
2.	COBOMAN	-	Name, image, status, distribution, habitat, ecology, conservation strategies implemented and suggested, biology and possible value, cultivation, description, and reference in the scientific community	SI, ENVIS ww.envisbsi.org/ FAMILY.html
3.	Encyclopedia of Indian Medicinal plant		Scientific name, trade information, distribution, agro technological propagation, seed storage, Eco map, pharmacology, and pharmaceutics	http://encyclopedia frlht.org.in/
4.	Garden Info	Garden Information	Details regarding plants conserved by TBGRI	TBGRI, Thiruvanthapuram, Kerala 695562
5.	MAPA	Medicinal and Aromatic Plant Abstract	Bimonthly computerized edition of the Inferencing journal on aromatic and medicinal plants	NISCAIR, Dr. K.S. Krishna Marg, NewDelhi 110012
6.	MAPI	Major Aromatic Plants of India	information gathered and assembled about India's aromatic plants	CIMAP, Lucknow 226015, UP
7.	Medicinal Plants		Phytography, distribution, ecology, cultivation, image, synonyms, English, vernacular, trade name, traditional and modern use, and chemical content	BSI, ENVIS www.envisbsi.org/ medi.htm
8.	Medicinal Plants of Conservation Concern		Vernacular name, trade name, taxon, data sheet, scientific name, family, habit, threat status, use, and distribution	FRLHT, Bangalore http://envis. frlht.org.in /iucnlist.php

TABLE 12.1 *(Continued)*

Sr. No.	Database	Full Data Base Name	Special Features of Database	Organizations and Institutions Developing Databases
9.	NAATS	Natural Products Database	Phytochemical information, collection specifics, pharmacological information, the chemical makeup of the active ingredients, applications in conventional and folk medicine, and a detailed review of the literature	CDRI Chattar Manzil Palace Lucknow 226001 UP
10.	Plant Info	Plant Information Series	Scientific name, English, Sanskrit, Malayalam name, uses in medical systems	TBGRI, www.tbgri.org / plantinfo/ plant01042003_ Local/index.asp
11.	REFMAP	References on Medicinal and Aromatic Plants	Details about references in the literature on aromatic and medicinal plants	CIMAP
12.	Seed Pack		TBGRI's seed bank data	TBGRI

Primarily focused on preserving propagules from conventional seed-bearing wild vascular plants, the Millennium Seed Bank (MSB) Partnership is overseen and developed by the RBG Kew. With 96 countries and territories currently participating, it is the world's biggest *ex situ* conservation program. Seeds are collected and conserved in their country of origin whenever possible, with duplicates being sent to the MSB at RBG Kew for storage. Using both quantitative and qualitative techniques, we evaluate the conservation value of the germplasm kept at the MSB in this chapter. The MSB holdings are a rich and superior biological resource. About 350 families, 6100 genera, 40000 species, and over 2.4 billion seeds are represented in the collections, which have a significant and distinctive taxonomic diversity. All 35 rich diversity zones and all nine biogeographic zones are included in the collections, and they come from 189 different countries and territories. The collections have significant population value and natural capital. At least one recognized usage for humans exists for 32% of taxa or 49% of collections; significant ecological, social, cultural, scientific, or economic importance is attributed to 74% of taxa, which are either noticed in a defined zone or near to extinction (locally or internationally). About 20% of taxa, or 13% of collections, are endemic at the national or territorial level, whereas 10% of taxa, or more than 8% of collections, are either rare, extinct, or threatened with extinction on a global and/or national scale.

12.7.2 NATIONAL MEDICINAL PLANTS BOARD (NMPB), INDIA

The NMPB makes a substantial contribution to efforts aimed at seed conservation and concentrates on the preservation, growth, and sustainable use of medicinal plants.

Many different types of medicinal plants can be found in India that the locals have been using for centuries to treat crops and domesticated animals' illnesses (Pashuayurveda and Vrikshayurveda) in addition to their own primary healthcare needs. Our indigenous healthcare traditions primarily rely on medicinal plants as their primary source of resources. Even though the cultivation of medicinal plants has gained popularity recently, a sizeable portion of our needs are still satisfied by foraging in the wild. Enhancing increasing through afforestation the amount of medicinally significant aromatic species and native medicinal plants in accordance with silvicultural principles and management prescriptions will help us meet the growing demand for medicinal plants from both domestic and foreign markets. We also need to concentrate on *ex situ* cultivation of medicinal plants as well as *in situ* conservation efforts through systematic surveys.

Aware of this, the cabinet first approved the "Medicinal Plant Conservation, Development, and Sustainable Management Scheme for the Central Sector" on June 26, 2008, under the reference number CCEA/21/2008. During the 12th Plan period, which ran from March 31, 2021, to December 1, 2014, the EFC meeting approved the scheme with an outlay of Rs. 450.00 crores. In the SFC meeting on March 17, 2021, it was decided to extend the Central Sector Plan for Medicinal Plant Development, Preservation, and Sustainable Handling beyond March 31, 2021, at a cost of 322.41 crores rupees for a five-year period from 2021–2022 to 2025–2026. The proposed activities under the scheme will also assist the nation in fulfilling its international commitments regarding the biodiversity of medicinal plants and will foster bilateral and international cooperation, both of which are essential for the sector's future growth and for positioning India as a global leader in it.

12.8 CONCLUSION

Sustainable biodiversity conservation heavily relies on the preservation of MAP seeds. To overcome the obstacles related to seed conservation, a combination of *ex situ* and *in situ* conservation tactics, solid research, and community involvement are required. Future generations can benefit from the genetic diversity of MAPs being preserved if these strategies are put into practice.

KEYWORDS

- **medicinal plants**
- **seed, conservation**
- **field gene bank**
- **seed gene bank**
- **cryopreservation**

REFERENCES

Google. (n.d.). NMPB focuses on the conservation, cultivation, and sustainable utilization of medicinal plants, contributing significantly to seed conservation efforts. https://www.google.com/search?q=NMPB+focuses+on+the+conservation%2C+cultivation%2C+and+sustainable+utilization+of+medicinal+plants%2C+contributing+significantly+to+seed+conservation+efforts.&oq=NMPB+focuses+on+the+conservation%2C+cultivation%2C+and+sustainable+utilization+of+medicinal+plants%2C+contributing+significantly+to+seed+conservation+efforts.&aqs=chrome..69i57.915j0j15&sourceid=chrome&ie=UTF-8 (accessed on 12th August 2025).

Heywood, V. H. (2011). Conservation and sustainable use of medicinal plants. In *Medicinal and Aromatic Plants* (pp. 441–459). Springer.

Lakshman, C. D. (2016). Bio-diversity and conservation of medicinal and aromatic plants. *Advances in Plants and Agriculture Research, 5*(4), 561–566. https://doi.org/10.15406/apar.2016.05.00186 (accessed on 12th August 2025).

Liu, U., Breman, E., Cossu, T. A., et al. (2018). The conservation value of germplasm stored at the Millennium Seed Bank, Royal Botanic Gardens, Kew, UK. *Biodiversity and Conservation, 27*, 1347–1386. https://doi.org/10.1007/s10531-018-1497-y (accessed on 12th August 2025).

Padulosi, S., Leaman, D., & Quek, P. (2002). Challenges and opportunities in enhancing the conservation and use of medicinal and aromatic plants. *Journal of Herbs, Spices & Medicinal Plants, 9*(4), 243–267.

Rao, R. R., & Bhatt, B. P. (2006). Seed storage behavior of some important medicinal plants. *Seed Science and Technology, 34*(2), 259–268.

Simmonds, M. S., & Blaney, W. M. (2011). *Medicinal and Aromatic Plants–Industrial Profiles*. CRC Press.

Vieira, R. F. (1999). Conservation of medicinal and aromatic plants in Brazil. In *Perspectives on New Crops and New Uses* (pp. 152–159).

CHAPTER 13

Packaging: Principles, Practices, and Materials, Bagging and Labeling of Medicinal and Aromatic Plant Seeds

ANASUYA SIL,[1] TANMOY MONDAL,[2,*] and AMEDA SWARNALATHA[1]

[1]*Department of Plantation Spices, Medicinal and Aromatic Crops, Bidhan Chandra Krishi Viswavidyalaya, Mohanpur, Nadia, West Bengal, India*

[2]*Department of Horticulture, MITS Institute of Professional Studies, Rayagada, Odisha, India*

*Corresponding author

ABSTRACT

Medicinal and aromatic plants (MAPs) are gaining popularity in contemporary culture due to their incredible healthcare properties in both preventive and curative measures. The seeds of MAP decay rapidly and lose their viability after several months of harvest and storage. Thus, implementation of suitable post-harvest management techniques are required to maintain the seed viability of MAPs. Drying, storage, packaging, and labeling are some important post-harvest operations that help to maintain the physical and chemical properties of seeds. Storage duration and temperature play a significant role in retaining seed viability and improving the germination ability of different MAPs species, whereas packaging and labeling are essential components of product marketing plans. Packaging plays an important sales strategy that motivates customers to purchase the item. It serves a variety of functions, including storage, protection, simplicity of handling, and ease of use. It also helps with product density, selecting

Seed Technology for Medicinal and Aromatic Plants: Production, Testing, Storage, and Marketing. Priyanka Sharma, Kishore Kumar Sharma, and Gopal Shukla (Eds.)

DOI: 10.1201/9781779641199-13

the best packaging material, and optimizing the use of available space. Labeling also serves to provide all relevant information about the product, such as quality, origin, and authenticity, while also giving the product a competitive advantage in the market. On the other hand, transportation greatly facilitates marketing activities such as product purchasing, assembly, sale, storage, and warehousing, and so forth.

13.1 INTRODUCTION

Medicinal and aromatic plants (MAPs) are defined as plants that are rich in secondary compounds such as alkaloids, glycosides, coumarins, flavonoids, steroids, terpenes, and low-boiling-point phenylpropanes that act as potential sources of drugs and are widely used in aromatherapy and healthcare. Besides this, they are also used in various other industries such as cosmetics, spices, flavors and fragrances, pesticides and repellents, and herbal beverages (Devika, 2021; Samarth et al., 2017). MAPs form the main base for the manufacture of drugs in Indian Systems of Medicine like Ayurveda, Unani, Siddha, and Homeopathy (Ranjith and Rohini, 2019). Essential oils, derived from various parts of these plants like flowers, seeds, leaves, and roots, are highly concentrated, volatile, hydrophobic compounds known for their distinct odors (Devika, 2021; Samarth et al., 2017). MAPs are found in various parts of the country in different environmental and climatic conditions (Devika, 2021). Natural products made from medicinal plants are increasingly prescribed, and this has led to a growing pharmaceutical industry and small-scale processing for the manufacturing of herbal goods. Thus, MAP are used in nutrition as well as the pharmaceutical and cosmetics industries (Dzoyem et al., 2013). The fundamental ingredients of many contemporary medications are still derived from medicinal plants, which have gained acceptance as viable treatments for a variety of reasons, including their accessibility, low cost, minimal side effects, environmental friendliness, and long-lasting healing potential (Chowti et al., 2018). Based on projections by the International Trade Centre (ITC), global trade in MAP resources is expected to increase each year by approximately 10 to 12% (Chandra and Sharma, 2019). Additionally, according to the World Health Organization (WHO), approximately 75 to 80% of the global population depends on herbal medicines for their primary healthcare necessities. Furthermore, with the ongoing growth in the world population, it is foreseen that the demand for medicinal plants worldwide will persistently increase (Bareetseng, 2022). India is regarded as a treasure trove of priceless aromatic and medicinal plant species. It is believed that this group of plants has the potential to generate over $3 billion in foreign exchange earnings annually. Approximately 1500 species of native plants are recognized for their flavor and aroma, while over 9000 have been shown to have medicinal qualities (Chowti et al., 2018). It is necessary to protect important MAP and conserve them through effective packaging methods. Packaging helps to prevent damage to the product, which can enhance the purity, security, and efficacy of MAP. The following section presents best practices based on recommendations from the World Health Organization (WHO, 2003) and the European Herb Growers Association (EUROPAM, 2016, 2022).

13.2 IMPORTANCE OF MEDICINAL AND AROMATIC CROPS

- Since ancient history, MAP have been utilized for therapeutic and cosmetic purposes because they are essential to human welfare and have a significant role in human wellbeing (Tiwari et al., 2022). Apart from helping with general health problems, these plants allow building bridges between different medical systems and even offer better ways to cure complex epidemics like Covid 19 (Benarba and Pandiella, 2020).
- Herbal remedies are still quite effective in preventing and treating viral respiratory infections. Many plant families have been thoroughly studied for their therapeutic value, especially Zingiberaceae, Lamiaceae, and Apiaceae, because of their strong antioxidant qualities. A few examples of medicinal plants that have been successfully utilized to treat skin diseases are *Senegalia visco, Artemisia nilagirica, Vaccinium microcarpum, Hypericum perforatum,* and *Curcuma longa* (Devika, 2021).
- Essential oils of aromatic and medicinal plants are widely used in aromatherapy and healthcare. They also have distinctive flavor and aroma qualities, as well as biological activities. India is a producer of essential oils, including those of lemon grass, citronella, palmarosa, spearmint, patchouli, eucalyptus, ajwain, celery seed, cedar wood, and turpentine (Devika, 2021).

The forests act as a home to most MAP. On the other hand, MAP found in their native environment are affected by biodiversity loss due to over exploitation caused by human activities (Perinchery, 2020; Sharma et al., 2018; Tiwari et al., 2022). The rapid genetic loss of MAP requires conservation to safeguard their accessibility for posterity. Although seeds are an essential material for protecting the genetic resources, several valuable MAP have become extinct in their native habitats as the seeds of these plants degrade more rapidly and lose their potency after a few months of harvesting and storage, which causes significant economic losses (Tiwari and Das, 2014). For example, *Withania somnifera* seeds kept for two months storage had a noticeably decreased germination percentage (Himangini and Thakur, 2018).

To protect and conserve the genetic resources of MAP, seed viability needs to be maintained for a prolonged period of time (Pradhan and Badola, 2012). When seeds are stored under inappropriate conditions, such as room temperature and high moisture, they naturally lose viability, resulting in low seed germination and seed deterioration (Schmidt, 2000). A number of variables, such as the type of seed, temperature, moisture content, storage circumstances, and packing techniques, are crucial for preserving the quality and durability of seeds while they are being stored (Gordon, 1992; Pradhan and Badola, 2008; Roberts, 1972b; Yang et al., 2005). For example, storage of *Moringa oleifera* seed in compact sealed super bags at a moisture content of 8% maintained the seed viability for a long period (Afzal et al., 2020). Onion seeds (*Allium cepa*) kept in glass containers and cold storage shows a higher germination percentage (Manna et al., 2020). Storage of *Hibiscus sabdariffa* L. seeds in darkness with flushing nitrogen may increase the germination rate for up to three months (Juhari and Petersen, 2018). In the case of *Desmodium gangeticum,* storing seeds at 8% moisture condition, encapsulating them in polythene bags, and keeping them in a frozen environment contribute to increasing the germination rate (Gayathri and Kanakamany, 2018). So it is important to know about different post-harvest techniques and their importance on maintaining the seed viability of different MAP.

13.3 POST-HARVEST TECHNIQUES OF MAPs

Post-harvest processing of MAP has a significant impact on the quantity and quality of active ingredients, making it an important step in the manufacturing process. When it comes to maintaining the product's physical and chemical properties, then drying, storing, and packing are essential post-harvest procedures. To preserve the quality of MAP after drying, it is important to create the ideal temperature and relative humidity conditions and avoid the attack of insects, fungi, and microorganisms while they are being stored. It is important to lower metabolic activity when storing medicinal plants so they become less prone to deterioration. To prevent deterioration during transportation, the harvested materials need to be handled, stored, and consolidated carefully both during and just after harvest. Cross-contamination from foreign materials, insects or other pests, product shrinkage, exposure to weather, temperature rise, and overheating are some of the risks that can affect product quality. These issues are relevant to good collection practices as well as appropriate cultivation practices.

13.3.1 STORAGE

The primary aim of preserving the seeds of aromatic and medicinal plants is to avoid decay and maintain both their qualitative and quantitative characteristics after drying. This can be achieved by using a modified environment, by cooling the system in which the MAPs are kept, or by reducing the moisture content of the product to the appropriate levels as well as inhibiting the growth of fungi, insects, and microbes during seed storage (Masand et al., 2014; Silva et al., 2013).

13.3.1.1 PRINCIPLES OF STORAGE OF MEDICINAL AND AROMATIC PLANTS

- To enhance storage life and increase availability round the year.
- Prevention of seasonal market gluts, orderly marketing, better income to grower, and preservation of nutritional qualities.
- To control rate of respiration, transpiration, undesirable bio-chemical changes, and disease infection.
- To avoid entry of rodents, birds, and other animals.
- To protect from dampness, dirt, and dust.

Harvested seed samples of medicinal and aromatic crops should not be stored in the open conditions or near cattle sheds to avoid contamination and microbial attack. The container must be sealed and properly labeled and stored on wooden pallets, in a cool and dry place. Never stack containers (tie bags, jute bags, woven sacks, corrugated boxes, etc.) directly on the floor.

13.3.1.2 ROLE OF STORAGE DURATION

The recommended shelf life for the seeds of MAP is one year, but if the manufacturer provides stability test results that ensure the product's qualities will be maintained during the suggested period, a longer shelf life might be approved (Brasil, 2010). MAP are frequently kept in storage for a long time before being utilized as raw materials to make different kinds of products (Sourestani et al., 2018). But improper storage conditions can lead to alterations in the physical, chemical, and microbiological realms (Mayuoni-Kirshinbaum et al., 2013; Peter, 2006).

The germination of MAP is modified considerably by the seed storage times. While *W. somnifera* and *A. paniculata* benefited from medium-term ageing, *P. corylifolia* and *B. orelina* seeds germinated more readily when they were fresh. Conversely, after extended storage, the seeds of *A. moschatus*, *O. basilicum*, and *P. zeylanica* showed increased germination. This suggests that seeds from medicinal plant species need to be stored carefully. Tiwari and Das (2014) also noted that the confounding influence of variables like plant species, temperature, and storage materials resulted in varying patterns in the germination of seeds from various medicinal plants. Likewise, the seeds are categorized into three groups according to their differing storability: orthodox, recalcitrant, and intermediate (Ellis and Roberts, 1988; Roberts, 1972a). Regardless of the type of storage material, weight loss was a consistent result of longer seed storage times for all MAP.

13.3.1.3 ROLE OF STORAGE TEMPERATURE

Seed quality and storage temperature affect the duration of seed storage (Fabrizius et al., 1999; Heatherly and Elmore, 2004). In terms of temperature gradients, plant species showed significant variation in the intensity of germination and other characteristics of stored seed. *A. moschatus*, *B. olerina*, *P. corylifolia,* and *P. zeylanica* seeds germinate best in ambient temperature (25°C), whereas *A. paniculata*, *W. somnifera,* and *O. basilicum* seeds germinate best in low temperature (4–8°C) conditions

(Tiwari and Das, 2014). It is necessary to regulate the specific storage temperature for different plant species, as improper storage temperature can result in a significant loss of MAP seeds, especially those of *B. olerina*, *A. paniculata*, *W. somnifera*, and *P. corylifolia* (Tiwari et al., 2022).

The kind of packaging used to store aromatic and medicinal plants is another crucial consideration. Packing materials that are frequently used to store plant leaves and seeds include cardboard boxes, paper bags, polyethylene bags, and double kraft paper bags with a non-toxic polyethylene inner layer (Correa Junior et al., 2006).

13.3.2 PACKAGING

To ensure the quality of MAPs and preserve their chemical properties during storage, plant material must be dried, and appropriate packaging should be used. The drying process helps to "concentrate" the active ingredients by weight by removing water and inhibiting the growth of microorganisms and enzymatic reactions, which reduces degradation of the active ingredients during storage. Packaging plays an important role in maintaining the product quality from processing to customer consumption (Cooksey, 2006). Numerous packing methods have been developed and enhanced as science has progressed. These consist of controlled atmosphere, modified atmosphere, and vacuum packaging. A multitude of factors must be considered during packaging and storage. The product's physical and chemical characteristics play a major role in determining the best packaging and storage method for each condition (Lisboa et al., 2022).

13.3.3 IMPORTANCE OF PACKAGING

- Using suitable packaging supplies and equipment will improve the quality and storage conditions of the packaged seeds.
- To avoid product damage and safeguard it against possible pest attacks and other means of contamination, MAP seeds should be packaged as soon as possible (EMEA, 2006).
- To remove undesirable components prior to and during the last phase of packaging, continuous quality assurance processes must be carried out (EUROPAM, 2016; WHO, 2013).

- Batch packing records must be retained for three years or as long as required by national and/or the local authorities (WHO, 2013).
- Assembling multiple units into aggregates is made easier by packaging, which functions as a material handling tool by holding the required unit amount of seeds in a single container (Yadav et al., 2014).
- Packaging offers various economic advantages such as preventing product spills, simplifying transportation, preventing contamination, lowering labor costs, and so on (Yadav et al., 2014).
- Packages serve as marketing tools. For the sales and marketing divisions of agricultural companies in particular, the packaging's aspects of sales appeal and recognition of products are critical.

13.3.4 PRINCIPLES OF PACKAGING OF MEDICINAL AND AROMATIC PLANTS

The packaging material must be carefully selected, keeping in mind both functional and marketing requirements to protect the quality of aromatic and medicinal crop seeds during handling, transportation, storage, and distribution. Listed below are the packaging specifications for aromatic and medicinal plants.

- To shield the product from environmental elements like heat, light, humidity, and oxygen, the packaging materials should have strong oxygen and water vapor barriers.
- To safeguard the product from spills and spoiling.
- To avoid aroma or flavor loss and external odor infiltration, the packing material ought to possess a high barrier property.
- The volatile oil found in the seeds of aromatic and medicinal products has the ability to react with the inner/contact layer of the packaging material, sometimes resulting in a smudged and greasy package. As a result, the packaging material needs to be compatible with the product and resistant to grease and oil. In addition to the aforementioned functional specifications, the packaging material should be readily available, disposable, and have good machinability and printability (NIFTEM, 2023).

13.3.5 PACKAGING MATERIALS

The seeds of aromatic and medicinal plants should be placed in dry, clean bags, sacks, or bales following a thorough examination and removal of any foreign objects (EUROPAM, 2022). Packaging materials ought to be kept out of the reach of pets, livestock, and other potential sources of contamination. They should also be kept dry, clean, and pest-free. Furthermore, it is important to make sure the product is free of contamination from the packaging material (particularly if it is a fiber bag) in the form of lost plastic fibers, chemical residues, odd smells, or any type of household waste (EMEA, 2006; EUROPAM, 2016, 2022). Contamination of packaging materials is not acceptable (WHO, 2013). Before being used, the material must be thoroughly cleaned (disinfected) and dried if it is reusable. It is imperative to guarantee that reusing bags does not result in contamination (EUROPAM, 2022). Packing should follow SOPs and local, national, and/or international laws in the countries of origin and the final consumer. It is advised to package delicate medicinal plant materials in stiff containers (WHO, 2013).

13.3.6 IMPACT OF PACKAGING

The choice of appropriate packaging is crucial because storage materials have a significant impact on the germination of seeds (Owalade et al., 2011; Pradhan and Badola, 2012; Saisanthosh and Patil, 2018). The seeds of diverse medicinal and aromatic species require different types of storage materials. For instance, high-density polythene bags (20 micron) worked well for storing *A. moschatus* seeds, but semi-impervious materials like unburned earthen pots worked best for *W. somnifera* seeds (Tiwari et al., 2022). Since polythene prevents or minimizes moisture absorption, regulates weight loss, and lessens interaction with the storage environment, seeds stored in polythene are more unlikely to experience increased oxidation (Netra et al., 2015). Reduced respiration rates in moisture-proof storage materials contribute to increased seed germination (Khalequzzaman et al., 2012; Rahman and Rahman, 1997; Saisanthosh and Patil, 2018; Sultana et al., 2016). The variations in the dissolution and polarization of genetic material in seed may also create differences in seed germination of aromatic and medicinal plants (Dapkevicius et al., 1998).

Furthermore, the genetic makeup of a species influences its phenotypic expression, vigor, and quality of the seeds (Al-Yahya, 1995; Vieira et al., 2001), which can lead to variations in seed germination.

13.3.7 BULK PACKAGING AND LABELING

It is important to pack the seeds of MAP as soon as possible to avoid product deterioration and to shield them from unwanted exposure to possible pest attacks and other contamination sources. Allowing a controlled flow of processing from the time raw MAP materials arrive on site until the processed plant materials are shipped will facilitate efficient and hygienic operations.

13.3.7.1 LABELING

It is the process of placing a tag on a good or its container and providing the customer with helpful details that will assist them, like the product's price, value, manufacturing details, expiration date, and more.

13.3.7.2 IMPORTANCE OF LABELING

- labeling is a crucial component of product marketing as it helps to attract customer attention.
- It helps the product stand out in the market and identifies it as part of a particular brand.
- The label needs to be non-toxic, permanently affixed, and easily readable (EUROPAM, 2022).
- The label must contain information about a product on its container. Some national and international labeling laws are followed to do that. The labels should show the following particulars (EUROPAM, 2022; WHO, 2013):
    - Common and scientific names of plants.
    - Parts of the plant used.
    - Product details (name and address).

- Batch number of the production.
- Quantitative information of the product (like, package weight).
- Conservation strategy, if any.
- Place of origin of the product.
- Cultivation or collection date of the product.
- Hazardous information about the product.
- If the product is organic, then the Organic Certification Controlled Number should be mentioned.
- Information that indicates the standard approval and compliance with other national or regional labeling requirements should be mentioned.
- A separate certificate with a similar batch number can be attached to the package to provide additional details on production guidelines and the quality of aromatic and medicinal plants (Marcelino et al., 2023).

13.4 TRANSPORTATION

Transportation is an indispensable part of marketing. It serves as the physical means to move products from production sites to market locations, creating place utility and ensuring a consistent supply between different areas. It plays a crucial role in marketing activities such as purchasing, gathering, selling, preservation, and storage, and so forth. A well-integrated transportation system is essential for the economy and its development (Smith, 2020). The harvested seeds of the aromatic and medicinal crop should be transported as soon as possible in dry, clean conditions. To facilitate transportation, seeds should be placed in sanitized containers, airtight bags, buckets, or other well-ventilated boxes and moved to a central location. Every container used during harvesting needs to be kept clean and free of foreign objects and plant products from previous harvests. If plastic containers are used, special attention should be paid to retaining potential moisture that could encourage mold growth. Containers ought to be stored in a dry location away from animals and humans, and away from insects, rodents, and birds when they are not used. It is important to clean conveyors in between loads when moving large plant materials from the production site to storage for processing (Figure 13.1). When used appropriately, bulk transportation such as ship or rail cars should have adequate ventilation to drive moisture out of plant materials and avoid condensation (Douglas et al., 2005).

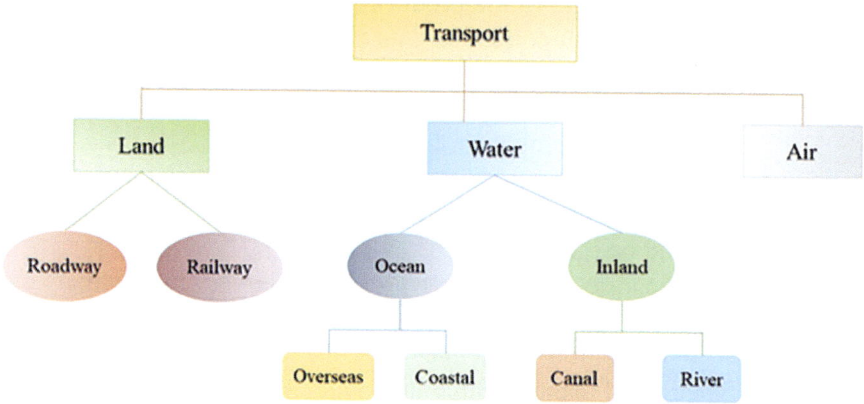

FIGURE 13.1 Different types of transport used for marketing.

13.5 CONCLUSION

Since ancient times, MAPs have had a significant role in many conventional medical systems and aromatherapy practices, including Ayurveda and Unani medicine and their derivatives. The natural forests provide nearly 90% of the resources of aromatic and medicinal plants. However, collection of wild species in natural environments poses unique challenges, particularly concerning misidentification with related plants, insufficient expertise among personnel, and so on. It is necessary to adhere to specific guidelines to develop Good Field Collection and Processing Practices for MAP, the majority of which are collected from the wild. The ability of aromatic and medicinal plants to germinate their seeds is maintained in part by a variety of post-harvest methods. The viability of both cultivated and wild species of MAPs can be restored with the aid of proper handling, storage, and packaging of seed materials. Certain species of MAPs can be kept for a long time with the aid of the right storage conditions, including the right temperature, length of time, and packaging materials. Hence, it is required to maintain optimum storage and packaging conditions for MAP for prolonged viability and quality.

KEYWORDS

- **labeling**
- **packaging materials**
- **processing**
- **seed germination**
- **storage**
- **transportation**

REFERENCES

Afzal, I., Jaffar, I., Zahid, S., Ur Rehman, H., & Basr, S. M. A. B. (2020). Physiological and biochemical changes during hermetic storage of *Moringa oleifera* seeds. *South African Journal of Botany, 129*, 435–441. https://doi.org/10.1016/j.sajb.2019.11.011 (accessed on 12th August 2025).

Al-Yahya, S. A. (1995). Losses of corn in the storage. *Arabian Gulf Journal of Scientific Research, 13*, 199–212.

Bareetseng, S. (2022). The worldwide herbal market: Trends and opportunities. *Journal of Biomedical Research and Environmental Sciences, 3*, 575–584.

Benarba, B., & Pandiella, A. (2020). Medicinal plants as sources of active molecules against COVID-19. *Frontiers in Pharmacology, 11*, e01189. https://doi.org/10.3389/fphar.2020.01189 (accessed on 12th August 2025).

BRASIL. (2021). *Agência Nacional de Vigilância Sanitária. Farmacopéia Brasileira* (5th ed.). Brasília.

Chandra, P., & Sharma, V. (2019). Marketing information system and strategies for sustainable and competitive medicinal and aromatic plants trade. *Information Development, 35*, 806–818.

Chowti, S. P., Rudrapur, S., & Naik, B. K. (2018). Production scenario of medicinal and aromatic crops in India. *Journal of Pharmacognosy and Phytochemistry, SP3*, 274–277.

Dapkevicius, A., Venskutonis, R., Van Beek, T. A., & Linssen, J. P. H. (1998). Antioxidant activity of extracts obtained by different isolation procedures from some aromatic herbs grown in Lithuania. *Journal of the Science of Food and Agriculture, 77*, 140–146.

Devika, M. (2021). An overview of medicinal and aromatic plants. *Medicinal & Aromatic Plants (Los Angeles), 10*, 387.

Douglas, M., Heyes, J., & Smallfield, B. (2005). *Herbs, spices and essential oils: Post-harvest operations in developing countries*. UNIDO and FAO. https://www.fao.org/3/ad420e/ad420e.pdf (accessed on 12th August 2025).

Dzoyem, J. P., Tshikalange, E., & Kuete, V. (2013). Medicinal plants market and industry in Africa. In V. Kuete (Ed.), *Medicinal Plant Research in Africa* (pp. 859–890). Elsevier.

EMEA. (2006). *Guideline on Good Agricultural and Collection Practice (GACP) for Starting Materials of Herbal Origin*. EMEA. http://www.ema.europa.eu/docs/en_GB/document_library/Scientific_guideline/2009/09/WC500003362.pdf-1 (accessed on 12th August 2025).

European Herb Growers Association (EUROPAM). (2016). *GACP Working Group: A Practical Implementation Guide to Good Agricultural and Wild Collection Practices (GACP)*. EUROPAM.

European Herb Growers Association (EUROPAM). (2022). *Guidelines for Good Agricultural and Wild Collection Practice (GACP) of Medicinal and Aromatic Plants* (Vol. 7, pp. 1–12). EUROPAM. https://www.europam.net/wp-content/uploads/2022/11/EUROPAM%20GACP-2022.pdf (accessed on 12th August 2025).

Fabrizius, E., Tekrony, D., Egli, D. B., & Rucker, M. (1999). Evaluation of a viability model for predicting soybean seed germination during warehouse storage. *Crop Science, 39*, 194–201.

Gayathri, P., & Kanakamany, M. T. (2018). Influence of storage environment and packing materials on seed germination and viability of *Desmodium gangeticum* (L.) DC. *Open Access Journal of Medicinal and Aromatic Plants, 9*, 23.

Gordon, G. A. (1992). *Seed Manual for Forest Trees*. Forestry Commission.

Heatherly, L. G., & Elmore, R. W. (2004). Managing inputs for peak production. In H. R. Boerma & J. E. Specht (Eds.), *Soybeans: Improvement, Production and Uses* (3rd ed., pp. 451–536). ASA, CSSA, and SSSA.

Himangini., & Thakur, A. (2018). Effect of seed storage conditions on seed germination and vigor of *Withania somnifera*. *Journal of Pharmacognosy and Phytochemistry, 7*, 1409–1413.

Juhari, N. H., & Petersen, M. A. (2018). Physicochemical properties and oxidative storage stability of milled roselle (*Hibiscus sabdariffa* L.) seeds. *Molecules, 23*, 385. https://doi.org/10.3390/molecules23020385 (accessed on 12th August 2025).

Khalequzzaman, K. M., Rashid, M. M., Hasan, M. A., & Reza, M. M. A. (2012). Effect of storage containers and storage periods on the seed quality of French bean (*Phaseolus vulgaris*). *Bangladesh Journal of Agricultural Research, 37*, 195–205.

Lisboa, C. F., Melo, E. de C., Sperotto, N. C. Z., Avila, M. B. R., Silva, L. C. da, Aleman, C. C., Carneiro, J. C. de S., Coelho, A. P. de F., & da Silva, C. S. (2022). Packaging and storage of medicinal plants. *Research, Society and Development, 11*(7), e50911724813.

Manna, D., Maity, T. K., & Basu, A. K. (2020). Comparison on condition and packaging materials for better storage of onion seeds. *Asian Journal of Horticulture, 15*, 15–25. https://doi.org/10.15740/HAS/TAJH/15.1/15-25 (accessed on 12th August 2025).

Marcelino, S., Hamdane, S., Gaspar, P. D., & Paço, A. (2023). Sustainable agricultural practices for the production of medicinal and aromatic plants: Evidence and recommendations. *Sustainability, 15*, 14095.

Mayuoni-Kirshinbaum, L., Daus, A., & Porat, R. (2013). Changes in sensory quality and aroma volatile composition during prolonged storage of 'Wonderful' pomegranate fruit. *International Journal of Food Science & Technology, 48*, 1569–1578.

National Institute of Food Technology Entrepreneurship and Management, Ministry of Food Processing Industries. (2023). *PM Formalization of Micro Food Processing Enterprises Scheme*. Sonipat, Haryana: National Institute of Food Technology Entrepreneurship and Management. http://www.niftem.ac.in (accessed on 12th August 2025).

Netra, N., Uma Rani, K., Gowda, R., Rajendra, P. S., & Narayanaswamy, S. (2015). Effect of packaging material and desiccant on storability of soybean seeds. *Seed Science and Technology, 44*, 207–211.

Owalade, O. F., Olasoji, J. O., & Afolabi, C. G. (2011). Effect of storage temperature and packaging materials on seed germination and seed-borne fungi of sorghum (*Sorghum bicolor*) in South West Nigeria. *African Journal of Plant Science, 5*, 873–877.

Perinchery, A. (2020). Forests that heal: Medicinal plants as an ecosystem service. *Mongabay: Environment and Health*. https://india.mongabay.com/2020/02/forests-that-heal-medicinalplants-as-an-ecosystem-service/ (accessed on 12th August 2025).

Peter, K. V. (2006). *Handbook of herbs and spices*. CRC Press.

Pradhan, B. K., & Badola, H. K. (2012). Effect of storage conditions and storage periods on seed germination in eleven populations of *Swertia chirayita*: A critically endangered medicinal herb in Himalaya. *The Scientific World Journal, 2012*, 128105. https://doi.org/10.1100/2012/128105 (accessed on 12th August 2025).

Rahman, M. M. K., & Rahman, G. M. M. (1997). Effect of containers and length of storage on germination and seed-borne fungi associated with jute seed. *Bangladesh Journal of Plant Pathology, 13*, 13–16.

Ranjith, K., Kannan, R., & Rohini, A. (2019). Economic analysis of production and marketing of *Gloriosa superba* L. in Tamil Nadu. *International Journal of Chemical Studies, 7*(3), 4275–4278.

Roberts, E. H. (1972a). *Viability of Seeds*. Chapman and Hall.

Roberts, E. H. (1972b). Storage environment and the control of viability. In E. H. Roberts (Ed.), *Viability of Seeds* (pp. 14–58). Chapman and Hall.

Roberts, E. H. (1988). Seed ageing–the genome and its expression. In L. D. Nooden & A. C. Leopold (Eds.), *Senescence and Ageing in Plants* (pp. 465–598). Academic Press.

Saisanthosh, K., & Patil, N. K. B. (2018). Effect of packaging materials and moisture content on seed storability of onion. *Journal of Pharmacognosy and Phytochemistry, 7*, 1745–1750.

Samarth, R. M., Samarth, M., & Matsumoto, Y. (2017). Medicinally important aromatic plants with radioprotective activity. *Future Science OA, 3*(4), FSO247. https://doi.org/10.4155/fsoa-2017-0061

Sarantópoulos, C. I. G. L., Alves, R. M. V., Coltro, L., Padula, M., Teixeira, F. G., & Moreira, C. Q. (2018). Propriedades de barreira. In C. I. G. L. Sarantopoulos & F. G. Teixeira (Eds.), *Embalagens plásticas flexíveis: Principais polímeros e avaliação de propriedades*. ITAL/CETEA.

Schmidt, L. (2000). *Guide to Handling of Tropical and Subtropical Forest Seed*. Danida Forest Seed Centre.

Sharma, T. V. R. S., Abirami, K., & Punnam, C. M. (2018). Medicinal plants used by tribes of Andaman and Nicobar Islands: A conservation appraisal. *Indian Journal of Plant Genetic Resources, 31*, 125–133. https://doi.org/10.5958/0976-1926.2018.00015.3 (accessed on 12th August 2025).

Smith, J. (2020). Essential functions of marketing: Transportation and storage. *Journal of Business Logistics, 25*(3), 45–58.

Sultana, R., Chowdhury, M. S. M., Islam, M. R., & Akhter, K. (2016). Effect of container and duration of storage on the quality of okra (*Abelmoschus esculentus*) seeds. *Agriculturists, 14*, 63–72.

Tiwari, R. K. S., & Das, K. (2014). Impact of different storage conditions on seed germination and viability of some medicinal plants. *African Journal of Agricultural Research, 20*, 1578–1585. https://doi.org/10.5897/AJAR2012.1758 (accessed on 12th August 2025).

Tiwari, R. S., Chandra, K. K., Dubey, S., & Tripathi, S. (2022). Influence of packaging materials and storage conditions on seed germination ability and biochemical changes in some medicinal plants of Indian forests. *Frontiers in Forests and Global Change, 5*, 868237. https://doi.org/10.3389/ffgc.2022.868237 (accessed on 12th August 2025).

Vieira, R. D., Te-Krony, D. M., Egli, D. B., & Rucker, M. (2001). Electrical conductivity of soybean seeds after storage in several environments. *Seed Science and Technology, 29*, 599–608.

World Health Organization. (2003). *WHO Guidelines on Good Agricultural and Collection Practices (GACP) For Medicinal Plants* (p. 80). World Health Organization. https://apps.who.int/iris/bitstream/handle/10665/42783/9241546271.pdf?sequence=1 (accessed on 12th August 2025).

Yadav, R. K., Goyal, R., & Singh, S. (2017). *Post Harvest Technology of Horticultural Crops*. https://doi.org/10.13140/RG.2.2.28507.98089 (accessed on 12th August 2025).

Yang, Q. H., Ye, W. H., Deng, X., Cao, H. L., Zhang, Y., & Xu, K. Y. (2005). Seed germination eco-physiology of *Mikania micrantha* HBK. *Botanical Bulletin of Academia Sinica, 46*, 293–299.

CHAPTER 14

Seed Germination Behavior of Hogweed (*Heracleum candicans* Wall.)

J. S. BUTOLA,[1,*] HIMSHIKHA GUSAIN,[1] PRIYA BANSAL,[1] A. R. MALIK,[2] and PANKAJ SHARMA[3]

[1]Department of Forestry and Natural Resources, Hemvati Nandan Bahuguna Garhwal University, Srinagar Garhwal, Uttarakhand, India

[2]Division of Forest Products and Utilization, Faculty of Forestry, Sher-e-Kashmir University of Agricultural Sciences and Technology of Kashmir (SKUAST-K), Benhama, Ganderbal, Jammu and Kashmir, India

[3]Himachal Pradesh Biodiversity Board, Paryavaran Bhawan, Shimla, Himachal Pradesh, India

*Corresponding author

ABSTRACT

Heracleum candicans (Wall.) is a native and endangered medicinal herb of the temperate Himalayan region. It is one of the prominent sources of Xanthotoxin, an active constituent of a commercial medicine for skin diseases like Psoriasis and Vitiligo. The chapter offers the results of an experimentation pertaining to the effect of different eco-physiological factors on the germinability of this species. An experimental trial was set up to understand the germination behavior of the species under different temperature regimes, *viz*., 5, 15, 25, 30°C and room temperature; photoperiodic conditions (continuous light, continuous dark, and alternate photoperiods – 16 hrs dark *vs* 8 hrs light); and sowing depths, that is, 0.5, 1.0, 1.5, and 2.0 cm. In the laboratory, seed germination was found under all temperature regimes, even at low temperature (5°C). Seeds stored at

Seed Technology for Medicinal and Aromatic Plants: Production, Testing, Storage, and Marketing.
Priyanka Sharma, Kishore Kumar Sharma, and Gopal Shukla (Eds.)

DOI: 10.1201/9781779641199-14

low temperature (4°C) showed a minor declination (up to 7%) in viability when tested after six months' storage. It offers further in-depth studies to understand the seed dormancy mechanism of this species. Significantly ($P < 0.05$) higher (50.67%) and the earliest germination (8 days) was recorded under alternate photoperiods at 25°C as compared to other conditions. In field conditions, seed sowing at the soil surface (only 0.5 cm depth) responded to maximum seedling emergence (36.93%) as compared to other sowing depths. Results of our previous studies on the germination behavior of the species have also been elaborated in this chapter. Seeds collected from different provenances have shown significant variation ($P < 0.05$) in germinability (14.67 to 82.33%). However, pre-sowing treatment of GA_3 (250 μM), KNO_3 (100 mM), and $NaHClO_3$ (15, 30, 45 minutes) were recommended to achieve better seed germination and mass multiplication of this species. Our studies suggest that, by adopting these simple, low-cost, and effective scientific interventions, mass-scale nursery and field production of *H. candicans* can easily be done in the temperate Himalayan region. Commercial-scale cultivation of this endangered species has been suggested as one of the strategies for conservation and consistent supply of raw material to pharmaceutical industries.

14.1 INTRODUCTION

The Indian Himalayan Region (IHR) is considered a goldmine for a great diversity of medicinal flora due to the altitude, topography gradient, and varying climatic conditions (Butola, 2011; Mehta, 2021; Samant et al., 1998). Such great diversity of medicinal plants not only supports the livelihood system for local rural communities (Butola, 2020) but also offers the most reliable solution to global human health issues (Ashraf et al., 2023). Genus *Heracleum* of the family Apiaceae is cosmopolitan and comprised of 120–125 species widely distributed in Asia (109 spp.), temperate Europe, North America, and Abyssinia (Pimenov and Leonov, 2004). In the Indian subcontinent, a total of 21 species have been reported in India and adjoining countries like Nepal, Bhutan, Sri Lanka, Burma, Pakistan, and Bangladesh (Hooker, 1879). Out of the total 15 species reported so far from IHR, 06 species from north-west Himalaya, 02 from Trans Himalaya, 03 in Central Himalaya, and the remaining 4 from Eastern Himalaya have been reported (Nasir, 1972). Many species are morphologically similar to each other and worldwide known for their therapeutic applications in medical science.

H. candicans is a Himalayan native species distributed from temperate to alpine zones of north India, west Pakistan, Nepal, Bhutan, Afghanistan, south-west China, and so forth. The species is an important source of Xanthotoxin, an active constituent of medicine used for the treatment of autoimmune conditions like Psoriasis and Vitiligo. Being a major source of Xanthotoxin, the species holds constant demand in pharmaceutical industries at domestic and global levels, which makes it one of the medicinal plant products being exported from India to abroad. In the absence of cultivation, the supply of its raw material in India is relying exclusively on wild harvest. Unsustainable *in situ* harvesting, deforestation, over-grazing, trampling, habitat loss, and fragmentation has threatened its natural populations to a great extent. Due to ruthless extraction, the species is very likely to become extinct in the near future. Considering the ever-increasing pressure on wild populations, it has been categorized as an Endangered species of north-west Himalaya and has been prioritized among the species of high conservation concern (Khan et al., 2005; Sastry and Chattergee, 2000).

Ex situ production is one of the globally accepted strategies for threatened species that mitigates harvesting pressure on their natural populations. Despite the fact that systematic cultivation of a very few threatened species, viz. *Saussurea lappa*, *Inula racemosa*, and *Rauvolfia serpentina*, has been initiated so far, the lack of standardized production technology can be considered one of the prominent reasons for this malady.

Temperature and light play a significant role in seed germination and seedling establishment, which are the most critical stages in a plant's life. Understanding these factors is extremely meaningful for the restoration of vegetation and commercial propagation through artificial regeneration (Chen et al., 2019; Kim et al., 2016). Development or standardization of a reliable protocol for improvement of seed germination and seedling establishment can help in achieving improved field production of high-value medicinal and aromatic crops (MAPs), including *H. candicans* (Khidrapure et al., 2018). Production technology of *H. candicans* has been standardized by some of the researchers (Badola and Butola, 2005; Butola and Badola, 2004, 2006; Butola et al., 2010a, b; Kaul, 1989). However, its germination behavior under different eco-physiological conditions has not been studied so far. Therefore, an experimental trial was set up to understand the germinability of this species under different temperature regimes, photoperiodic conditions, and sowing depths.

14.2 IDENTIFICATION OF SEED SOURCE AND COLLECTION OF SPECIMENS

Sample specimens of mature seeds of *Heracleum candicans* (Wall.) were collected from different locations at Parvati valley (1850–2300 m asl), Himachal Pradesh (HP), India. The samples were identified and matched with the specimens available in our laboratory. Details of the seed source are presented in Table 14.1. After drying at room temperature (20–25°C) for 10 days and retaining 12% moisture, which is acceptable for most of the orthodox species, the seeds were kept in hermetic plastic containers and stored in a refrigerator at 4°C for further estimation of viability and germination behavior.

TABLE 14.1 Description of Source of *Heracleum candicans* Seeds

Parameters	Description
Seed source	Bhandag Thach, Parvati valley, HP.
Altitudinal range (m amsl)	3040–3120
Aspect	north-east
Slope (°)	20–60
pH	6.7
Organic Carbon (%)	6.06 (0.06)
N (%)	0.44 (0.14)
Average plant density (ind./m^2)	1.40 (0.49)
No. of mature individuals	70
Habitat characteristics	Open grasslands, Grass slopes, humus rich area, moist area
Major associated species	*Cedrus deodara, Picea smithiana, Abies pindrow* and *Indigofera heterantha*

14.3 SEED VIABILITY TEST

Before conducting a germination test, seed viability was examined to determine whether individual seeds appeared to be dead or alive and metabolically active. Immediately after collection and during the time of germinability tests, seed viability was determined as per ISTA (1976). Before immersing in 0.5% tetrazolium solution, the wings of seeds were excised. The seeds were examined under an electronic microscope.

The embryos that turned dark red were considered viable. It is already a well-known phenomenon that during this process, hydrogen ions reduce the colorless TZ solution to red formazan, which stains live tissues red, while dead tissues remain unstained.

14.4 SEED GERMINATION AND SEEDLING EMERGENCE

Screenings of 3000 seeds were done for a germination test. Fifty seeds per replicate (in triplicate) were used for each experiment. For sterilization, all seeds were soaked in 0.04% $HgCl_2$ for 1 minute, followed by washing thoroughly with distilled water. Three experiments with CRD (Completely Randomized Design) were undertaken for the purpose. To evaluate germination behavior under different temperature regimes, seeds were placed in petri dishes (9 cm diameter) lined with Qualigens (615A) filter paper. The petri dishes were placed at different temperature regimes, *viz*., 5°C, 15°C, 25°C, and 30°C, inside a germinator and at room temperature (25–33°C) under continuous (24 hrs) light conditions. To evaluate germination behavior under different photoperiodic conditions, seeds were placed in petri dishes (9 cm diameter) having filter paper. Petri dishes were placed in a seed germinator at 25 ± 2°C under light (24 hrs), dark (24 hrs), and an alternate light regime (8 hrs light and 16 hrs dark). To evaluate germinability under different sowing depths, seeds were sown at 0.5 cm, 1.0 cm, 1.5 cm, and 2.0 cm depths in Styrofoam trays containing a mixture of soil (field soil), pure sand (coarse), and forest litter in equal ratio (1:1:1). The trays were placed inside a net house (temperature 24.71 ± 14.88°C max., 11.12 ± 5.05°C min; relative humidity 81.53 ± 1.93% max., 74.81 ± 4.09% min). Seed germination in the laboratory and seedling emergence in field conditions were monitored on a daily basis, and constant readings were recorded for two weeks. The seedling in nursery condition was considered emerged upon the appearance of the cotyledon above the soil.

Mean germination time was calculated by using the formula (Nichols and Heydecker, 1968):

$$MGT=\Sigma(fx)/\ \Sigma x,$$

where x is the number of newly germinated seeds on each day and,

f is the number of days after seeds were set to germinate. Germination Potential Index (GPI) was calculated as per Butola (2011) as:

GPI = (Days to start germination + Days to complete germination) / (Mean germination (%) X 100)

14.5 STATISTICAL ANALYSIS

Statistical analysis of the data was done with the help of MS Excel, 2011. One-way ANOVA and F-test ($P < 0.05$) were applied to calculate significant differences between the means (Snedecor and Cochran, 1967). Before ANOVA, the data in percentages were converted into the value through arcsine transformation.

14.6 RESULTS

14.6.1 SEED VIABILITY

Seed viability test is essential in germplasm conservation and research to determine the success of *ex* situ conservation efforts such as seed banking. In the present experiment, seed viability after two days of collection and six months' storage (40°C) was found to be 82% and 78%, respectively. Similarly, in our previous study, a minor declination (7%) in viability was recorded after six month's storage at low temperature (Butola et al., 2010). In many circumstances, seed lots under similar laboratory germinations may give widely differing field emergence values. Similarly, two seed lots having the same germination percentage in the laboratory may age differently when stored under ambient conditions.

14.6.2 GERMINABILITY UNDER DIFFERENT PHOTOPERIODIC CONDITIONS

Germinability under photoperiodic conditions is presented in Figures 14.1 and 14.2. At alternate light regime, germination took place earlier (8 days) in comparison to continuous light condition (13.33 days). Maximum mean germination percentage was found ($P < 0.05$; 50.67%) in alternate light regime and minimum (13.33%) in continuous light condition. These observations were in tune with the findings of Bisht et al. (2009) in case of Pyrethrum (*Chrysanthemum cinerariaefolium* vis.). Similar observations were found by Butola et al. (2014) in *Angelica glauca*, of family Apiaceae. Germination potential of seven alpine and sub-alpine species was observed by Nautiyal (1996) which concluded that almost all the species germinated better in light than in dark. In the present study, GPI was recorded maximum (283.5) in continuous dark condition and minimum (120.9) in alternate light regime whereas non-significant ($P > 0.05$) variation was observed in all conditions (Table 14.2).

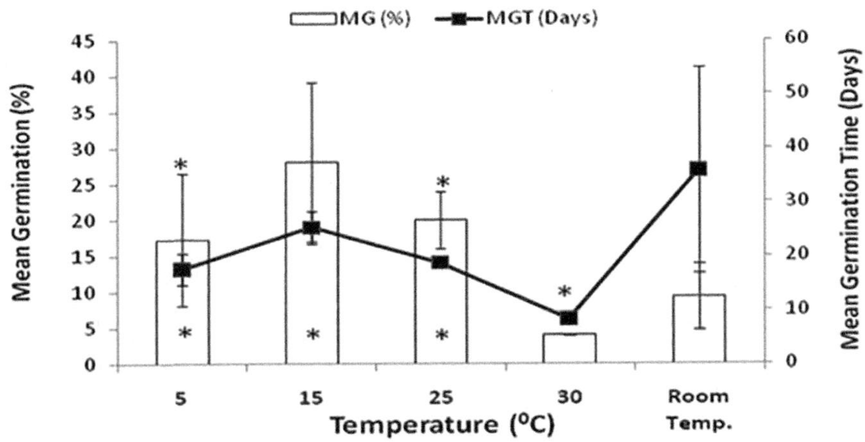

FIGURE 14.1 Germinability of *Heracleum candicans* seeds under different photoperiodic conditions; *indicates significant difference.

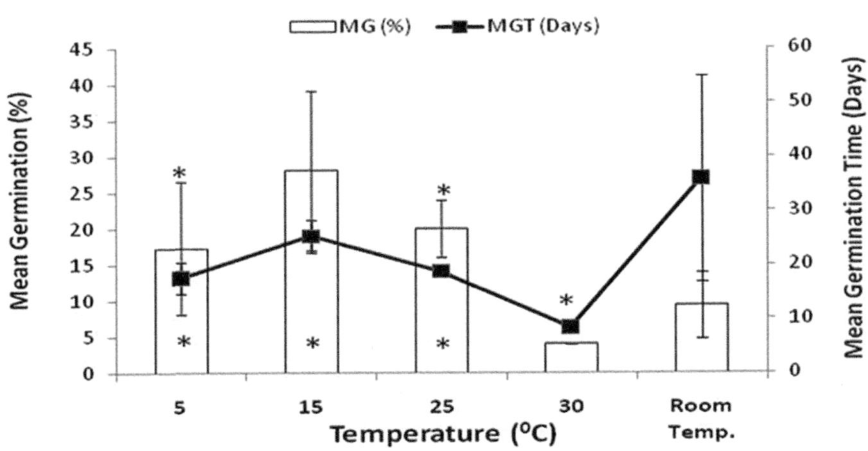

FIGURE 14.2 Germinability of *Heracleum candicans* seeds under different temperature regime; *indicates significant difference in mean germination.

TABLE 14.2 Germinability of *Heracleum candicans* Under Different Photoperiodic Conditions

Parameters	Photoperiodic Conditions (25°C)			LSD (P< 0.05)	F
	Light	Dark	Alternative regime (8L/16D)		
Days taken to first germination	13.3±4.04	9.3±1.15	8.00±0.00	4.85	3.92 ns
Days taken to 50% germination	20.3±5.51	12.0±0.00*	26.67±6.66	9.97	6.52*
Days taken to complete germination	35.7±13.3*	39.7±1.15*	56.67±2.89	15.77	5.98*
Mean Germination (%)	13.3±2.31	20.0±4.00	50.67±6.11*	5.68	59.35*
	21.37	26.49	45.38		
Mean Germination Time (Days)	21.00±2.02*	18.8±0.93*	27.40±4.10	5.39	8.30*
Germination Potential Index	284±108	213±39.7	121±13.6	133.4	4.45 ns

Abbreviations used: LSD = least significant difference; F = F ratio; *Indicate significant difference at P < 0.05 level; NS = non-significant; Data in bold phase are transformed values.

14.6.3 GERMINABILITY UNDER DIFFERENT TEMPERATURE REGIME

Germinability under different temperature regimes is presented in Figures 14.2 and 14.3. Maximum mean germination (28.0%) was found at 15°C and minimum (4.0%) at 30°C. In general, the effect of temperature on germination is quite complex because it affects each stage of the germination process in a different way and is not independent of other factors (Mayer and Poljakoff-Mayber, 1982). Nautiyal et al. (1985) observed poor seed germination under low (10°C) as well as relatively high temperature (30°C) in *Aconitum atrox*. The highest germination of alpine seeds was observed under 20–30°C temperature (Sayers and Ward, 1966). Alternate low and high moderate temperature treatments induce and enhance germination of dormant seeds (Bose and Sharma, 2000). Low and fluctuating temperatures hamper seed germination and seedling establishment in the high-altitude areas (Osburn, 1961). Therefore, most of the germination takes place soon after the snowmelt in early summers (Billings and Mooney, 1968) (Table 14.3).

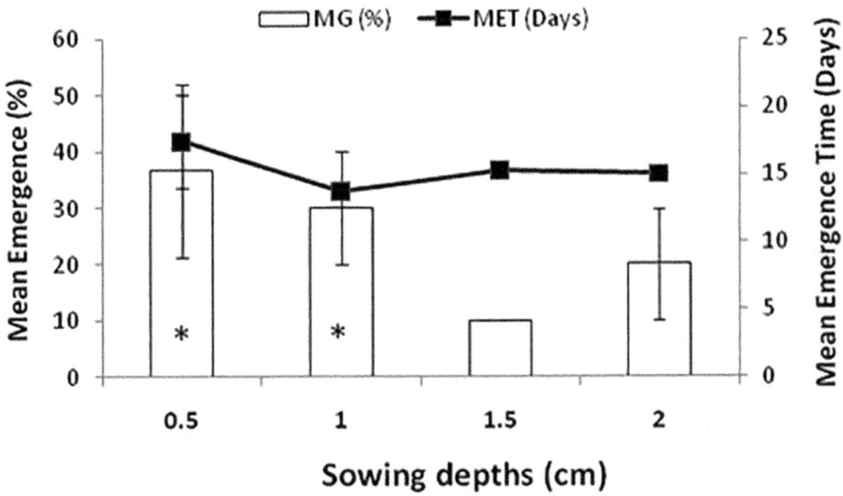

FIGURE 14.3 Seedling emergence of *Heracleum candicans* under different sowing depths; *indicates significant difference in mean emergence.

TABLE 14.3 Germinability of *Heracleum candicans* Under Different Temperature Regime

Parameters	**Temperature regime (0C)**					**LSD (P<0.05)**	**F**
	5	**15**	**25**	**30**	**RT (25–33)**		
Days taken to first germination	14.0±0.00	13.3±2.30	9.3±1.15	8.3±0.58	28.7±23.0	19.03	1.83 ns
Days taken to 50% germination	17.0±2.70*	22.3±1.50	12.0±0.0*	8.3±0.58*	35.3±19.0	15.69	4.48*
Days taken to complete germination	21.3±4.60*	43.7±4.00	39.7±1.20	8.3±0.58*	43.7±20.0	16.77	8.85*
Mean Germination (%)	17.3±9.2* 24.16	28.0±11.0* 31.71	20.0±4.0* 26.49	4.0±0.00 11.54	9.3±4.62 17.36	9.04	7.59*
Mean Germination Time (Days)	17.6±2.9*	25.2±3.0	18.77±0.9*	8.33±0.58*	35.89±19.0	16.12	3.96*
Germination Potential Index	154±57.0	183±56.0	213±40.0	217±15.0	698±51	534.5	1.81 ns

RT = room temperatures; LSD = least significant difference; F = F ratio; *Indicate significant difference at P < 0.05 level; NS = non-significant; Data in bold phase are transformed values

14.6.4 SEEDLING EMERGENCE UNDER DIFFERENT SOWING DEPTHS

Seedling emergence under different sowing depths is presented in Figure 14.3 and Table 14.4. First seedling emergence was recorded after 13 days at 1.0 cm depth; however, maximum days were taken to complete emergence (22 days) at 0.5 cm depth. Mean germination percentage was found significantly (P < 0.05) higher (36.70%) at 0.5 cm depth than 1.5 cm depth (10.0%). Overall, seedling emergence was found poor with increasing seed sowing depths. Similarly, Butola et al. (2014) revealed similar results in *Angelica glauca*. It is a very common understanding that the availability and requirement of light, moisture, and temperature at different sowing depths affect seed germinability.

TABLE 14.4 Seedling Emergence of *Heracleum candicans* Under Different Sowing Depths

Parameters	**Sowing Depths (cm)**				**LSD (P<0.05)**	**F**
	0.5	**1.0**	**1.5**	**2.0**		
Days taken to first emergence	15.0±5.20	13.0±0.00	15.3±0.58	14.7±0.6	4.95	0.5 ns
Days taken to 50% emergence	16.3±4.04	13.3±0.58	15.3±0.58	15.0±0.00	3.88	1.1 ns
Days taken to complete emergence	22.0±5.57	14.7±0.58*	15.3±0.58*	15.7±1.2*	5.41	4.2*
Mean Emergence (%)	36.7±15.3*	30.0±10.0*	10.0±0.00	20.0±10.0	12.8	4.32*
	36.93	33.00	18.43	26.07		
Mean Emergence Time (Days)	17.5±3.41	13.8±0.25	15.3±0.58	15.1±0.59	3.31	2.33 ns
Emergence Potential Index	84.7±40.1*	65.5±16.4*	169±6.35	109±49.6	62.30	5.50*

LSD = least significant difference; F = F ratio; *Indicate significant difference at P < 0.05 level; NS = Non-significant; Data in bold phase are transformed values

14.6.5 GERMINATION BEHAVIOR UNDER DIFFERENT PRE-SOWING CHEMICAL TREATMENTS

Apart from the present experimentation presented here, the results of previous studies of one of the authors (Butola and Badola, 2004, Butola et al., 2010b) on germination behavior of *H. candicans* have also been discussed in the present chapter. A total of 15 treatments were applied to stimulate seed germination (8 for PGRs, 4 for nitrogenous compounds, and 3 for $NaHClO_3$), among which, 5 were found significantly effective over control. For the PGRs (Table 14.5), a higher concentration of GA_3 (250 μM) significantly enhanced (P < 0.05) seed germination (76.7%, compared to 58.7% in control), which is similar to that for *Podophyllum hexandrum* (Nadeem et al., 2000). Promoting effects of GA_3 on seed germination by lowering the endogenous abscisic acid (ABA) content of seed (an inhibitor to germination) have been suggested. Similar results were obtained by Joshi and Dhar (2003) as promoting effect of GA_3 in one population of *H. candicans*. Hence, the finding supports the idea that GA_3 is generally effective for germination in this species.

Amongst nitrogenous compounds, only a higher concentration of KNO_3 (100 mM) significantly ($P < 0.001$) enhanced germination (74.7%) over control (Table 14.5). Stimulatory effects of KNO_3 at similar concentrations were observed in *H. candicans* (Joshi and Dhar, 2003). In such cases, it is reported that the germination is stimulated through oxidized forms of nitrogen causing a shift in respiratory metabolism to the pentose phosphate pathway (Roberts and Smith, 1977).

In the case of $NaHClO_3$, all three concentrations significantly enhanced seed germination ($P < 0.001$) over control (Table 14.5). The highest germination (81.3%) was achieved by treating seeds for 30 minutes. The same treatment significantly ($P < 0.05$) reduced MGT (11.4 days) over control (16.6 days). These results indicated that any modification of the seed coat by different chemicals may generate stimulatory effects to seed germination. These may help early germinating seedlings by providing them a higher competitive ability and then reducing mortality in the field condition. Owing to low cost and simple applicability, KNO_3 and $NaHClO_3$ can be easily adopted by the farmers.

TABLE 14.5 Germination Behavior of *Heracleum candicans* Under Different Pre-Sowing Treatments

Treatments	Mean Seed Germination (%)	Mean Germination Time (Days)
Control	58.7±3.1	16.6±1.6
GA_3 25 µM	68.0±7.2	17.1±3.0
GA_3 100 µM	55.3±8.3	17.0±3.4
GA_3 250 µM	76.7±16.3[a]	16.0±2.0
BAP 25 µM	44.7±16.7	16.7±2.0
BAP 100 µM	21.3±3.1	19.7±2.1
BAP 250 µM	23.3±7.6	25.8±3.6
BAP 25 µM +GA_3 25 µM	35.3±6.1	17.0±1.8
BAP 25 µM +GA_3 250 µM	29.3±8.1	20.0±1.6

TABLE 14.5 *(Continued)*

Treatments	Mean Seed Germination (%)	Mean Germination Time (Days)
Thiourea 65 mM	18.7±6.1	21.1±5.8
Thiourea 130 mM	1.3±1.2	13.0±12.1
KNO_3 50 mM	61.3±3.1	16.3±1.3
KNO_3 100 mM	74.7±2.3[c]	13.5±1.6
$NaHClO_3$ 15 minute	80.0±5.3[c]	12.4±0.8
$NaHClO_3$ 30 minute	81.3±4.2[c]	11.4±2.1
$NaHClO_3$ 45 minute	76.7±4.6[b]	17.3±3.9[a]

Source: Butola and Badola (2004).

14.6.7 INTER-POPULATION VARIABILITY IN GERMINABILITY

Germinability amongst 6 populations of *H. candicans* varied significantly ($P < 0.05$; 14.67% to 82.33%, Figure 14.4). While testing with Tz solution, 30% of ungerminated seeds were found viable. It can be envisaged that the low germination in this species is due to dormancy. Different dormancy levels, sizes, and weight of seeds from different populations are the most possible reason for such variability in germination (Baskin and Baskin, 1998). Variation in fruit mass and germinability is reported in *Heracleum mantegazzianum* (Moravcova et al., 2005). Germinability not only determines the success of natural regeneration but also indicates the superiority of germplasm desirable for mass multiplication.

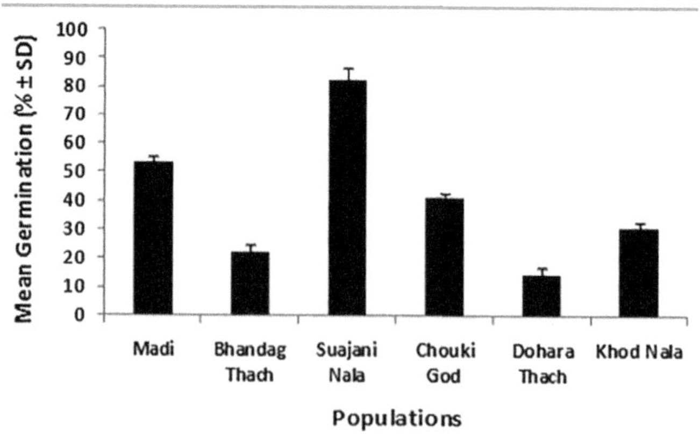

FIGURE 14.4 Inter-population variability in seed germination of *Heracleum candicans* ($P < 0.05$ level; LSD = 6.15, F = 59.55).

14.7 CONCLUSION

Findings of the present study give a better understanding of the seed germination behavior of *H. candicans*. Low-temperature (40°C) storage conditions were found suitable for maintaining long-term seed viability in the species. Moreover, there was a significant difference in the germinability of seeds collected from different provenances, which indicates that different levels of interpopulation dormancy and viability exist in this species. In laboratory conditions, higher germinability was observed under alternate photoperiodic conditions (8 hrs light/16 hrs dark) and at 25°C temperature. Pre-sowing treatment of GA3 (250 μM), KNO_3 (100 mM), and $NaHClO_3$ (15, 30, 45 minutes) were observed to be very effective in achieving better seed germination and mass multiplication of this species. In field conditions, seed sowing at the soil surface (only 0.5 cm depth) resulted in maximum seedling emergence, which indicates the requirement of light for seedling emergence in the species. By adopting these simple and effective scientific interventions, a commercial-scale nursery can be raised in temperate Himalayan regions. Commercial-scale cultivation of the species has been suggested as one of the strategies for conservation and a consistent supply of raw materials to pharmaceutical industries.

ACKNOWLEDGMENT

We are grateful to the Director of GBPNIHE, HP, for providing necessary facilities and consistent encouragement.

KEYWORDS

- **germination**
- ***heracleum candicans***
- **medicinal plant**
- **pre-sowing treatments**
- **photoperiod**
- **sowing depth**

REFERENCES

Ashraf, M. V., Pant, S., Khan, M. A. H., Shah, A. A., Siddiqui, S., Jeridi, M., Alhamdi, H. W. S., & Ahmad, S. (2023). Phytochemicals as antimicrobials: Prospecting Himalayan medicinal plants as source of alternate medicine to combat antimicrobial resistance. *Pharmaceuticals, 16*, 881. https://doi.org/10.3390/ph16070881 (accessed on 12 August 2025).

Badola, H. K., & Butola, J. S. (2005). Effect of ploughing depth on the growth and yield of *Heracleum candicans*: A threatened medicinal herb and a less-explored potential crop of the Himalayan region. *Journal of Mountain Science, 2*(2), 173–180.

Baskin, C. C., & Baskin, J. M. (1998). *Seed: Ecology, Biogeography and Evolution of Dormancy and Germination*. Academic Press.

Billings, W. D., & Mooney, H. A. (1968). The ecology of arctic and alpine plants. *Biological Reviews of the Cambridge Philosophical Society, 43*, 481–529.

Bisht, C., Badoni, A., & Vashistha, R. K. (2009). Photoperiodic effect on seed germination in pyrethrum (*Chrysanthemum cinerariaefolium* Vis.). *The Journal of American Science, 5*(4), 147–150.

Bose, B., & Sharma, M. C. S. (2000). Influence of maize seed treatment with nitrates on nitrogen economy and nitrogen pollution. *Indian Journal of Plant Physiology, 5*, 93–95.

Butola, J. S. (2011). Post-cultivation evaluation of germplasm in Himalayan threatened medicinal herbs: Implication for *ex-situ* cultivation and conservation. *National Academy Science Letters, 34*(1, 2), 49–58.

Butola, J. S. (2020). Value chain analysis of four important ABS potential bioresources in Garhwal region. Report submitted to GIZ, New Delhi, p. 90.

Butola, J. S., & Badola, H. K. (2004). Seed germination improvement using chemicals in *Heracleum candicans* Wall, a threatened medicinal herb of Himalaya. *Indian Forester, 130*(5), 565–572.

Butola, J. S., & Badola, H. K. (2006). Chemical treatments to improve seedling emergence, vigour and survival in *Heracleum candicans* Wall. (Apiaceae): A high value threatened medicinal and edible herb of Himalaya. *Journal of Plant Biology, 33*(3), 215–220.

Butola, J. S., Samant, S. S., & Vashistha, R. K., & Malik, A. R. (2010a). Propagation and cultivation techniques of *Heracleum candicans* Wall.: A Himalayan medicinal resource under peril. In *Medicinal Plants and Sustainable Development*. Nova Publishers.

Butola, J. S., Vashistha, R. K., Kuniyal, C. P., Malik, A. R., Ali, A., & Samant, S. S. (2014). Germination eco-physiology of *Angelica glauca* Edgew seeds. *European Journal of Medicinal Plants, 4*(4), 405–412.

Butola, J. S., Vashistha, R. K., Malik, A. R., & Samant, S. S. (2010b). Assessment of inter-population variation in *Heracleum candicans* Wall. with emphasis on seed characteristics and germination behavior. *Journal of Medicinal Plants Research, 4*(3), 1523–1534.

Chen, Y., Cao, Q., Li, D., Liu, H., & Zhang, D. (2019). Effects of temperature and light on seed germination of ephemeral plants in the Gurbantunggut Desert, China: Implications for vegetation restoration. *Journal of Arid Land, 11*, 916–927. https://doi.org/10.1007/s40333-019-0027-4 (accessed on 12 August 2025).

Hooker, J. D. (1879). *Flora of British India* (Vol. 2). L. Reeve and Co.

ISTA. (1976). International rules for seed testing. *Seed Science and Technology, 4*(1), 1–180.

Joshi, M., & Dhar, U. (2003). Effect of various pre-sowing treatments on seed germination of *Heracleum candicans* Wall. Ex DC.: A high value medicinal plant. *Seed Science and Technology, 31*, 737–743.

Kaul, M. K. (1989). *Himalayan Heracleum* Linn. (Hogweed)–A review. CSIR.

Khan, S. K., Karnat, N. M., & Shankar, D. (2005). India's Foundation for the Revitalization of Local Health Traditions: Pioneering *in situ* conservation strategies for medicinal plants and local cultures. *HerbalGram, 68*, 34–48.

Khidrapure, G., Lakshmana, D., Hanumanthappa, M., Badalingappanavar, R., & Chandana, B. C. (2018). Seed quality enhancement techniques in medicinal and aromatic crops: A review. In *National Conference on Conservation, Cultivation and Utilization of Medicinal and Aromatic Plants* (pp. 59–66).

Kim, H. J., Choi, S. H., & Kim, Y. J. (2016). Effects of temperature on seed germination and photoperiod on seedling growth of *Hosta yingeri* S. B. Jones. *The Horticulture Journal, 85*(3). https://www.researchgate.net/publication/286522978_ (accessed on 12 August 2025).

Mayer, A. M., & Poljakoff-Mayber, A. (1982). *The Germination of Seeds* (Vol. III). Pergamon Press.

Mehta, P., Bisht, K., & Chandra Sekar, K. (2021). Diversity of threatened medicinal plants of Indian Himalayan region. *Plant Biosystems–An International Journal, 155*(6), 1121–1132.

Moravcova, L., Perglova, I., Pysek, P., Jarosik, V., & Pergl, J. (2005). Effects of fruit position on fruit mass and seed germination in the alien species *Heracleum mantegazzianum* (Apiaceae) and the implications for its invasion. *Acta Oecologica, 28*, 1–10.

Nadeem, M., Palni, L. M. S., Purohit, A. N., Pandey, H., & Nandi, S. K. (2000). Propagation and conservation of *Podophyllum hexandrum* Royle: An important medicinal herb. *Biological Conservation, 92*, 121–129.

Nasir, E. (1972). Umbelliferae. In E. Nasir & S. I. Ali (Eds.), *Flora of West Pakistan.* Rawalpindi, Pakistan.

Nautiyal, M. C. (1996). Cultivation of medicinal plants and biosphere reserve management in alpine zone. In Ramakrishna et al. (Eds.), *Conservation and Management in Alpine Zone* (pp. 569–582). Oxford and IBH.

Nautiyal, M. C., Rawat, A. S., & Bhadula, S. K. (1985). Germination in two *Aconitum* species. *Seed Research, 14*(1), 133–139.

Nichols, M. A., & Heydecker, W. (1968). Two approaches to the study of germination data. *Proceedings of International Seed Testing Association, 33*, 531–540.

Osburn, W. S., Jr. (1961). Successional potential from differential seedling establishment in alpine tundra stand. *Bulletin of the Ecological Society of America, 42*, 146–147.

Pimenov, M. G., & Leonov, M. V. (2004). The Asian umbelliferae biodiversity database (ASIUM) with particular reference to South-West Asian taxa. *Turkish Journal of Botany, 28*, 139–145.

Roberts, E., & Smith, R. D. (1977). Dormancy and the pentose phosphate pathway. In A. Khan (Ed.), *The Physiology and Biochemistry of Seed Dormancy and Germination* (pp. 385–411). North-Holland Publishing Co.

Samant, S. S., Dhar, U., & Palni, L. M. S. (1998). *Medicinal Plants of Indian Himalaya: Diversity, Distribution, Potential Values*. HIMAVIKAS Publ. No. 13, Gyan Prakash, Nainital.

Sastry, A. R. K., & Chatterjee, S. (2000). Prioritization of medicinal plants of India. In S. Singh, A. R. K. Sastry, R. Mehta, & V. Uppal (Eds.), *Setting Biodiversity Conservation Priorities for India* (Vol. II, pp. 467–473). World Wildlife Fund.

Sayers, R. L., & Ward, R. T. (1966). Germination responses in alpine species. *Botanical Gazette, 127*, 11–16.

Snedecor, G. W., & Cochran, W. G. (1967). *Statistical Methods*. Oxford and IBH Publishing.

CHAPTER 15

Drying Temperature and Duration on Seed Germination and Storability with Particular Reference to Seeds of Medicinal Plants

ANCHAL BISHT[1,*] and GOPAL SHUKLA[2]

[1]*Institute of Wood Science and Technology, Bangalore, Karnataka, India*

[2]*Department of Forestry, North-Eastern Hill University, Tura Campus, Tura, Meghalaya, India*

*Corresponding author

ABSTRACT

The Indian subcontinent is characterized by its abundant diversity and ancient healthcare practices, where medicinal plants play a pivotal role not just in healthcare but also serve as valuable economic resources for marginalized communities. The enduring popularity of these plants is attributed to a belief in herbal products and concerns over adverse effects of allopathic medicine. Beyond healthcare, these plants play a multifaceted role, contributing to economic empowerment, cultural preservation, and environmental sustainability. Recognizing these traditions provides an opportunity for comprehensive well-being and community advancement. In the field of medicinal plants, the crucial practice of drying seeds for preservation is essential to maintain their medicinal attributes. While natural drying suffices for small quantities, large-scale production requires technical drying methods. Nevertheless, it is advisable to maintain low

Seed Technology for Medicinal and Aromatic Plants: Production, Testing, Storage, and Marketing.
Priyanka Sharma, Kishore Kumar Sharma, and Gopal Shukla (Eds.)

DOI: 10.1201/9781779641199-15

drying temperatures to safeguard active ingredients, even though this may prolong the drying process. Seed storage, vital for preserving genetic diversity, involves distinct behaviors, with some seeds categorized as orthodox and others as recalcitrant. This chapter delves into the intricacies of seed storage in medicinal plants, emphasizing the influence of factors such as duration and temperature on seed germination and storability. Understanding seed storage dynamics encompasses processes from collection to germination studies. Timely seed collection, diverse drying techniques, controlled storage conditions, and regular germination tests collectively contribute to efficient storage. The chapter highlights disparities in seed characteristics between medicinal and non-medicinal plants, underscoring the importance of comprehending species-specific requirements. It emphasizes the impact of drying temperatures on seed viability, stressing the necessity for a balanced approach to temperature and duration, given that medicinal plant seeds, enriched with bioactive compounds, often benefit from low-temperature drying to preserve their chemical composition and medicinal properties. The chapter concludes by elucidating the consequences of seed drying on storage stability, emphasizing prolonged longevity, inhibition of microbial growth, prevention of germination, and the preservation of genetic integrity. The conservation of secondary metabolites, vital for the therapeutic properties of medicinal plants, underscores the significance of employing low-temperature drying methods for optimal seed storage.

15.1 INTRODUCTION

The Indian subcontinent is renowned for its rich diversity and ancient indigenous healthcare traditions. Beyond its healthcare applications, medicinal plants also serve as an additional source of income for underprivileged communities. The continued popularity of medicinal plants can be attributed to beliefs in herbal products and concerns about the adverse side effects of allelopathic medicine. The multifaceted role of medicinal plants in the Indian subcontinent goes beyond healthcare, encompassing economic empowerment, cultural preservation, and environmental sustainability. As society increasingly recognizes the value of these ancient traditions, there is a growing opportunity to leverage the potential of medicinal plants for holistic well-being and community development.

In the realm of medicinal plants, a crucial aspect is the drying of seeds for storage, as it is one of the most common methods for preserving medicinal qualities. This method allows for the quick conservation of the plant materials in the form of medicinal attributes. For small quantities, natural drying (without auxiliary energy, either in the field or in sheds) is recommended. However, for mass production, the use of technical drying applications is essential. To preserve the active ingredients of medicinal plant materials, lower drying temperatures are recommended, albeit at the cost of a longer drying duration. Drying constitutes a significant portion, ranging from 30 to 50%, of the total costs in medicinal plant production (Qaas and Schiele, 2001). Seed drying is typically performed for the extended storage of seeds. "Seed storage" involves preserving viable seeds from the time of collection until they are needed for sowing. The purpose of seed storage is to maintain seeds in good physical and physiological condition from harvesting until planting. Factors affecting seed longevity in storage include moisture content, relative humidity, storage temperature, oxygen pressure, fluctuating environmental conditions, and seed treatment. These factors need optimization for effective seed storage and subsequent activities. This chapter explores the storage potential of medicinal plants, the effects of factors such as duration and temperature on seed germination and storability, and potential future technologies.

15.2 UNVEILING THE SEED STORAGE DYNAMICS IN MEDICINAL PLANTS

Seed storage behavior pertains to the capability of seeds to endure the processes of drying, cooling, and storage while retaining their viability and potential for germination. Seeds demonstrate diverse storage behaviors, typically categorized into two main groups: orthodox seeds and recalcitrant seeds. Comprehending the storage behavior of seeds from medicinal plants holds significant importance for seed banks, conservation initiatives, and enhancement programs. Examining the seed storage behavior of medicinal plants is essential for formulating effective storage strategies crucial for preserving plant genetic diversity.

The seed storage behavior of medicinal plants can vary widely among species. Some medicinal plant seeds exhibit orthodox storage behavior, allowing them to withstand desiccation and be suitable for conventional storage methods. On the other hand, some may display recalcitrant or intermediate storage behavior, necessitating specific conditions to maintain their viability. Orthodox seeds can endure drying to extremely low moisture content without losing viability, making them tolerant to water removal during storage. They are often stored in airtight containers at low temperatures to prevent deterioration. Examples of plants with orthodox seeds include common crops and certain trees.

Intermediate seeds, however, have a limited ability to withstand desiccation and cannot tolerate as much drying as orthodox seeds. They are typically stored under cool and moderately moist conditions, requiring careful temperature and humidity management to prevent dehydration or excess moisture. Non-orthodox seeds cannot be successfully stored for long-term using conventional genebank protocols, involving drying and storage at low temperatures. Instead, cryopreservation, usually in liquid nitrogen at –196°C, is recommended. Advances in cryopreservation technology have reached a point where it should become a routine practice for the *ex situ* conservation of non-orthodox species (Hay and Sershen, 2021). The seed storage behavior of selected medicinal plants is summarized in Table 15.1.

Considering the widely recognized importance of medicinal plants in both traditional and modern medicine, the storage of seeds becomes crucial for safeguarding their genetic diversity. The intricacies of seed storage involve various processes, each exerting an influence on the viability and longevity of the seeds (Figure 15.1).

15.3 SEED COLLECTION

During seed collection, it is crucial to consider how and when the harvesting occurs. The techniques and timing employed in the harvesting process have a direct influence on the dynamics of seed storage. Employing gentle harvesting methods that prevent damage to seeds is essential for preserving their structural integrity and viability (Dickie et al., 1990). Careful handling at this stage is necessary to prevent premature deterioration. Timing plays a critical role in achieving optimal seed quality. The collection time is contingent on the seed maturation period, a pivotal stage where seeds reach their maximum quality and viability. Properly timed seed collection during this phase is vital to ensure that the seeds possess

TABLE 15.1 Seed Storage Behavior of Some Medicinal Plants

Scientific Name	Family	Storage Behavior	Temperature Range (°C)	Moisture Content Range	Reference
Withania somnifera	Solanaceae	Orthodox	5–10	Low moisture content, cool conditions	Tiwari et al. (2022)
Abelmoschus moschatus	Malvaceae				
Andrographis paniculata	Acanthaceae				
Bixa orellana	Bixaceae				
Ocimum basilicum	Lamiaceae				
Plumbago zeylanica	Plumbaginaceae				
Psoralea corylifolia	Fabaceae				
Azadirachta indica	Meliaceae		0–5		Kumar (2015)
Terminalia arjuna	Combretaceae		0–5		Kundu and Tiwari (2019)
Abrus precatorius	Fabaceae		20–24		Das (2014)
Costus speciosus	Costaceae				
Desmodium motorium	Fabaceae				
Coleus forskohlii	Lamiaceae				
Centratherum anthelminticum	Asteraceae				
Saraca asoca	Fabaceae	Recalcitrant	15	Shedding moisture content	Kundu et al. (2020)
Rauvolfia serpentina	Apocynaceae	Orthodox	0–5	Dry storage, cool conditions	Choudhary and Chauhan (2015)
Salacia gambleana	Celastraceae	Recalcitrant	–	–	Anaz et al. (2019)
Swertia chirayita	Gentianaceae	Orthodox	4	Dry storage, cool conditions	Pradhan and Badola (2012)
Embelia tsjeriam	Primulaceae	Orthodox	5	Dry storage, cool conditions	Chauhan and Chauhan (2020)

TABLE 15.1 *(Continued)*

Scientific Name	Family	Storage Behavior	Temperature Range (°C)	Moisture Content Range	Reference
Mesua ferrea	Calophyllaceae	Recalcitrant	10	Air tight container, dry	Mithun et al. (2021)
Holostemma annulare	Apocynaceae		–196	cryopreservation	Gupta et al. (2022)
Dioscorea bulbifera	Dioscoreaceae				
Emblica officinalis	Phyllanthaceae	Orthodox	0–5	Dry storage, cool conditions	
Asparagus racemosus	Asparagaceae	Orthodox	0–4	Dry storage, cool conditions	Bittencourt et al. (2005)
Berberis aristata	Berberidaceae	Orthodox	5–7	Dry storage, cool conditions	Mukhopadhyay et al. (2013)
Terminalia chebula	Combretaceae	Orthodox	25–30	Dry storage, room temperature	Chauhan et al. (2002)

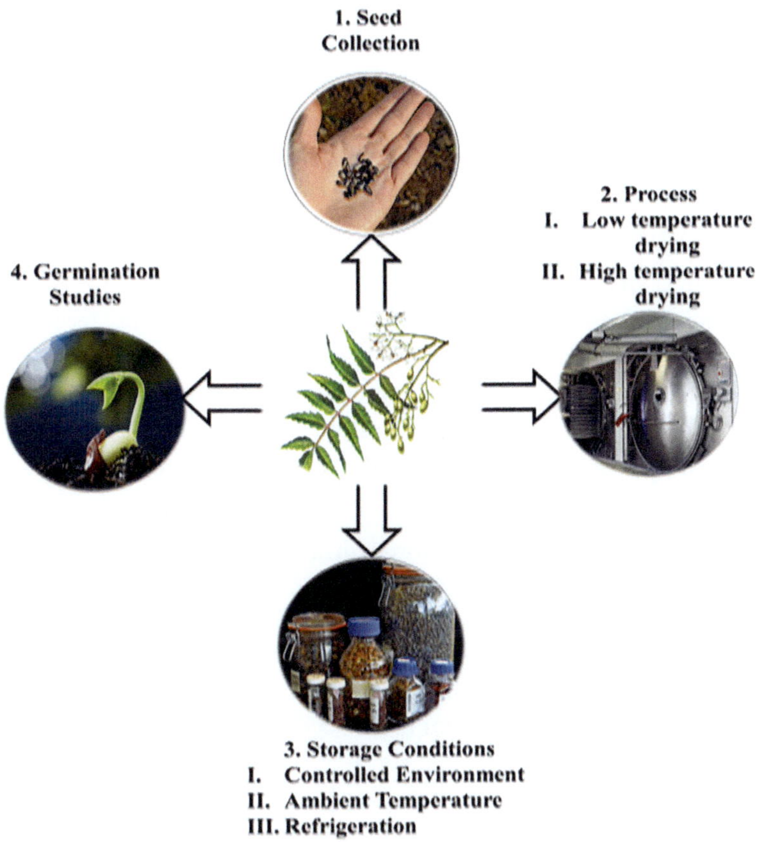

FIGURE 15.1 Seed storage dynamics in medicinal plants.

optimal attributes for storage and subsequent germination (Bewley and Black, 1994). The timing of seed collection significantly affects the success of storage protocols.

15.4 DRYING METHODS

The process of drying plays a pivotal role in the dynamics of seed storage, affecting moisture content and, consequently, the longevity of seeds. Various drying methods, including air drying and low-temperature drying, have diverse impacts on seed quality (Copeland and McDonald, 2012). Employing suitable drying techniques helps decrease microbial activity and sustains seed viability. The proper execution of drying processes is essential for the effective storage of seeds from medicinal plants. By adhering to these procedures, researchers and seed conservators can prolong the longevity and viability of medicinal plant seeds, safeguarding their genetic diversity and medicinal properties for future utilization.

15.5 STORAGE CONDITIONS

The duration of seed viability is significantly influenced by the storage environment. It is crucial to control storage conditions, including the regulation of temperature and humidity, to preserve seed quality (Nagel and Börner, 2010). Creating storage conditions that replicate the natural habitat of seeds can enhance their ability to endure prolonged storage periods. When storing medicinal plant seeds, certain considerations should be kept in mind:

- Ensure that seeds are stored in a clean and dry environment to prevent the growth of fungi and bacteria.
- Choose storage areas that are not prone to high humidity.
- Maintain a cool storage temperature to slow down metabolic processes and minimize the risk of seed aging; optimal temperatures for most medicinal plant seeds typically range between 5°C to 15°C.
- Preferentially opt for dark storage for seeds, as exposure to light can trigger premature germination and compromise seed quality.
- Store seeds in airtight containers to prevent moisture from entering and affecting seed viability.
- Avoid storage areas with fluctuating temperatures, as rapid changes can lead to condensation and increase the risk of fungal contamination.
- Conduct regular germination tests to evaluate seed viability.

15.6 GERMINATION STUDIES

Examining the germination process provides a deeper understanding of how storage conditions impact seed vigor and overall quality. The observation and analysis of germination rates under various storage scenarios yield valuable insights, aiding in the refinement of storage strategies (Dagnachew et al., 2023).

In summary, a comprehensive approach to understanding the dynamics of seed storage in medicinal plants necessitates careful consideration of seed maturation, harvesting methods, drying procedures, storage conditions, viability assessment, and germination studies. By addressing these factors and taking into account the influence of environmental variables, researchers and conservationists can formulate effective strategies to preserve the genetic diversity of medicinal plants. This approach ensures the continued contribution of these plants to human health and well-being.

15.7 ENHANCING GERMINATION: IMPACT OF DRYING TEMPERATURE ON SEED VIABILITY

Understanding the drying temperature and duration for seed germination is crucial, especially in conservation programs (Bewley and Black, 1994). The drying temperature notably affects physiological and biochemical processes within seeds, and high temperatures can adversely impact seed viability and germination potential (McDonald and Copeland, 1997). Another critical aspect of storage conditions is the duration of drying, determining the exposure time of seeds to drying conditions (Ellis and Roberts, 1981). Prolonged drying can lead to over-drying, affecting seed germination negatively (Vertucci and Roos, 1990). Hence, finding the optimal balance of temperature and duration is essential for preserving seed quality and germination potential during storage (Nadarajan et al., 2023).

Species exhibit varied responses to drying conditions, emphasizing the importance of understanding species-specific requirements (Baskin and Baskin, 2014). Several factors influence seed drying conditions, and comprehending these factors is crucial for maintaining seed quality during storage. Factors like initial moisture content, air temperature, relative humidity, ventilation, airflow, seed species and variety, seed coat characteristics, and even seed size and structure can impact the drying rate and overall effectiveness of the drying process (Bradford, 2002; Copeland and McDonald, 2012; ISTA, 2020).

The drying process, particularly at high temperatures, can damage cellular structures, enzymes, and proteins within seeds, which are sensitive to temperature. High-temperature drying may deteriorate seed viability and germination rates, especially for medicinal plant seeds rich in bioactive compounds. Low drying temperatures are generally preferred to minimize damage, as extremely low temperatures can slow down the process and damage cell structures, leading to membrane damage, metabolic changes, and oxidative stress.

Drying duration is another critical aspect affecting germination in the seed storage process. Inadequate drying can result in higher moisture content, making seeds susceptible to fungal growth and premature germination during storage (Chala and Bekana, 2017). Prolonged exposure to over-drying, particularly at higher temperatures, can reduce seed viability and germination rates (Huang et al., 2021).

The impact of drying temperature on seed germination, especially concerning medicinal plants, is vital for seed preservation. Optimal drying conditions can vary among different medicinal plant species. Medicinal plant seeds often benefit from low-temperature drying to minimize potential damage to delicate structures and bioactive compounds. Low temperatures are less likely to cause denaturation of proteins and degradation of secondary metabolites. The characteristics of medicinal plant seeds, including their bioactive compounds such as alkaloids, flavonoids, and essential oils, make low-temperature drying preferable to preserve their chemical composition and medicinal properties.

Lastly, the moisture content of seeds is a crucial factor affecting germination. Low-temperature drying may require a longer duration to achieve the desired moisture content without compromising seed viability. While there may not be distinct and universal differences between the seeds of medicinal plants and non-medicinal plants, certain characteristics may be more emphasized or specific to medicinal plant seeds (Figure 15.2).

- **Bioactive Compounds:** The seeds of medicinal plants frequently possess elevated levels of bioactive compounds, including alkaloids, flavonoids, and essential oils, contributing to their therapeutic qualities. In contrast, seeds of non-medicinal plants may also contain bioactive compounds, but these may be less specifically tailored for medicinal purposes (Duke et al., 2002).
- **Secondary Metabolites Diversity:** Seeds from medicinal plants may display a greater variety of secondary metabolites, which frequently play a role in the plant's pharmacological effects. In contrast, non-medicinal plant seeds may contain secondary metabolites, but the diversity and concentration might be less prominent (Wink, 2010).

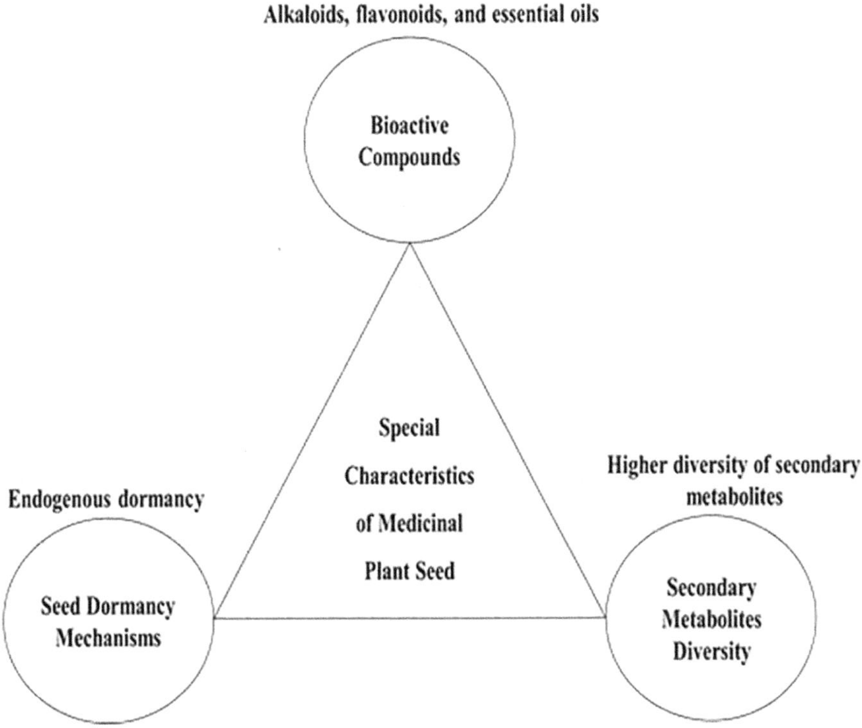

FIGURE 15.2 Some specialized characters of seeds of medicinal plants.

- **Seed Dormancy Mechanisms:** The seeds of medicinal plants may showcase particular dormancy mechanisms that contribute to synchronizing germination with environmental conditions favorable for plant growth. Due to the elevated concentrations of bioactive and secondary metabolites in medicinal plants, they might have inherent dormancy (Baskin and Baskin, 2014).

15.8 PRESERVING POTENTIAL: UNDERSTANDING THE IMPACT OF DRYING ON SEED STORAGE STABILITY

An essential component of seed preservation methods is seed drying, a prerequisite for seed storage. Significantly, the moisture content of seeds is pivotal in determining their longevity, and the drying process is employed to decrease this moisture content. The significant impacts of seed drying on storage are illustrated in Figure 15.3.

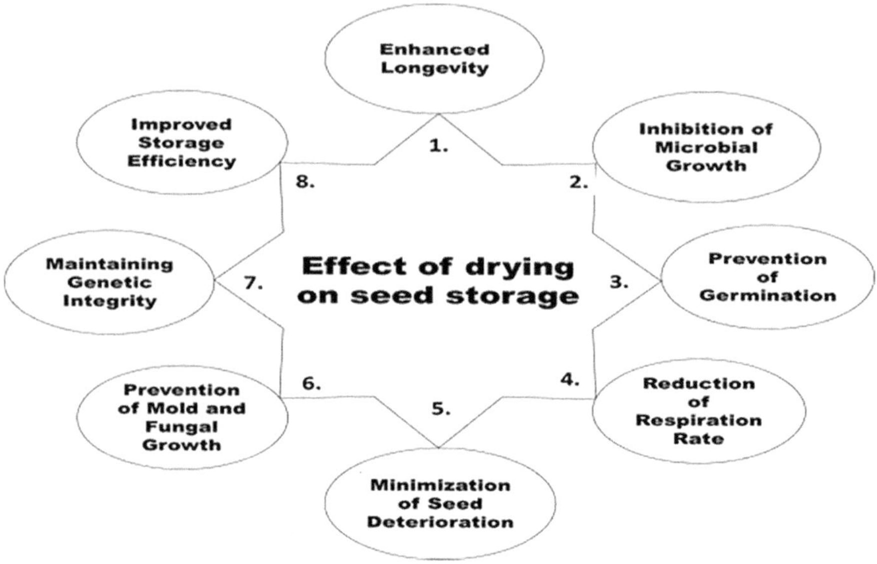

FIGURE 15.3 Effect of drying on seed storage.

- **Enhanced Longevity:** The longevity of seeds is influenced by their moisture content. Drying aids in lowering the moisture levels, thereby inhibiting germination and microbial growth throughout the storage period. Studies suggest that lower moisture levels are linked to prolonged seed lifespan.
- **Inhibition of Microbial Growth:** Significantly, microorganisms like fungi and bacteria rely on moisture for their proliferation, leading to the degradation of seeds. Drying seeds diminishes the moisture content, creating an environment less favorable for microbial growth and contamination (Martín et al., 2019).
- **Prevention of Germination:** While seed germination is a natural process that can deplete stored energy reserves, drying seeds to adequately low moisture content hinders germination, ensuring that seeds maintain a dormant state during storage (Saha et al., 2022).

- **Reduction of Respiration Rate:** Given that the metabolic activities of seeds, such as the respiration rate, are contingent on optimal moisture content, the drying process reduces these metabolic activities. Crucially, a high respiration rate necessitates elevated energy reserves, leading to the consumption of significant stored materials. Lower respiration rates play a role in preserving stored energy reserves and contribute to maintaining seed quality over time (El-Maarouf-Bouteau, 2022).
- **Minimization of Seed Deterioration:** Crucially, in optimal moisture conditions, there is an increase in enzyme activity and oxidation rates, contributing to seed deterioration. These physiological activities are often slowed down or inhibited at lower moisture levels. The drying process plays a crucial role in minimizing various forms of seed deterioration, including enzymatic activity and oxidation (Rao et al., 2023).
- **Prevention of Mold and Fungal Growth:** Elevated moisture content creates favorable conditions for fungi, molds, and other pathogens, leading to a loss of seed viability. Drying prevents the growth of these microorganisms, thereby decreasing the risk of fungal infections during storage (Martin et al., 2019).
- **Maintaining Genetic Integrity:** The primary method employed for preserving plant genetic resources (PGR) is seed banking through the establishment of a gene bank. Since the majority of seeds tend to lose viability quickly, appropriate drying is essential to uphold the genetic integrity of seeds. This is vital for safeguarding the desirable traits and characteristics of plant varieties over prolonged storage periods (Panis et al., 2020).
- **Improved Storage Efficiency:** Dried seeds are generally lighter and occupy less space compared to seeds with higher moisture content. This makes storage and transportation more efficient, as less space and energy are required. Apart from this, drying is often a prelude to storing seeds in cold storage conditions. Low-temperature storage is effective in further slowing down metabolic processes and extending the viability of seeds (Martins et al., 2019).

In the case of medicinal plant seeds containing secondary metabolites like alkaloids, the presence and characteristics of these compounds can vary among plant species. Preserving the viability of seeds while maintaining secondary metabolites is crucial, as these compounds, including alkaloids, often play roles in diverse biological functions within plants, such as defense mechanisms against herbivores and pathogens. The impact of drying on secondary metabolites depends on the sensitivity of these compounds to heat and other factors during the drying process. Some secondary metabolites remain stable during drying, while others may degrade or undergo chemical changes.

For the preservation of seed quality and biochemical constituents, low-temperature drying methods, such as freeze-drying, are preferable, as high temperatures may lead to the degradation of specific secondary metabolites. Additionally, storing seeds in cool, dark, and dry conditions minimizes the risk of deterioration and loss of secondary metabolites.

15.9 CONCLUSION

This chapter emphasizes the need for a comprehensive approach to effective seed storage dynamics in medicinal plants. Meticulous attention to seed maturation, harvesting techniques, drying methods, storage conditions, viability assessment, and germination studies is essential. By addressing these aspects, researchers and conservationists can develop strategies ensuring the ongoing contribution of medicinal plants to human health and well-being.

KEYWORDS

- **genetic integrity**
- **medicinal plants**
- **secondary metabolites**
- **seed germination**
- **seed maturation**
- **seed viability**

REFERENCES

Anaz, M., Sasidharan, N., Dilsha, M. V., & Deepa, K. (2019). Seed storage and germination of *Whiting & Kaul* (Celastraceae), an endemic and endangered medicinal plant of Western Ghats. *Journal of Horticulture, Forestry and Biotechnology, 23*(2), 8–13.

Baskin, J. M., & Baskin, C. C. (2014). *Seeds: Ecology, Biogeography, and Evolution of Dormancy and Germination.* Academic Press.

Bewley, J. D., & Black, M. (1994). *Seeds: Physiology of Development and Germination.* Springer Science & Business Media.

Bittencourt, M. L. D. C., Dias, D. C. F. D. S., Dias, L. A. D. S., & Araújo, E. F. (2005). Germination and vigor of primed asparagus seeds. *Scientia Agricola, 62*, 319–324.

Bradford, K. J. (2002). Applications of hydrothermal time to quantifying and modeling seed germination and dormancy. *Weed Science, 50*(2), 248–260.

Chala, M., & Bekana, G. (2017). Review on seed process and storage condition in relation to seed moisture and ecological factor. *Journal of Natural Sciences Research, 7*, 84–90.

Chauhan, R., & Chauhan, J. S. (2020). Seed storage behavior of *Embelia tsjeriam-cottam* (An important medicinal plant of India). *Journal of Pharmacognosy and Phytochemistry, 9*(2), 761–765.

Chauhan, S., Bhatt, B. P., & Todaria, N. P. (2002). Germination behavior of *Terminalia* species as influenced by temperature, light and seed maturity in Garhwal Himalaya. *Journal of Plant Biology, 29*(2), 203–208.

Choudhary, T., & Chauhan, J. S. (2015). Seed storage behavior in *Rauvolfia serpentina. International Journal of Latest Technology in Engineering, Management & Applied Science, 4*(12).

Copeland, L. O., & McDonald, M. B. (2001). *Principles of Seed Science and Technology.* Kluwer Academic Publishers.

Copeland, L. O., & McDonald, M. B. (2012). *Principles of Seed Science and Technology.* Springer Science & Business Media.

Dagnachew, S., Teketay, D., Demissew, S., & Awas, T. (2023). Evaluation of germination under different storage conditions of four endemic plant species from Ethiopia: Implications for *ex situ* conservation in seed banks. *Seeds, 2*(1), 45–59.

Dickie, J. B., Ellis, R. H., Kraak, H. L., Ryder, K., & Tompsett, P. B. (1990). Temperature and seed storage longevity. *Annals of Botany, 65*(2), 197–204.

Duke, J. A. (2002). *Handbook of Medicinal Herbs.* CRC Press.

Duke, J. A., Bogenschutz-Godwin, M. J., duCellier, J., & Duke, P. A. K. (2002). *Handbook of Medicinal Herbs* (2nd ed.). CRC Press.

Ellis, R. H., & Roberts, E. H. (1981). The quantification of ageing and survival in orthodox seeds. *Seed Science and Technology, 9*(2).

El-Maarouf-Bouteau, H. (2022). The seed and the metabolism regulation. *Biology, 11*(2), 168. https://doi.org/10.3390/biology11020168 (accessed on 12 August 2025).

Gupta, R., Nigam, A., & Kapila, R. (2022). Cultivation and conservation of underutilized medicinal and agricultural plants in India. *Proceedings of the National Academy of Sciences, India Section B: Biological Sciences, 92*(4), 741–745.

Hay, F. R., & Sershen, D. (2021). New technologies to improve the *ex situ* conservation of plant genetic resources.

Huang, Y., Lu, M., Wu, H., Zhao, T., Wu, P., & Cao, D. (2021). High drying temperature accelerates sunflower seed deterioration by regulating the fatty acid metabolism, glycometabolism, and abscisic acid/gibberellin balance. *Frontiers in Plant Science, 12*, 628251. https://doi.org/10.3389/fpls.2021.628251 (accessed on 12 August 2025).

International Seed Testing Association. (2020). International Seed Testing Association: Tetrazolium testing handbook. *Seed Science and Technology, 48*(3), 337–381.

Kumar, D. (2015). Storage response of neem (*Azadirachta indica* A. Juss.) seed under different moisture and temperature regime. *Global Journal of Science Frontier Research: C Biological Science, 15*(6), 1–18.

Kundu, M., & Tiwari, S. (2019). Germination and *ex-situ* conservation of *Terminalia arjuna* (Roxb.) seeds. *International Journal of Forestry and Crop Improvement, 10*(1), 9–13.

Kundu, M., Tiwari, S., & Haldkar, M. (2020). Collection, germination and storage of seeds of *Saraca asoca* (Roxb.) Willd. *Journal of Applied Research on Medicinal and Aromatic Plants, 16*, 100231. https://doi.org/10.1016/j.jarmap.2020.100231 (accessed on 12 August 2025).

Kuntal, D. (2014). Impact of differential storage conditions on seed germination and viability of some medicinal plants. *African Journal of Agricultural Research, 9*(20), 1578–1585.

Martins, A. B. N., Carvalho, I. R., Costa, C. J., Marques, F. S., Marques, R. L. L., Xavier, F. D. M., Cavalcante, J. A., Vergara, R. D. O., Teixeira, S. B., Bohn, A., & Tunnes, L. V. M. D. (2019). Storage of amaranth seeds: Reflex in physiologic potential. *Journal of Agricultural Science, 11*(12), 79.

McDonald, M. B., & Copeland, L. O. (1997). *Seed Production: Principles and Practices*. Chapman & Hall.

Mithun, V., Pradeep, N. S., & Krishnan, P. N. (2021). Seed storage studies in *Mesua ferrea* L. a medicinal tree of Indo-Malayan region. *Plant Science Today, 8*(3), 712–716.

Mukhopadhyay, D., Chauhan, J. C., Parihar, S. S., & Dadlani, M. (2013). Seed storage behavior in *Berberis aristata. The Indian Journal of Agricultural Sciences, 83*(3).

Nadarajan, J., Walters, C., Pritchard, H. W., Ballesteros, D., & Colville, L. (2023). Seed longevity—The evolution of knowledge and a conceptual framework. *Plants, 12*(3), 471. https://doi.org/10.3390/plants12030471 (accessed on 12 August 2025).

Nagel, M., & Börner, A. (2010). The longevity of crop seeds stored under ambient conditions. *Seed Science Research, 20*(1), 1–12.

Panis, B., Nagel, M., & Van den Houwe, I. (2020). Challenges and prospects for the conservation of crop genetic resources in field genebanks, in *in vitro* collections and/or in liquid nitrogen. *Plants, 9*(12), 1634. https://doi.org/10.3390/plants9121634 (accessed on 12 August 2025).

Pradhan, B. K., & Badola, H. K. (2012). Effect of storage conditions and storage periods on seed germination in eleven populations of *Swertia chirayita*: A critically endangered medicinal herb in Himalaya. *The Scientific World Journal*. https://doi.org/10.1100/2012/128105 (accessed on 12 August 2025).

Qaas, F., & Schiele, E. (2001). Einfluss der Energiekosten auf die Rentabilität im Trocknungsbetrieb. *Zeitschrift für Arznei- und Gewürzpflanzen, 6*(3), 144–145.

Rao, P. J. M., Pallavi, M., Bharathi, Y., Priya, P. B., Sujatha, P., & Prabhavathi, K. (2023). Insights into mechanisms of seed longevity in soybean: A review. *Frontiers in Plant Science, 14*. https://doi.org/10.3389/fpls.2023.1143216 (accessed on 12 August 2025).

Saha, D., Choyal, P., Mishra, U. N., Dey, P., Bose, B., Prathibha, M. D., Gupta, N. K., Mehta, B. K., Kumar, P., Pandey, S., & Chauhan, J. (2022). Drought stress responses and inducing tolerance by seed priming approach in plants. *Plant Stress, 4*, 100066. https://doi.org/10.1016/j.stress.2022.100066 (accessed on 12 August 2025).

Tiwari, R. S., Chandra, K. K., Dubey, S., & Tripathi, S. (2022). Influence of packaging materials and storage conditions on seed germination ability and biochemical changes in some medicinal plants of Indian forests. *Frontiers in Forests and Global Change, 5*, 868237. https://doi.org/10.3389/ffgc.2022.868237 (accessed on 12 August 2025).

Vertucci, C. W., & Roos, E. E. (1990). Theoretical basis of protocols for seed storage. *Plant Physiology, 94*(3), 1019–1023.

Wink, M. (2010). Introduction: Biochemistry, physiology, and ecological functions of secondary metabolites. In *Annual Plant Reviews: Biochemistry of Plant Secondary Metabolism* (Vol. 40, pp. 1–19). Wiley.

CHAPTER 16

Detection of Seed Storage Fungi, Bacteria, and Viruses and Identification of Storage Fungi

KODIDHALA VAISHNAVI,[1] HEMANTH KUMAR MANNE,[2,*] KUMARI PUNAM SINHA,[3] PARTHA MONDAL,[4] NISHA KUMARI,[2] and SUNAYANA RATHI[5]

[1]*Department of Plant Pathology, Uttar Banga Krishi Viswavidyalaya, Cooch Behar, West Bengal, India*

[2]*Department of Biochemistry, Chaudhary Charan Singh Haryana Agricultural University, Hisar, Haryana, India*

[3]*Department of Seed Science and Technology, Uttar Banga Krishi Viswavidyalaya, Cooch Behar, West Bengal, India*

[4]*Department of Agricultural Biochemistry, Bidhan Chandra Krishi Viswavidyalaya, Mohanpur, West Bengal, India*

[5]*Department of Biochemistry and Agricultural Chemistry, Assam Agricultural University, Jorhat, Assam, India*

*Corresponding author

ABSTRACT

Plant infections can be effectively introduced into new areas and given a method of survival from one farming season to the next by using seeds, which are miniature embryonic plants. One of the major biotic barriers to seed production globally is the presence of fungi that are carried by seeds. In addition to producing mycotoxins, certain fungus related to seeds have

Seed Technology for Medicinal and Aromatic Plants: Production, Testing, Storage, and Marketing.
Priyanka Sharma, Kishore Kumar Sharma, and Gopal Shukla (Eds.)

DOI: 10.1201/9781779641199-16

the potential to seriously harm people, animals, and even the seeds themselves. Fungi that infect seeds either before or after harvest have the potential to significantly degrade both seed quality and seed lifetime. Diseases can spread to new places through the movement of seeds. The genetic potential of a cultivar is safeguarded by seed, which also establishes the production thresholds for each cropping system. In today's agricultural science, a healthy seed bank is essential for a desired plant population and a healthy yield. Thus, in recent years, producers, dealers, and consumers have become more conscious of the importance of seed health. The most effective and popular technique for preserving plant genetic resources (PGR) is seed storage. Low temperature and low moisture content seed storage, which are utilized in gene banks, lengthen seed shelf life and are often conducive to the survival of seed-borne mycoflora. A seed-borne inoculum ensures that a virulent strain of the pathogen is present with the seed and promotes the earliest infection of seedlings. It also raises the possibility of the introduction of new physiological races or stains with the seed. An essential step in managing crop diseases is seed health testing, which looks for infections that are transmitted through seeds. To preserve premium seeds and stop the spread of pathogens, it is crucial to identify and diagnose seed fungus early on. The degree of pathogen-free seed is a measure of seed health. Standardized seed health test methods are published by three main organizations: ISHI, NSHS, and ISTA. The primary criteria for choosing test methods for seed health are specificity, sensitivity, speed, simplicity, cost-effectiveness, and reliability. Polymerase chain reaction (PCR) is very useful for detecting seed-borne diseases because of its many advantageous features. It is essential to examine seed before utilizing it as planting material because seeds allow plant infections to spread and survive. The initial step in treating plant illnesses that are borne from seeds is to test for and detect seed health issues.

16.1 INTRODUCTION

A seed is a miniature plant that is still in the embryonic stage and is covered in a layer known as the seed coat. The most crucial ingredient for producing crops is seed. It is the byproduct of the mature ovule, which develops in both gymnosperm and angiosperm plants following fertilization and some internal growth. The fundamental unit of production for the world's food crops is the seed. In recent times, seeds have evolved into a global commodity that is traded for germplasm. Since the Neolithic era, seeds have served as the foundation for agriculture and food production. The majority of crops are grown from seeds, and the primary source of raw materials for both human and animal feed is cereal and legume grains. But seed can also be a useful tool for spreading plant diseases to new regions and ensuring their survival from one cropping season to the next (Walcott, 2003). Many different microorganisms, particularly fungi, can be found in seeds and pose a serious threat to agricultural growth, human and animal health, and the sanitary conditions of the seeds themselves. The primary detrimental impacts of fungi associated with seeds include the spread and transmission of plant diseases, a decrease in the longevity and quality of seeds, and the generation of mycotoxins, which when ingested by humans or animals can be extremely toxic. In the right circumstances, seeds can act as passive carriers of diseases, which can then spread to developing plants. Since seed is the guardian of a cultivar's genetic potential, its quality dictates the maximum productivity that may be achieved in a particular cropping system.

Any microorganism carried with, on, or inside the seed is referred to as seed-borne. The act of infecting seedlings with seed-borne inoculums is referred to as "seed-transmission." The earliest known reports of seed-borne illnesses are from the mid-1800s, when bunt and smut diseases of cereals were described. It was discovered that the fungus causing bean anthracnose is transmitted by seeds a century later (Maude, 1996). Many new disease outbreaks at the start of the 20th century were most likely caused by the expansion of the seed industry and the global seed market (Elmer, 2001). Many instances of harmful infections spread by seeds were documented in the years that followed (Neergaard, 1977). The risk of seed-borne diseases spreading into new areas has grown in the last few decades due to the growing international exchange of plant germplasm for breeding program and research objectives, however to a lesser level (Kumar et al., 2021; Plucknett, 1989). The germination of seeds kept for plant reproduction or genetic preservation may be significantly reduced by a variety of seed-borne fungi (Magan et al., 2004). It has long been recognized that certain fungi can lower seed quality when they infect seeds in the field prior to harvest. Nevertheless, the consequences of additional fungi, referred to as "storage fungi," which mostly infiltrate seeds following harvest, were not documented until the 1940s (Christensen and Meronuck, 1986).

According to Rahman et al. (2008), seed health is a well-known component in contemporary agricultural science that determines the desired plant population and successful harvest. The production and sale of organic seed have increased recently, which has increased scrutiny of organic seed quality and raised awareness of concerns about contamination from seed-borne diseases. Many plant diseases spread by seeds, which can result in significant losses to crops. Pathogens like bacteria, fungi, or viruses that reside on the outside or inside of seeds and have the ability to infect following crops are known as seedborne diseases. The most significant of these is fungus contamination (Colley, 2009). There is a good chance that fungal infection has both direct and indirect effects on the quality. The development and spread of the fungus throughout the kernel, as well as the generation of metabolites that may change the composition or metabolism of the grain or make it unfit for human or animal feeding, may have direct consequences on grain quality. Reduced yields linked to pollution are referred to as indirect consequences. A significant portion of the spread of economically significant plant diseases in the field that significantly lower crop output is caused by contaminated seeds. Additionally, poor germination and field establishment are caused by decreased seed quality in infected seed, which lowers its market value (Akranuchat et al., 2007). Seed fungus can cause losses in the stages of seed growth, storage, or germination. Loss of seed viability or infection of the seedlings after germination causes damage. The incidence of seed-borne fungus appears to be on the rise in general when seed-collecting techniques such as letting the seed fall on nets or the ground before harvesting are used. According to Gitaitis and Walcott (2007), seedborne infections are still a problem and may even be to blame for the spread of diseases into new regions and the resurgence of old illnesses. Seedborne infections pose a significant risk to the establishment of seedlings (Walcott, 2003). More than ever in the global economy of today, plant diseases spread by seed across great distances, natural barriers, and political borders (Gitaitis and Walcott, 2007). Crops' capacity to establish themselves and reach their maximum yield and value potential is significantly impacted by the quality of the seeds that are sown (McGee, 1995). Fungi that spread through seeds are one of the major biotic barriers to seed production in the world. They cause changes in plant shape, pre- and post-emergence death of grains, and vigor of seedlings, which in turn reduces germination to some extent (Niaz and Dawar, 2009; Rajput et al., 2005). The introduction of new strains or physiologic races of the pathogen along with new germplasm from other nations, loss of germination, discoloration and shriveling, development of plant diseases, and toxin production in infected seed are all possible outcomes of seedborne pathogens (Agarwal and Graur, 1995). Due to their capacity to produce diseases in cultivated crops that result in significant yield losses, fungi exceed all other types of pathogens that attack plants and have a very major negative economic impact on agricultural production (Paplomatas, 2006).

Attacks can also occur when handling seeds during extraction. Cone-processing techniques that produce high cone temperatures and moisture content frequently lead to an increase in seed fungus. In addition to harming seeds, they create an environment that encourages the growth of seed fungal, allowing the fungus to infiltrate the seed. Inadequate maintenance of temperatures and seed moisture levels might lead to the accumulation of fungi during storage. According to Mehrotra and Aggarwel (2003), cool, damp conditions throughout the nursery's planting process appear to favor these fungi and the damage they do later on. In order to eradicate diseases, conventional seeds are frequently treated or covered with chemical fungicides. Seed-borne illnesses are not common in all crops, though, and certain diseases that are known to be seed-borne and that provide a risk in a particular producing zone should receive the majority of attention. If high-quality seed is utilized, using organic seed does not come with any greater risks than using conventional seed (Colley, 2009). A seed-borne illness's risk of infection varies greatly depending on the crop, the disease, and the area. Many diseases only become an issue if they are grown in an area or environment that supports the sickness. The eventual fate of the crop may depend as much on crop management techniques and variety resistance as it does on the presence of seed-borne inoculums and common illnesses that are present on seed. These diseases can also be soil- or air-borne (Colley, 2009).

In addition, low percentages of seed infection can cause significant crop losses (Pellegrino et al., 2010), especially if the pathogens are quarantine bacteria. Infected seeds, unlike infected vegetative plant tissues, are typically symptomless. However, it is challenging to screen seeds for diseases because of low inoculum levels and the frequently uneven distribution within seed batches. Several methods have been used to create the detection strategies used for fungal seed pathogens. Few of these, meanwhile, are able to meet the minimal standards needed for sufficient seed testing.

Since serological and molecular tools have become more widely available, efforts have been made to develop assays that meet both technical and financial requirements. Examples of these demands include short diagnosis times, high throughput, minimal taxonomic expertise, and low costs (Lievens and Thomma, 2005; Walcott, 2003). Given that timely and precise identification of these organisms is necessary to comply with phytosanitary regulations and, in turn, to effectively control disease, this chapter critically examines and summarizes the conventional and novel diagnostic methods that have been used to identify seedborne fungal pathogens.

Growers, dealers, and consumers have been more conscious of seed health in recent times, particularly in the aftermath of the GATT era. With the establishment of the WTO, seed health has become a worldwide concern. Seed pathologists have been prompted to take notice of many instances of seed-borne illnesses, such as *Tilletia indica*, which causes karnal bunt, a type of wheat that is unfit for human consumption because it contains a tri-methylamine chemical that causes noticeable discoloration, an unpleasant fishy odor, and incapacity to be consumed. The mycotoxin ergotin, which is present in ergot disease of pearl millet, sorghum, and triticale produced by *Claviceps* spp., renders the plant poisonous to eat for both humans and animals. Early in 1956, an epidemic of ergot illness in pearl millet and sorghum emerged in western areas of India. It has been determined that the Marathawada region of Maharashtra is endemic for the sunflower downy mildew disease (Upadhyay and Singh, 2019). In Punjab, Haryana, and the surrounding areas of Uttar Pradesh, wheat bunt (also known as Karnal bunt) and rice bunt are currently the biggest concerns. Potato scabs manifest in a severe form in the Indo-Gangetic region. The ear cockle is native to several areas of Rajasthan and Madhya Pradesh's northwest. The outbreaks caused many to reconsider using clean seeds.

A successful food crop production requires healthy and high-quality seeds, which are a crucial component of the crop. Seed-borne infections have the potential to cause up to 90% of production loss, increase morphological variety in plants, and reduce germination (Gebeyaw, 2020). Consequently, in order to preserve seed health, various scientific assessments and interventions are crucial. Plant pathogen detection and identification, together with the availability of quick, sensitive, and precise techniques to enhance disease control decision-making, are some of the most crucial tactics for sustainable plant disease management (Belete and Boyraz, 2018). As important as seed vigor, purity, and germination capacity, one aspect of seed quality that cannot be overlooked is seed health (Thomson, 1970). A pathogen-free seed is one that is considered healthy. The presence or lack of illness resulting from bacteria, viruses, nematodes, fungus, or insects in a seed lot is a sign of the health of the seed. A contaminated seed lot may bring new diseases or pests to other areas, and pathogen development in the field may lower the crop's market value. Healthy seed is crucial to agricultural productivity (Misra et al., 1994). Additionally, seed infection poses a risk, particularly when it comes to highly pathogenic or quickly spreading pathogens in agricultural settings (Neergaard, 1977). In order to ensure that the seeds are free from infection and pathogenic contamination, it is crucial to assess the health quality of the seeds.

Since infected seeds frequently show no symptoms, early detection of seedborne fungal pathogens is crucial to preventing unchecked pathogen spread through long-distance interchange of such material. This will lower expenses and the release of hazardous materials into the environment by preventing financial losses and the needless use of fungicides. Incubation and grow-out procedures form the foundation of conventional approaches for identifying seedborne fungus. These are commonly employed due to their ease of application; nonetheless, they are laborious, necessitate mycological expertise, and can lack sensitivity to low levels of seed infection. Due to their great sensitivity and specificity, new identification techniques based on DNA analysis have been used recently and are quite effective. The most widely used method is traditional PCR; more recently, nested PCR has been used to avoid low target pathogen levels; multiplex PCR has been used to detect many pathogens at once; real-time PCR has been used to quantify fungus on seeds; and magnetic-capture hybridization (MCH) PCR has been used to detect numerous infections simultaneously. The incapacity to differentiate between essential and non-essential inocula and the challenge of getting high-quality DNA templates because of PCR inhibitors in seeds are the primary limitations of molecular techniques. Numerous modified PCR processes, including loop-mediated isothermal amplification (LAMP), and non-destructive testing techniques have been developed to lessen inhibitory effects. Given the many benefits they offer, LAMP and next-generation sequencing have found extensive use in nucleic acid analysis. In the future, their use for the identification of fungal infections in seeds may be greatly expanded (Mishra et al., 1994).

When seed pathologists from various laboratories and nations examined their own data, the first issues that emerged were the standardization, unification, and enhancement of diagnostic methodologies. Traditional techniques have been centered on the identification of fungi, primarily through the use of incubation and grow-out techniques. Traditional approaches, although useful, can be time- and space-consuming, and they may not be sufficient in cases of low seed infection levels. When new diagnostic techniques, such as ELISA, ISEM, and others, were successfully used on seeds in the 1980s, the results showed that the seeds could detect viruses, bacteria, and fungi at extremely low contamination levels. Recently, new, developing DNA-based diagnostic methods have shown great promise due to their high sensitivity and specificity. Examples of these methods include PCR, or PCR. DNA analysis results have advantages, but they also have limitations. Specifically, they cannot identify vital from non-vital inoculum, nor can they pinpoint the precise position of the inoculum within the seed. False positive or false negative test findings could represent another issue. However, a comparison of the outcomes obtained on the sub-samples in various laboratories is required prior to the large-scale implementation of the novel molecular approaches.

16.2 SEED STORAGE FUNGI

The field fungus does not appear to withstand the drying process once the grain is lifted, and a second type of fungi, known as storage fungi—species of saprophytic and weakly parasitic fungi that multiply at lower seed moisture levels—become prominent during the drying, picking, and storing processes. According to Mashilla (2004), the main storage fungus is a member of the genera *Aspergillus, Penicillium, Rhizopus, Sclerotium,* and *Fusarium*. Temperature, the composition of the intergranular gas, and the availability of water in the substrate all affect the capacity of storage fungi to germinate, develop, and sporulate in grain that has been stored. These elements may combine to have a significant impact on when fungi begin to deteriorate grain that has been stored. Improved and safer grain storage techniques could be developed with an awareness of the ecological factors that influence the establishment of mold (Magan and Lacey, 1988). It is recognized that a wide variety of fungus, many of which are plant pathogens, can colonize seeds. According to Mehrotra and Aggarwel (2003), the fungus may cause seed rot, shrunken seeds, reduced sizes, sclerotization or somatization of the seed, necrosis, discoloration, reduced or whole loss of germinability, and certain physiological abnormalities.

It was found that the watermelon seedling growth and germination were significantly reduced by the seed-borne fungus *Rhizopus nigricans* and *Mucor racemosus* (Anjorin and Mohammed, 2009). No matter the age of the seed, *Tilletia indica*-infected wheat had very little effect on viability, according to Mehrotra and Aggarwel (2003). Conversely, it seemed that the wheat cultivars and the seeds themselves determined whether the contaminated seeds germinated. Comparing infected seed to healthy seed from the same lots, infected seed showed a lower survival rate in storage. In addition, soybean seeds soaked in culture filtrates of *Fusarium solani, F. oxysporum, Aspergillus flavus, A. niger, Alternaria tenuis*, and *A. alternata* for 24 hours demonstrated a reduction in percentage seed germination (Haikal, 2008). Soybean seeds soaked in *Phomopsis phaseoli* filtrates also showed a reduction in percentage seed germination.

A number of variables, including the type of fungus, where it is located within the seed, the degree of infection, and the existence of hostile microbiota, affect how long seed-borne fungi live. On the other hand, the primary factor influencing seed-borne fungus survival is storage conditions. Many viruses inoculum can be significantly reduced by a few months or years of storage under standard room or warehouse settings; most of them have been seen to vanish after 10 to 15 years (Agarwal and Sinclair, 1997; Neergaard, 1977). However, because seed viability also decreases, extending the commercial storage duration is typically not thought to be a workable way to free seeds from seed-borne fungus. The lifespan of *Alternaria tenuis* (*A. alternata*) and Helminthophobia sativus (*Bipolaris sorokiniana*, sexual morph *Cochliobolus sativus*) on wheat seed kept in room temperature was examined for 16 years in a paper by Russell (1958). Over time, both fungi's vitality gradually declined. But although Helminthophobia was discovered in a tiny percentage of seeds after 15 years, when the germination rate was less than 30%, Alternaria died off more quickly and vanished after 7 years. After 3–5 years of unsupervised storage, fungus from the *Fusarium* genus were not found on barley grains in a Polish experiment, although *Alternaria* and *Bipolaris sorokiniana* were still present (Wiewióra, 2009).

16.3 SEED HEALTH DETECTION METHODS

The goal of detection is to determine whether a certain target organism is present in a sample, with a focus on asymptomatic persons (Lopez et al., 2003). Generally speaking, there are two main types of seed health detection assays: traditional seed detection assays and PCR-based seed detection assays.

16.3.1 DETECTION OF SEED BORNE FUNGI

Seed fungus can be detected in the gametophyte, embryo, and seed coat. Certain fungi, like molds, cannot be identified by evaluating the external surface of seeds. When a seed is cut open, some internal seed fungus can be identified by looking for visible mycelium. However, putting sample seeds on the right culture media is the only reliable method to determine the incidence of seed fungus. Mycelium or fruiting bodies on the surface of the seed and the growing media indicate the presence of fungus following incubation. An expert is typically required to identify the numerous fungi that infect seed due to their huge quantity (Walcott, 2003). The following are some methods for identifying fungal pathogens linked to seeds.

16.3.2 CONVENTIONAL FUNGAL DETECTION ASSAYS

16.3.2.1 DIRECT INSPECTION

Seed or seed washings may need to be visually inspected for this. Generally, dried seeds are analyzed using the former method; for example, dry seed samples can be analyzed with a stereomicroscope or without one to check for the presence of ergots, other sclerotia, and smut balls. Fungal structures like sclerotia, encrusted mycelium, discoloration, pigmentation, and so forth, can be identified with the naked eye using optical lenses and stereomicroscopes (De Tempe and Binnerts, 1979). For example, ear rot or tundu disease of wheat seeds, which is caused by the nematode *Anguina tritici* (Steinbuch) Chitwood, can be identified visually by looking for black galls. By visually examining the seeds, morphological anomalies and seed discoloration are also examined. Certain fungal species can occasionally be identified by their symptoms. For example, *Phomopsis spp.* and *Cercospora kikuchii* are known to produce discoloration, shriveling, and care (Li, 2011; Murakishi, 2002), while *Cylindrocladium parasiticum* is responsible for the speckles on peanut seeds (Randall-Schadel et al., 2001). Since these symptoms might be shared by a number of fungi, it is generally not advised to identify a pathogen only on the basis of them (Andersen and Leach, 1961). When seeds are examined under a stereoscopic microscope, fungal reproductive or overwintering structures can be found. Examples of these structures include the pycnidia of *Septoria apii* embedded in the seed coat of celery seeds (Horst, 2008), the acervuli and microsclerotia of *Colletotrichum dematium* on chili seeds (Kumar et al., 2020), and the crust of oospores of *Peronospora manshurica* on soybean seeds (Agarwal et al., 2006). But no kind of visual examination has a valid basis as a detection assay when fungal-infected seeds do not exhibit macroscopic symptoms (Walcott, 2003). When surface-borne pathogenic fungi are suspected, or when seedborne fungi are detected as microscopic spores on the seed surface, seed washing techniques can be employed in conjunction with visual assessment (Agarwal et al., 2006; Rao et al., 2006). By washing and agitating the seeds in distilled water, these methods enable the removal of spores, hyphal components, or other fungal structures (Reeves, 1998). After washing, the particle that is left over after centrifugation can be inspected using a compound microscope, or the resultant suspension can be decanted and viewed under a stereomicroscope. Using a hemocytometer counting chamber, the quantity of spores for a given amount of seed can also be ascertained microscopically (Andersen and Leach, 1961).

16.3.2.2 INCUBATION METHOD

The most popular approach for identifying seed-transmitted diseases is incubation, particularly for high-incidence fungi (*Alternaria spp.*, *Drechslera spp.*, *Fusarium spp.*, *Stemphylium spp.*, etc.) that can be found on seed samples at a level greater than 1%. This method also tells us about the viability of the inoculum on the seeds (Marcinkowska, 2002). This method confirms the presence of the pathogen by keeping seeds in a particular medium for a predetermined amount of time (De Tempe, 1961; De Tempe and Binnerts, 1979). Following the period of incubation, the fungi that have grown on each seed are analyzed using a stereomicroscope or a compound microscope to determine the morphological characteristics of the conidia, spores, and so forth. In petri plates with medium-impregnated filter paper (blotting) or various agar media, seeds can be cultured under conditions that encourage fungal growth and sporulation.

These samples are typically subjected to 12-hour cycles of light and dark. Depending on the kind of fungus species, the duration and temperature of plate incubations can vary, but typically they last between two and 10 days at a temperature of 20–28°C (Elwakil et al., 2009; Naqvi et al., 2013; Shovan et al., 2008). It is common for seed surfaces to harbor saprophytic bacteria. If the pathogen to be identified is within the seed, surface sterilization of the seed can be used to curtail or eliminate it (Du Toit et al., 2005; El-Nagerabi and Elshafie, 2000; Rodrigues and Menezes, 2005). Various surface disinfectants, including the most widely used 0.5–1.0% sodium hypochlorite (Peres et al., 2002; Roy et al., 2000), 0.1% mercuric chloride (El-Nagerabi and Elshafie, 2000; Ora et al., 2011), or 10–90% ethanol (Vallad et al., 2005), can be used for this purpose.

16.3.2.3 TESTING ON AGAR MEDIA

Viable fungal material can be found using Agar Plate Incubation techniques, even in the early stages of the fungus's development. In order to promote the growth of seed-borne fungus, this is typically accomplished by seeding sterile agar medium (malt or potato dextrose agars are the most popular types). When a blotter test is unable to produce sufficient conditions for the establishment of mycelial growth, sporulation, or pathogen symptoms on seedlings and seeds, but on nutrient agar typical colonies are formed, the approach is preferred. To reduce or free surface microorganisms, surface-sterilized seeds are typically plated into petri dishes on sterile standard media such as PDA (Blaszkowski, 1994; Kuaemierz and Gorajczyk, 1991; Pyndji et al., 1987; Osorio and McGee, 1992; Tseng et al., 1995; Wiwart and Korona, 1997), MEA (Ali et al., 1982, Bathgate et al., 1989), or specific media, for example, Coons (Marcinkowska, 1998), SNA (Wakuliñski and Chelkowski, 1993). To encourage the growth of seedborne necrotrophs, the seeds are kept on the medium for a few days (5–8), usually at a temperature of about 20°C. Time, temperature, and light for plate incubation differ since there are various versions of the agar test depending on species requirement (Yeh and Sinclair, 1982). A standard sample of 400 seeds is often used for testing. When it comes to identifying high-incidence infections like *Ascochyta pisi* on peas, *Botrytis allii* on onions, Septoria (Stagonospora) nodorum on wheat, *Colletotrichum lindemuthianum* on beans, *C. linicola*, and *B. cinerea* on flax, agar tests are the most effective (Marcinkowska, 2002). It can be used to either qualitatively ascertain the species composition (direct plate approach) or quantitatively ascertain the fungal burden (e.g., CFU/gm of seed). The dilution plate method is the most widely used technique to quantify the amount of fungal tissue present in kernels. This approach has two variations: the spread plate method, which is preferable at low sample contamination, and the pour plate method, which is used more commonly. However, one of the best ways to ascertain the genus and species composition of the grain fungus is through the use of the direct plating method. Using this technique, entire kernels are positioned on the culture medium's surface after being thoroughly cleaned on the outside. According to Trojanowska (1991), the direct plating technique is a highly valuable instrument for assessing the quality of bulk grain because it is an effective procedure for measuring the interior colonization of kernels by fungi.

16.3.2.4 BLOTTER TESTING

The blotter test is unquestionably one of the most crucial techniques for detecting fungal infections in seeds and assessing their health (Limonard, 1966). The blotter test finds seed infection and, in rare cases, seedling infection as well. When routine testing for seed health is conducted and the agar test is impractical, the blotter method is frequently utilized. The test combines the benefits of both *in vivo* observations and *in vitro* analysis. For the expression of several microorganisms and their incidence on either small or big seeds of *Lagestroemia microcarpa* and *Pterocarpus marsupium*, respectively, the standard blotter method was the most effective (Sharma et al., 1997). For some studies, seed germination is decreased by temperature (deep freezing) or chemicals (2,4-D) which destroy seed tissues, and seedborne fungus proliferate easier (Jorgensen, 1977; Limonard, 1966; Maguire et al., 1978). Blotter tests are similar to germination tests in that seeds are placed on moistened layers of blotter paper and incubated under conditions that promote fungal growth. The seed may then be allowed to germinate, and fungal seed-borne infections may manifest themselves by any pertinent signs or symptoms. The blotter test indicates whether the seed is infected based on the presence of mycelium and fruiting bodies; in certain cases, it also indicates whether the seedlings are infected based on symptoms on the young plants. In certain experiments, seeds are incubated, allowing them to develop and exhibiting symptoms (such as black patches on the cotyledons of bean seeds harbouring the anthracnose disease). In other tests, the germination of seeds is purposefully inhibited to facilitate the development of seed-borne infection (for example, the herbicide 2,4-D is applied prior to incubation to enable the pycnidia of seedborne *Phoma lingam* to develop on *Brassica* seeds, allowing a larger number of seeds to be tested) (Limonard, 1966). In blotting tests, chemicals or low temperatures can be used to stop seeds from germinating, which can hinder the growth of fungus that are carried by seeds. These treatments cause the seed tissues to become less vital (Ha et al., 2009). For instance, in blotting tests, the filter paper is frequently moistened with the sodium salt of the herbicide 2,4- dichlorophenoxy) acetic acid monohydrate (2,4-D) (Alves and Pozza, 2009; Kumar et al., 2020). In the freeze blotter test, there is an extra step where seeds are soaked for 24 hours and then kept at 20°C for 24 hours to kill the embryo (Du Toit et al., 2005; Elwakil et al., 2009).

According to Rao et al. (2006), the blotter and agar-plate procedures are easy and affordable strategies to identify fungi that are spread by seeds and that sporulate. But sometimes, the blotter test cannot create the right environment for mycelial growth to flourish, sporulation to occur, or for the pathogen's symptoms to manifest on the seeds. It is best to incubate the seeds on agar media for these traits, as more fungal colonies can form there and they will also grow into distinctive colonies (Marcinkowska, 2002; Nasir, 2003).

16.3.2.5 BIOASSAYS

Grow-out is arguably the oldest technique for assessing seed health. Because it depends on the uniqueness of the host-pathogen relationship, this process is very selective. Its sensitivity is less certain because inoculation thresholds can change based on the plant cultivar under test, changes in the surrounding environment, fertility, and other elements. Furthermore, as it is frequently required to physically identify individual lesions among 1000s of plants, the capacity of plant inspectors to consistently identify low occurrences of disease is a crucial component. However, grow-outs are commonly utilized and acknowledged as the gold standard for identifying a seed lot's level of infection. Because grow-outs rely on symptom expression, a positive result usually represents indisputable evidence that the bacteria was present, live, and harmful (Gitaitis and Walcott, 2007).

One example of a commonly used grow-out test is looking for the bacterial fruit blotch disease, *Acidovorax avenae* subsp. *citrulli* (Aac). For the assay to be trustworthy and effective, ideal environmental conditions for the development of the disease must be met. Cross-contamination from other inoculum sources must also be avoided. Using commercial greenhouse potting mixture or steam-sterilized soil, selecting an appropriate sample size of at least 30,000 seeds or 10% seedlot, and creating ideal conditions for seed germination and seedling emergence are additional factors. Frequent disinfestations of planting trays, greenhouse floors, walls, and benches are also important. Relative humidity should be kept at least 70% throughout the test, and it should not be permitted to drop below 50% for longer than 12 hours. In an ideal world, grow-out temperatures would be between 30 and 32°C, never dropping below 25° or rising over 38° for longer than 12 hours. In an isolated section of the greenhouse, an Aac-infested seed sample should be included as a check to confirm further that the conditions are favorable for symptom expression. The control seedlot should be infested with an Aac strain that carries an antibiotic resistance marker or a distinct DNA fingerprint in order to mitigate the possibility of cross-contamination from these inoculum sources. Ample spacing between seedlots and the construction of plastic barriers 60–90 cm high can help prevent potential issues related to the bacterium's splash dissemination during irrigation. Also important are good management measures to restrict the passive dissemination of diseases by insects, mites, or other vectors. These precautions are important to prevent contamination across seedlots, but secondary spread inside a seedlot is desirable since it results in the establishment of infection foci that improve the odds of visual detection (Gitaitis and Walcott, 2007). Up to the assay's conclusion, cotyledons and true leaves should be inspected every day following germination. Every seedling should be checked for symptoms in well-lit natural light settings. Certain plants frequently experience natural water congestion; this is not to be confused with the water soaking that leads to the development of diseases. Apart from removing sick seedlings for testing and isolation, all contact with seedlings should be avoided until the final inspection. Workers should either sanitize their hands before handling seedlings or wear disposable gloves that can be replaced between seedlots. If there are no visible symptoms after three weeks, the seedling grow-out assay can be stopped; if symptoms are present, bacteria should be isolated and further diagnostic tests (such as immuno-strips, PCR, and/or pathogenicity) should be performed to confirm the pathogen's identity (Gitaitis and Walcott, 2007).

16.3.2.6 GROWING ON TEST (SEEDLING SYMPTOM TEST)

In greenhouses, seeds are sown in appropriate substrates (such as agar tubes, sand, or other sterile substrate) under varying environmental conditions. This technique is very helpful when seedlings begin to exhibit symptoms quickly.

16.3.2.7 STAINING METHODS

Use of staining compounds specific to fungal hyphae (such as trypan blue, aniline blue, rose bengale, etc.) is employed for pathogen detection when the inoculum is situated in the internal tissues of the seed, and it is not feasible to detect the fungi in a reasonable amount of time using the previously mentioned methods. These techniques are used to identify seedborne infections that are biotrophs, or that grow on artificial substrates but do so very slowly. Most of these pathogens, like Ustilagineous, are unable to form fruiting structures, or they take a long time to develop—typically more than 10 days—like Septoria. One common technique for determining the health of seeds is to stain barley embryos to check for the presence of *Ustilago nuda* mycelium (Anonymous, 1993). For the identification of Septoria (Stagonospora) nodorum, the fluorescent technique is used. Following incubation of the wheat seed, the percentage of kernels with fluorescent mycelium is calculated (Scheel, 1997). Stereomicroscopes and microscopes are the primary required instruments for seedborne fungal detection when using the incubation and staining procedures. However, one must be skilled in identifying etiological indicators, particularly fruiting bodies and/or sporulation on infected seeds, or when a distinctive pure culture of a fungus on medium is well developed, in order to identify species using incubation tests. Keys and descriptions of organism morphology are important for identification (Arxvon, 1974; Hawksworth et al., 1995). These techniques are simple to use for someone with training in fungal diagnostics, but because different species have varying capacities for sporulation, it may take several days to get results (for example, for *Colletotrichum lindemuthianum* or *Alternaria spp.*) or even longer if additional mycelium structure characteristics (such as chlamydospores from *Phoma spp.* or microsclerotia from *Verticillium spp.*) are required for identification.

16.4 MOLECULAR METHODS

Prior to serological testing in 1918, serology was used as a technique for the first time to detect microorganisms. The older serological techniques based on polyclonal antibodies that are still in use for some bacteria include agglutination, precipitation, and immunodiffusion testing (Ball and Reeves, 1992). A specific antigen clumps or agglutinates in the first test due to the antigen/antibody reaction (Lyons and Tylor, 1990). Antiserum diffuses from a central well across the agar in an immunodiffusion test, which is typically double-diffusion, to precipitate against particular antigens diffusing from nearby wells, or not. Antigens are regarded as same when comparable precipitation lines appear (Guthrie et al., 1965). In the precipitation test, a particular antibody precipitates an antigen out of solution. The methods discussed have been applied to the identification of *Pseudomonas syringae* (savastanoi) pv. phaseolus and *Xanthomonas campestris pv. phaseoli* in bean and pea seeds, respectively, as well as *P. syringae pv. pisi* (Trujillo and Seattler, 1979; van Vuurde and van den Bovenkamp, 1981). A representative fungal isolate was selected for the diagnosis of the molecular properties of *A. alternata* and *Arthrinium arundinis* from lemon balm seeds and *C. sphaerospermum* isolated from fennel seeds. Identity confirmation of *A. alternata* isolate TR-MeAa3 and *C. sphaerospermum* isolate TR-ReCs13 and *A. arundinis* TRMeArtA5 isolate were realized by the PCR (Kaygusuz and Coşkuntuna, 2022).

Immunological techniques for identifying seed-transmitted infections rely on the use of monoclonal or polyclonal antibodies that attach selectively to a target antigen. This enables the pathogen to be identified later on by substrate conversion enzymatically or by employing fluorescent tags. The double antibody sandwich–enzyme linked immunosorbent assay (Ward et al., 2004) is the most widely utilized immunological technique. Serological assays can be used with both necrotrophic and biotrophic seedborne diseases since they do not require the pathogen to be completely isolated.

Due to the lack of species-specific antibodies, these assays are less frequently used to detect fungal plant diseases than seedborne viruses. Moreover, non-viable infections might be found using serology-based assays, which may lead to incorrect interpretations (Afouda et al., 2009; Walcott, 2003). Serological seed assays rely on antibodies (polyclonal or monoclonal) produced against specific antigens on the surfaces of plant diseases (Hampton et al., 1990). There are various formats for serology-based seed tests, such as immunofluorescence microscopy (Franken, 1992) and the commonly used enzyme linked immunosorbent assay (ELISA) (McLaughlin and Chen, 1990). Agglutination tests, flow cytometry, immunomagnetic separation (IMS), immunofluorescence microscopy (IF), immunofluorescence colony-staining (IFC), enzyme-linked immunosorbent assays (ELISA), Western blot, lateral flow devices (e.g., immunostrips), immunocapture techniques, and lateral flow devices are among the serological methods used to detect and identify bacterial pathogens (Munkvold, 2009).

16.5 ENZYME-LINKED IMMUNOSORBENT ASSAY (ELISA)

Clark and Adams (1977) were the first to adapt the enzyme-linked immunosorbent test, or ELISA, for the identification of plant viruses. Clark (1981) provided a description of this method's development and guiding ideas. For ELISA, both polyclonal and monoclonal antibodies could be used; however, monoclonal antibodies increased ELISA's repeatability when it came to seed pathogen detection (Fox, 1993). There are many different variants of ELISA (Ball and Reeves, 1992), but three of them are main procedures: direct (double antibody sandwich DAS-ELISA), indirect, and competitive (Lange, 1986). The most used test in the seed health tests is the direct test (Kazinczi and Horvath, 1998; Lange, 1986). This technique uses a solid-phase-specific antibody to selectively capture and immobilize a seed (or seedling) extract, or antigen, in microtiter wells. After the enzyme-labeled antibody has reacted with the immobilized antigen and the unreacted antibody has been removed, the bound enzyme is tested by adding an appropriate substrate. While visual assays for qualitative analysis can be carried out, adding substrates that result in colored hydrolysates can enhance visualization. Colorimetric or spectrophotometric equipment is used to perform quantitative tests (Clark, 1981).

Diagrams and descriptions of the steps required in both direct and indirect ELISA tests were provided by Fox (1993). Polystyrene microtiter plates, which have up to 96 wells apiece and are ideal for indexing several seed samples and sub-samples, provide the foundation for ELISA seed testing methods. ELISA methods are mostly employed using monoclonal antibodies to identify microorganisms. To differentiate between strains of *Pseudomonas syringae pv. pisi* in indirect and competitive ELISA assays, Candlish et al. (1988) generated highly specific monoclonal antibodies. A sensitive, specific, and quick ELISA method was recently established by Rajeshwari et al. (1998) for the detection of isolates of *Ralstonia solanacearum* from tomato seeds. Fluorescence *in situ* microscopy, one highly suggested immunodiagnostic technique for detecting bacteria, is immunofluorescence microscopy (Van Vuurde, 1997). ELISA results could be easily automated, standardized, and could suit for screening large numbers of seeds (Sheppard et al., 1986), according to Vuurde (1990) out of the most widely used serological methods for seedborne bacteria detection and identification (agglutination, double diffusion, ELISA, and immunofluorescence microscopy). However, immunofluorescence is a second important assay.

The utilization of infrared fluorescence (IF) in Georgia's black rot seed certification program was discouraged due to the challenge of determining a threshold of positive fluorescent cells that can result in crop disease development. Researchers and regulatory staff came to the conclusion that IF produced an excessive number of false positives, possibly as a result of the bacteria attaching to nonviable cells or exposed antigenic determinants (Munkvold, 2009). Techniques such as IFC were developed to solve the problem of probable false positives (Glynn et al., 2008). The IFC Seed Health Assay uses seed extract and an equivalent volume of agar media. Target bacterium-specific antibodies coupled with a fluorescent dye are added to the mixture after it has been dried and incubated. With the use of a glass capillary tube, bacteria within colonies stained with the antibody-dye conjugate can be extracted and moved to an appropriate growth medium. Colonies can be seen using fluorescence microscopy.

Another interesting serological method for application in seed health assessments is flow cytometry (Broders et al., 2007), which automatically sorts and analyses bacterial cells tagged with dye-conjugated antibodies while they are in suspension. Thus, by measuring the amount of light scattering and fluorescence emitted by 1000s of individual cells retrieved from an infected seed sample, several parameters can be ascertained in a matter of minutes. As a stream containing tagged bacterial cells passes through a flow cell, a laser beam lights the cells and activates the fluorescent tags bound to the antibodies. Multiple characteristics, including cell size, granularity, and cell roughness, can be measured at the same time by utilizing various fluorescence detectors. Flow cytometry can also be utilized to assess the vitality of target cells when combined with fluorescent probes that target membrane potential, respiratory activities, or important enzymes.

Using antibody-antigen specificity is also possible with IMS PCR. IMS separates target cells from heterogeneous mixtures using tiny magnetic polystyrene beads covered with antibodies (Ojeda and Verdier, 2000). Following their immobilization using a magnet particle concentrator, the immunomagnetic beads are rinsed to get rid of any nontarget bacteria and inhibitory substances. Captured cells can be plated onto a semi-selective medium, or template DNA can be extracted by boiling the cells and utilized for PCR. Walcott et al. (2003) demonstrated that IMS-PCR, as opposed to traditional PCR, increased the detection threshold of *A. avenae* subsp. *citrulli* by a factor of 100 in watermelon seed as saying. IMS, essentially a concentration step to efficiently enhance the amount of target DNA in the PCR sample, was also successfully utilized to detect *P. ananatis* in onion seeds. By removing seed components that can impede PCR and increasing the effectiveness and dependability of PCR-quality pathogen DNA extraction from seeds, IMS-PCR can lower false-negative results (Munkvold, 2009).

16.6 NUCLEIC ACID BASED DETECTION METHODS

The potential of molecular nucleic acid tests in practical seed-borne organism detection approaches has not yet been completely realized. Because of their versatility, molecular methods are used either in place of or in addition to more traditional morphological, biochemical, and serological techniques. Their very short development and validation times compared to more time-consuming biochemical and serological diagnostics have contributed to their introduction. Molecular assays can offer straightforward, rapid, high-throughput diagnosis that is reasonably priced and simple to transfer to other laboratories once they are created and validated. Nonamplification tests were first employed, but their poor reproducibility and requirement for pristine cultures rendered them unsuitable for everyday diagnosis. The advancement of PCR-based *in vitro* amplification is yielding robust, sensitive, and specific assays for pure cultures and, more crucially, for the direct identification and detection of organisms from seed and plant material. Nucleic acid-based approaches tend to be relatively expensive to apply (Rao et al., 2006). It is also possible to utilized nucleic acid hybridization assays, often known as Southern and Northern blotting. These techniques involve transferring DNA or RNA from an electrophoresis gel onto a membrane and then detecting the nucleic acids using a labelled probe. Another method that can be utilized without going through the PCR stage is nucleic acid spot hybridization (NASH), in which a labelled DNA pathogen hybridizes directly to the pathogen DNA immobilized on a nylon membrane (Rao et al., 2006). Although nucleic acid-based techniques are more costly, they are quicker and need less time to identify a disease. Long-term costs for nucleic acid testing may be lower, but a major drawback of these seed health tests is quantification. Since PCR is only accurate when pure DNA is amplified, seed material presents another issue that could prevent DNA amplification (Reeves, 1995). In order to ensure seed health quality standards and meet phytosanitary requirements globally in the new millennium, research into nucleic acid-based methods will be essential. This is because the application of advanced DNA amplification techniques allows for the detection of seedborne pathogens that may go unnoticed by traditional methods (Nameth, 1998).

16.7 POLYMERASE CHAIN REACTION (PCR)

The technique known as PCR (Mullis and Faloona, 1987) allows the target DNA sequence to be amplified and multiplied up to a million times (Oliver, 1993; Saiki et al., 1988). As a result, previously undetectable amounts become detectable ones. Amplification involves the use of two primers. They are placed in front of the DNA segment that is going to be amplified and treated to several cycles of heat denaturation. Selecting the right primers is crucial (Fox, 1993). The primer-directed, *in vitro* enzymatic amplification of nucleic acids is known as PCR (Erlich et al., 1988; Saiki et al., 1988). Plant disease diagnostics is one of the many various applications for this technology. In PCR, primers are tiny oligonucleotide probes that anneal to particular DNA sequences in the chromosomal DNA or RNA of the target organism. These primers hybridize with the target sequence to drive the amplification of millions of copies of that sequence. This amplified DNA can be visualized following electrophoresis in ethidium bromide-stained agarose gels (Walcott, 2003). The process produces a concentration increase that is so great that it improves the efficiency of using a unique probe or gel electrophoresis for DNA sequence detection (Old and Primrose, 1985). Small tubes arranged in a block within a programmed thermal cycling machine are used to observe reactions. Typically, 30 cycles are used, with each cycle lasting between two and five minutes (Ball and Reeves, 1991; Fox, 1993). There have been multiple attempts to create plant pathogen-specific PCR primers (Hensen and French, 1993). The technique for identifying seed pathogens may be applied to seed washings (Vivian, 1992), negating the necessity for the organism to be extracted and isolated (Ball and Reeves, 1991). However, a significant quantity of non-specific hybridization with primer DNA may arise from a substantial amount of extraneous DNA in seed samples (Vivian, 1992). A few bacterial species have been identified using PCR techniques, such as Pantoea (*Erwinia*) stewartii in maize seeds (Blakemore and Reeves, 1993) and *Pseudomonas syringae* pv. phaseolicola and pisi in bean and pea seeds, respectively (Prosen et al., 1991; Rasmussen and Wulff, 1991; Reeves et al., 1994). A representative fungal isolate was selected for the diagnosis of the molecular properties of *A. alternata* and *Arthrinium arundinis* from lemon balm seeds and *C. sphaerospermum* isolated from fennel seeds. Identity confirmation of *A. alternata* isolate TR-MeAa3 and *C. sphaerospermum* isolate TR-ReCs13 and *A. arundinis* TRMeArtA5 isolate were realized by the PCR (Kaygusuz and Coşkuntuna, 2022). In the vegetable seed sector and in certain official seed testing laboratories for quality control testing, the astonishing expansion of PCR-based technologies for detecting pathogens in seeds has produced highly helpful tools that are available and have started to be utilized. There are various hurdles that have hampered the introduction of PCR-based technologies for seed health testing (Walcott, 2003). The financial outlay and technological know-how required to set up PCR capacity can be challenging in developing nations. Poor quality DNA and PCR inhibitors from seed extracts can be technical obstacles that result in misleading negative results, even in cases when price and knowledge are not the primary obstacles. For PCR-based techniques, low sample intensity can also lead to poor sensitivity. One of the greatest hurdles to the implementation of nucleic acid–based seed health diagnostics has been the possibility for false positives due to the detection of residual DNA from nonviable disease propagules (Agarwal et al., 2006).

16.8 BIO-PCR

Very low levels of fungal infestation frequently occur in seeds; therefore, the pathogen's DNA is insufficient for the reactions that follow (De Boer et al., 1995). The very sensitive PCR method known as BIO-PCR was created by Schaad et al. (1995) to solve this issue. To enhance the biomass of the fungal pathogen on the seeds, a pre-assay incubation phase is required. This is followed by DNA extraction and PCR amplification. Initially, this technique was utilized largely for phytopathogenic bacteria, as these are easily and rapidly cultivated over 2–3 days in a growth medium (Weller et al., 2000); however, this was also demonstrated to be useful for fungi (Munkvold, 2009). One of the incubation techniques that has been evaluated to promote uniform and quick fungal development is soaking seeds in liquid fungal growth media (Phan et al., 2002). The PCR assay's dependability is impacted by the possibility that the presence of bacteria will restrict or repress fungal development, and that adding antibiotics to the substrate will not always stop bacterial growth (Mbofung and Pryor, 2010; Pryor and Gilbertson, 2001). As demonstrated by *L. maculans* and *A. rabiei* from rapeseeds (Taylor, 1993) and chickpea seeds (Phan et al., 2002), respectively, the use of particular medium can prevent this issue and increase the biomass of fungal pathogens. To increase the low levels of target *F. oxysporum f. sp. lactucae* on lettuce seeds and *A. radicina* on carrot seeds, other pre-assay methods involve incubating the samples on moistened filter paper or simply on the surface of a plastic Petri dish under high humidity (Mbofung and Pryor, 2010; Pryor and Gilbertson, 2001). In order to find a balance between the target microorganism's growth and the restriction of other saprophytic fungi, it's also critical to accurately define the incubation period (Schaad and Frederick, 2002). BIO-PCR provides several advantages over regular PCR: enhanced sensitivity, elimination of PCR inhibitors, and identification of live cells exclusively, thereby minimizing false positives due to detection of DNA from dead cells (Marcinkowska, 2002). The drawbacks of BIOPCR include its higher cost compared to traditional PCR, particularly when selective media are employed (Schena et al., 2004), and the fact that the assays take longer to complete because the fungal growth typically takes five to seven days (Walcott, 2003).

16.9 NESTED PCR

Nested PCR is an additional method of circumventing the low concentrations of the target pathogen on seeds. By enhancing the assay's sensitivity and specificity, this technique can detect target DNA at levels that are several times lower than those of traditional PCR (Chiocchetti et al., 2001). In fact, DNA levels of 10 g for *Colletotrichum lindemuthianum* in bean seeds and 1 g for *F. oxysporum* f. sp. Lactuca in lettuce seeds have been detected using nested PCR (Chen et al., 2007; Mbofung and Pryor, 2010). However, this molecular technique is more labor intensive, more costly, and more prone to contamination than standard PCR (McCartney et al., 2003; Tomlinson et al., 2005).

16.10 REAL-TIME PCR

Conventional PCR has the advantages of speed, relative sensitivity, and specificity, but its primary drawback is that it produces only qualitative data. Real-time PCR is able to provide a specific fluorescent signal detected by an integrated fluorometer to give real-time study of reaction kinetics and so permits quantification of specific DNA targets (Atkins and Clark, 2004; Schena et al., 2004; Ward et al., 2004). With minimum sample modification, real-time PCR considerably minimizes the possibility of false positives due to cross contamination of the reaction mixes (Cullen et al., 2001; Gachon et al., 2004; Tomlinson et al., 2005). This method has a high sensitivity on the order of scale of a few femtograms of DNA and requires less time than other assays (Chilvers et al., 2007; Guillemette et al., 2004). (Chen et al., 2013). Using TaqMan-labelled probes and SYBR Green dye, quantitative real-time PCR tests have demonstrated high sensitivity in detecting and quantifying seedborne diseases when applied to diseased seeds. To detect and quantify, for instance, *Verticillium dahliae* on spinach seeds (Duressa et al., 2012), *A. brassicae* on cruciferous seeds (Guillemette et al., 2004), *Botrytis* spp. on onion seeds (Chilvers et al., 2007), and *C. lindemuthianum* on dry bean seeds (Chen et al., 2013), quantitative real-time PCR has been employed in these and other instances. Several primers and TaqMan test probes are used in a single reaction during multiplex real-time PCR to cut expenses and labor consumption. In real-time PCR experiments, selecting the right target DNA fragments for primer and probe design is essential to guaranteeing sufficient specificity, sensitivity, and equivalent amplification efficiency of various pathogens.

16.11 MAGNETIC-CAPTURE HYBRIDIZATION PCR

The process of MCH-PCR involves extracting DNA first, purifying it using a single-stranded DNA probe on magnetic beads, and then amplifying the appropriate DNA region using PCR. By capturing particular target DNA, this method makes it easier for PCR to detect particular DNA sequences (Jacobsen, 1995; Reeves, 1998). In materials containing PCR inhibitory chemicals, such as *Botrytis aclada* in onion seeds, MCH-PCR has been successfully employed to detect harmful microbes (Walcott et al., 2004). Also, an MCH multiplex real-time PCR assay has been developed to detect two distinct pathogens, *Didymella bryoniae* and *Acidovorax avenae* subsp. citrulli, from watermelon and melon seeds (Ha et al., 2009). Notwithstanding the aforementioned benefits of this application, MCH-PCR use is still restricted. However, its enhanced pathogen identification will enable further research.

16.12 IMS-PCR

This technique involves the selective binding of target cells from solutions comprising a mixture of cells using small magnetic beads coated with antibodies specific to a given pathogen (Olsvik et al., 1994; Walcott and Gitaitis, 2000). After being captured, cells can either be directly used for DNA extraction and PCR testing, or they can be cultured on specific media to produce more of the target pathogen. Despite being more sensitive and efficient than traditional PCR, IMSPCR depends on certain antibodies, which are not always readily available. Furthermore, beads can only be successfully utilized for bacteria because they are difficult to employ for catching the cells of filamentous fungi. MCH-PCR: This technique uses magnetic beads coated in single-stranded DNA probes to collect DNA fragments for PCR. Using materials containing PCR inhibiting chemicals, this approach has been utilized to detect viruses, bacteria, and fungus (Jacobsen, 1995). Although this method is quick and can get past the inhibitory effects of seed chemicals, it is unable to discriminate between cells that are viable and those that are not.

16.13 LOOP-MEDIATED ISOTHERMAL AMPLIFICATION

A relatively new technique for the quick, easy, and economical detection of genomic DNA is called LAMP of DNA, which was created by Notomi et al. (2000). According to Mumford et al. (2006), LAMP amplifies DNA with great specificity in less than an hour using a set of four or six primers and a thermophilic DNA polymerase from *Geobacillus stearothermophilus* that possesses strand displacement activity. Compared to conventional PCR and real-time PCR experiments, this approach provides a number of advantages, including the ability to perform all reactions under isothermal conditions, the lack of expensive equipment requirements, and a reduced number of preparatory steps. Moreover, LAMP has a very high amplification efficiency in addition to being highly specific, which improves sensitivity and robustness against inhibitors that typically negatively impact PCR techniques (Fu et al., 2011). The visualization of LAMP products can be achieved through various methods such as gel electrophoresis, magnesium pyrophosphate (Fukuta et al., 2003; Nie, 2005), real-time turbidity reader (Fukuta et al., 2004; Mori et al., 2004; Thai et al., 2004), or the addition of an intercalating dye (SYBR Green I), which results in a color change when the LAMP reaction is positive (Iwamoto et al., 2003; Mumford et al., 2006). The LAMP technique has been used in a number of tests to date to identify plant pathogenic bacteria and viruses in host tissues and in insect vectors, as well as to diagnose bacterial and viral infections in people and animals (Fukuta et al., 2003; Nie, 2005; Okuda et al., 2005). It was not until lately that the first LAMP uses for fungus were documented. The use of LAMP for the early detection of toxic Fusarium species in cereals was demonstrated by its high effectiveness in detecting *Fusarium graminearum* in total genomic DNA isolated from ground wheat grains (Niessen and Vogel, 2010) and from contaminated and germinated wheat seeds (Abd-Elsalam et al., 2011). The LAMP approach has been extensively used in nucleic acid analysis over the past 10 years. As a result, LAMP's applicability can be expanded to include a wider range of plant tissues, including seeds, due to its many benefits.

Conventional techniques for screening seeds: researchers initially concentrated on finding fungi on seeds, but as their understanding of pathogen transmission in seeds grew, they also began to discover viruses and bacteria. The following techniques were part of these conventional approaches.

16.14 DETECTION OF SEED-BORNE BACTERIA AND VIRUSES

16.14.1 GROWING-ON RESTS FOR BACTERIA AND VIRUSES

Under test conditions, these tests indicate the overall potential transmission of the seed sample, both from internal and external sources of inoculum. To stop the spread of inoculum, seed samples are typically sown in a greenhouse or a separate field away from other similar plant species. Isolation and identification can be done if the sprouting seedlings show symptoms. It was in this manner that numerous seedborne bacteria (Schaad, 1989) were discovered, such as *Pseudomonas syringae* (savastanoi) *pv. phaseolicola* (Grogan and Kimble, 1967). In order to find the lettuce mosaic virus (LMV) in commercial seed stocks, the *Chenopodium quinoa* inoculation test was employed (Marrou and Messiaen, 1967). The tests are simple to do, but they take time and space.

16.14.2 LABORATORY TESTS FOR BACTERIA DETECTION

Standard methods for detection of bacteria in seeds involve: Extraction from seeds: The goal of extraction procedures is to maximize the target bacterium's recovery. The saprophytic organisms on seeds and the inhibitory chemicals in seeds are the primary agents responsible for the release of bacteria from a seed sample (Roth, 1989). In a liquid medium, bacteria are extracted from seeds or seed flour. Although sterile buffered saline is typically used, the type of bacteria being soaked will determine the temperature and duration of soaking.

Isolation in culture: Small volumes of the extraction medium are transferred with the extracted microorganisms onto either semi-selective or general media (Schaad, 1989).

16.15 IDENTIFICATION BY DIFFERENT METHODS

16.15.1 STANDARD TESTS

Morphological, physiological, nutritional, and traditional procedures (Stead, 1992). Agar plating methods using nutrient-starch cycloheximide agar (NSCA) or King's medium B, where *Pseudomonas syringae* (savastanoi) *pv. phaseolicola* produced a characteristic fluorescent pigment (Taylor, 1970) or *Xanthomonas campestris pv. campestris* formed typical starch-hydrolyzing colonies (Mguni et al., 1999; Schaad and Donaldson, 1980), are two general plating media that can be used to identify certain seedborne bacteria. When isolating pathogenic bacteria that are abundant in extracts with minimal saprophytic bacterial incidence, agar plating techniques prove to be highly efficacious. In the event of the opposite circumstance, semi-selective medium enhanced with chemicals are advised in order to inhibit saprophyte growth. Certain bacteria from seeds directly plated on agar can also be found using semi-selective medium (Schaad, 1989). This procedure is often not entirely effective since the incidence of germs that are transmitted through seeds is typically quite low. Mguni et al. (1999) pointed out that additional testing, such as the host pathogenicity test, is necessary to confirm the identity of *X. campestris* pv. campestris colonies isolated on three semi-selective medium (FS, NSCA, and NSCAA).

16.15.2 HOST PATHOGENICITY TESTS

To identify a pathovar and race, pure cultures of bacteria must inoculate a distinct series of cultivars of the suitable host (Mguni et al., 1999; Taylor et al., 1989).

16.15.2.1 BACTERIOPHAGE TESTS

Bacteriophages are used to detect and identify bacteria since the interaction between a phage and its bacterial host can be precise (Taylor, 1970). The absence of real species specificity and bacterial resistance to phages limit the use of phage tests, despite the initiation of multiple trials (Sheppard et al., 1989). Laboratory tests for the detection and identification of bacteria don't take a lot of room or time to conduct, but they do require highly qualified individuals who are proficient in bacteriological techniques.

16.15.3 OTHER NEWLY DEVELOPED DIAGNOSTIC TECHNIQUES

1. A technology called "biospeckle" laser has been used recently to detect the presence of harmful fungus on seedlings. The optical phenomenon of interference, which is produced when a laser light interacts with the seed coat, is the basis for this approach. By examining seeds under a laser beam, distinct activity zones can be identified (Braga et al., 2005; Rabelo et al., 2011). Due to their biological activity, fungus on seeds can be detected using this method.
2. A multispectral vision system, such as the Videometer Lab device, is another tool that can be used to detect healthy seeds from sick seeds. It is useful in determining the color, texture, and chemical content of seed surfaces. The identification of healthy spinach seeds from those infected with *A. alternata, Fusarium spp., Cladosporium spp., Stemphylium botryosum*, or *Verticillium spp.* was made possible by combining features from images taken at visible light and near-infrared wavelengths (Olesen et al., 2011). A similar methodology has also been applied to identify Fusarium spp. in a malting barley study (personal communication, S. Bodevin, Carlsberg Research Centre, Copenhagen, Denmark). The ability to differentiate between healthy and sick seeds, as well as the color, texture, and chemical makeup of seed surfaces, is made possible by recently created instruments based on multispectral vision systems. According to the reflectance, color, and texture aspects of the seed pictures, multispectral imaging (MSI) in conjunction with machine vision was utilized as a non-invasive alternative way to detect *Drechslera avenae* in black oat seeds.
3. Luminex has created innovative technology that has been used to diagnose plant pathogens since 2008 (Ishii et al., 2008). The foundation of the Luminex PCR system is the hybridization of an oligonucleotide probe attached to separate beads that uniquely

detect variations in nucleotide sequences with a biotin-labelled PCR product. This technology has great potential and could be useful in the area of seed diagnosis because it does not involve time-consuming and arduous processes. An alternative to ELISA is the Luminex xMAP technique, which uses immuneobeads, which are paramagnetic microspheres coated in antibodies and internally fluorescently labelled. These beads function as tiny ELISA wells. These methods are currently utilized for plant pathology applications, such as multiplex detection of bacteria and viruses transmitted by seeds (Bergervoet et al., 2008; Peters et al., 2007), after first being employed in the clinical context (De Boer and Lopez, 2012). Although fungal pathogens in seeds have not yet been tested using this technology, its large multiplexing capacity—which allows for the simultaneous analysis of up to 100 pathogens in a single run—as well as its quick analysis make it a desirable option for laboratory-based sample testing (De Boer and Lopez, 2012).

16.16 CONCLUSION

As microorganism detection typically requires the destruction of pertinent quantities of material, the quantity of seeds stored is a common constraint for determining the sanitary status of the samples, even though the general guidelines of seed health testing of commercial seed lots apply to germplasm material in gene banks. When there are few seeds accessible, which happens frequently, particularly in large-seeded accessions, this presents an issue. In these circumstances, routine germination tests for seed viability monitoring to identify specific pathogens may be preferable, or it may be desired to recover viable seeds after testing (from blotter tests or washing tests). Additionally, a few nondestructive tests have been suggested to preserve genetic material. The final choice of analytic method will be determined by these criteria as well as the availability of specialized laboratories or experienced staff, albeit it appears that a combination of several approaches is the best option.

KEYWORDS

- **agar plating methods**
- **bacteriophage detection**
- **host pathogenicity tests**
- **luminex PCR system**
- **multispectral imaging**
- **seedborne pathogens**

REFERENCES

Abd-Elsalam, K., Bahkali, A., Moslem, M., Amin, O. E., & Niessen, L. (2011). An optimized protocol for DNA extraction from wheat seeds and loop mediated isothermal amplification (LAMP) to detect *Fusarium graminearum* contamination of wheat grain. *Int. J. Mol. Sci., 12*, 3459–3472.

Afouda, L., Wolf, G., & Wydra, K. (2009). Development of a sensitive serological method for specific detection of latent infection of *Macrophomina phaseolina* in cowpea. *J. Phytopathol., 157*, 15–23.

Agarwal, P., Dev, U., Singh, B., Indra, R., & Khetarpal, R. (2006). Seed-borne fungi detected in consignments of soybean seeds (*Glycine max*) imported into India. *EPPO Bulletin, 36*, 53–58.

Agarwal, V. K. S., & Sinclair, J. B. (1997). *Principles of Seed Pathology* (2nd ed.). CRC Press: Boca Raton, FL, USA.

Agarwal, V. K., & Graur, A. (1995). Seed health tests. *Seed-borne Diseases, 9*, 121–130.

Akranuchat, P., Noimanee, P., Krittigamas, N., Horsten, D. V., & Vearasilp, S. (2007). Control of seed-borne fungi by radio frequency heat treatment as alternative seed treatment in barley (*Hordeum vulgare*). *Conference on International Agricultural Research for Development*, University of Gottingen, October 9–11.

Ali, S. M., Paterson, J., & Crosby, J. (1982). A standard technique for detecting seed-borne pathogens in peas, chemical control, and testing commercial seed in South Australia. *Aust. J. Exp. Agric. Anim. Husb., 22*, 348–352.

Alves, M. D. C., & Pozza, E. A. (2009). Scanning electron microscopy applied to seed-borne fungi examination. *Microsc. Res. Tech., 72*, 482–488.

Andersen, A. M., & Leach, C. M. (1961). Testing seeds for seedborne organisms. In A. Stefferud (Ed.), *Seeds: The Yearbook of Agriculture* (pp. 453–457). Washington, DC, USA: US Department of Agriculture.

Anjorin, S. T., & Mohammed, M. (2000). Effects of seed-borne fungi on germination and seedling growth of watermelon (*Citrullus lanatus*). *JASR, 5*, 77–80.

Arx, J. A. von. (1974). *The Genera of Fungi Sporulating in Pure Culture*. Cramer: Vaduz, Germany.

Atkins, S. D., & Clark, I. M. (2004). Fungal molecular diagnostics: A mini review. *J. Appl. Genet.*, *45*, 3–15.

Ball, S. F. L., & Reeves, J. C. (1991). The application of new techniques in the rapid testing for seed-borne pathogens. *Plant Varieties and Seeds*, *4*, 169–176.

Ball, S. F. L., & Reeves, J. C. (1992). Application of rapid techniques to seed health testing: Prospects and potential. Blackwell Scientific Publications: Oxford.

Bathgate, J. A., Sivasithamparam, K., & Khan, T. N. (1989). Identity and recovery of seed-borne fungal pathogens of field peas in Western Australia. *N. Zeal. J. Crop Hort. Sci.*, *17*, 97–101.

Belete, T., & Boyraz, N. (2018). Biotechnological tools for detection, identification and management of plant diseases. *Afr. J. Biotechnology.*, *18*, 797–807.

Bergervoet, J. H. W., Peters, J., van Beckhoven, J. R. C. M., van den Bovenkamp, G. W., Jacobson, J. W., & van der Wolf, J. M. (2008). Multiplex microsphere immuno-detection of potato virus Y, X and PLRV. *J. Virola. Methods*, *149*, 63–68.

Blakemore, E. J. A., & Reeves, J. C. (1993). PCA used in the development of a new seed health test to identify *Erwinia stewartii*, a bacterial pathogen of maize. In *Proc. First ISTA Plant Disease Committee Symp. on Seed Health Testing* (pp. 19–22). Agric. Canada Central Seed Laboratory: Ottawa, Ontario.

Blaszkowski, J. (1994). The occurrence of *Septoria nodorum* Berk. and associated mycoflora in seeds of wheat cultivated in the Szczecin voivodship. *Acta Mycol.*, *29*, 43–52.

Braga, R. A., Jr., Rabelo, G. F., Granato, L. R., Santos, E. F., Machado, J. C., Arizaga, R., Rabal, H. J., & Trivi, M. (2005). Detection of fungi in beans by the laser biospeckle technique. *Biosyst. Eng.*, *91*, 465–469.

Broders, K. D., Lipps, P. E., Paul, P. A., & Dorrance, A. E. (2007). Evaluation of *Fusarium graminearum* associated with corn and soybean seed and seedling disease in Ohio. *Plant Dis.*, *91*, 1155–1160.

Candlish, A. A. G., Taylor, J. D., & Cameron, J. (1988). Immunological methods as applied to bacteria pea blight. *Proc. Brighton Crop Prot. Conf. –Pests and Diseases*, *2*, 787–794.

Chen, Y. Y., Conner, R. L., Gillard, C. L., Boland, G. J., Babcock, C., Chang, K. F., Hwang, S. F., & Balasubramanian, P. M. A. (2007). Specific and sensitive method for the detection of *Colletotrichum lindemuthianum* in dry bean tissue. *Plant Dis.*, *91*, 1271–1276.

Chen, Y. Y., Conner, R. L., Gillard, C. L., McLaren, D. L., Boland, G. J., Balasubramanian, P. M., Stasolla, C., Zhou, Q. X., Hwang, S. F., Chang, K. F., & Babcock, C. (2013). A quantitative real-time PCR assay for detection of *Colletotrichum lindemuthianum* in navy bean seeds. *Plant Pathol.*, *62*, 900–907.

Chilvers, M., Du Toit, L., Akamatsu, H., & Peever, T. (2007). A real-time, quantitative PCR seed assay for *Botrytis* spp. that cause neck rot of onion. *Plant Dis.*, *91*, 599–608.

Chiocchetti, A., Sciaudone, L., Durando, F., Garibaldi, A., & Migheli, Q. (2001). PCR detection of *Fusarium oxysporum* f. sp. *basilici* on basil. *Plant Dis.*, *85*, 607–611.

Christensen, C. M., & Meronuck, R. A. (1986). *Quality Maintenance in Stored Grains and Seeds*. University of Minnesota Press: Minneapolis, MN, USA.

Chun, S., Gopal, J., & Muthu, M. (2022). A consolidative synopsis of the MALDI-TOF MS accomplishments for the rapid diagnosis of microbial plant disease pathogens. *Trends Anal. Chem.*, *156*, 116713.

Clark, M. F. (1981). Immunosorbent assays in plant pathology. *Annu. Rev. Phytopathol.*, *19*, 83–106.

Clark, M. F., & Adams, A. N. (1977). Characteristics of the microplate method of enzyme-linked immunosorbent assay. *J. Gen. Virol.*, *34*, 475–483.

Colley, M. (2009). Field report: Seed-borne disease and organics.

Cullen, D. W., Lees, A. K., Toth, I. K., & Duncan, J. M. (2001). Conventional PCR and real-time quantitative PCR detection of *Helminthosporium solani* in soil and potato tubers. *Eur. J. Plant Pathol.*, *107*, 387–398.

De Boer, S. H., & Lopez, M. M. (2012). New grower-friendly methods for plant pathogen monitoring. *Annu. Rev. Phytopathol.*, *50*, 197–218.

De Boer, S. H., Ward, L. J., Li, X., & Chittaranjan, S. (1995). Attenuation of PCR inhibition in the presence of plant compounds by addition of BLOTTO. *Nucleic Acids Res.*, *23*, 2567–2568.

De Tempe, J. (1961). Routine methods for determining the health condition of seeds in the seed testing station. *Proc. Int. Seed Test. Assoc.*, *26*, 27–60.

De Tempe, J., & Binnerts, J. (1979). Introduction to methods of seed health testing. *Seed Sci. Technol.*, *7*, 601–636.

Diekmann, M. (1988). Seed health testing and treatment of germplasm at the International Center for Agricultural Research in the Dry Areas (ICARDA). *Seed Sci. Technol.*, *16*, 405–417.

Du Toit, L. J., Derie, M. L., & Hernandez-Perez, P. (2005). Verticillium wilt in spinach seed production. *Plant Dis.*, *89*, 4–11.

Duressa, D., Rauscher, G., Koike, S. T., Mou, B., Hayes, R. J., Maruthachalam, K., Subbarao, K. V., & Klosterman, S. J. (2012). A real-time PCR assay for detection and quantification of *Verticillium dahliae* in spinach seed. *Phytopathology*, *102*, 443–451.

Elmer, W. H. (2001). Seeds as vehicles for pathogen importation. *Biol. Invasions*, *3*, 263–271.

El-Nagerabi, S., & Elshafie, E. (2000). Incidence of seed-borne fungi and aflatoxins in Sudanese lentil seeds. *Mycopathologia*, *149*, 151–156.

Elwakil, M. A., El-Refai, I. M., Awadallah, O. A., El-Metwally, M. A., & Mohammed, M. S. (2009). Seed-borne pathogens of faba bean in Egypt: Detection and pathogenicity. *Plant Pathol. J.*, *8*, 90–97.

Erlich, H. A., Gelfand, D. H., & Saiki, R. K. (1988). Specific DNA amplification. *Nature*, *331*, 461–462.

Forster, M. K., Sedaghatjoo, S., Maier, W., Killermann, B., & Niessen, L. (2022). Discrimination of *Tilletia controversa* from the *T. caries/T. laevis* complex by MALDI-TOF MS analysis of teliospores. *Appl. Microbiol. Biotechnol.*, *106*, 1257–1278.

Fox, R. T. V. (1993). *Principles of Diagnostic Techniques in Plant Pathology*. CAB International: Wallingford, UK.

França-Silva, F., Rego, C. H. Q., Gomes-Junior, F. G., Moraes, M. H. D., Medeiros, A. D., & Silva, C. B. (2020). Detection of *Drechslera avenae* (Eidam) Sharif [*Helminthosporium avenae* (Eidam)] in black oat seeds (*Avena strigosa* Schreb) using multispectral imaging. *Sensors*, *20*, 3343.

Franken, A. A. J. M. (1992). Comparison of immunofluorescence microscopy and dilution plating for the detection of *Xanthomonas campestris* pv. *campestris* in crucifer seeds. *Neth. J. Plant Pathol.*, *98*, 169–178.

Franken, A. A. J. M., & Van Vuurde, J. W. L. (1990). Problems and new approaches in the use of serology for seedborne bacteria. *Seed Sci. Technol.*, *18*, 415–426.

Fu, S., Qu, G., Guo, S., Ma, L., Zhang, N., Zhang, S., Gao, S., & Shen, Z. (2011). Applications of loop-mediated isothermal DNA amplification. *Appl. Biochem. Biotechnol.*, *163*, 845–850.

Fukuta, S., Iida, T., Mizukami, Y., Ishida, A., Ueda, J., Kanbe, M., & Ishimoto, Y. (2003). Detection of Japanese yam mosaic virus by RT-LAMP. *Arch. Virol.*, *148*, 1713–1720.

Fukuta, S., Ohishi, K., Yoshida, K., Mizukami, Y., Ishida, A., & Kanbe, M. (2004). Development of immunocapture reverse transcription loop-mediated isothermal amplification for the detection of tomato spotted wilt virus from chrysanthemum. *J. Virol. Methods*, *121*, 49–55.

Gachon, C., Mingam, A., & Charrier, B. (2004). Real-time PCR: What relevance to plant studies? *Journal of Experimental Botany*, *55*, 1445–1454.

Gebeyaw, M. (2020). Impact of seed-borne pathogens on seed quality. *Journal of Plant Biology*, *5*, 77–81.

Gitaitis, R., & Walcott, R. (2007). The epidemiology and management of seedborne bacterial diseases. *Annual Review of Phytopathology*, *45*, 371–397.

Glynn, N. C., Hare, M. C., & Edwards, S. G. (2008). Fungicide seed treatment efficacy against *Microdochium nivale* and *M. majus in vitro* and *in vivo*. *Pest Management Science*, *64*, 793–799.

Grogan, R. G., & Kimble, K. A. (1967). The role of seed contamination in the transmission of *Pseudomonas phaseolicola* in *Phaseolus vulgaris*. *Phytopathology*, *57*, 28–31.

Guillemette, T., Iacomi-Vasilescu, B., & Simoneau, P. (2004). Conventional and real-time PCR-based assay for detecting pathogenic *Alternaria brassicae* in cruciferous seed. *Plant Disease*, *88*, 490–496.

Guthrie, J. W., Huber, D. M., & Fenwick, H. S. (1965). Serological detection of halo-blight. *Plant Disease Reporter*, *49*, 297–299.

Ha, Y., Fessehaie, A., Ling, K., Wechter, W., Keinath, A., & Walcott, R. (2009). Simultaneous detection of *Acidovorax avenae* subsp. *citrulli* and *Didymella bryoniae* in cucurbit seedlots using magnetic capture hybridization and real-time polymerase chain reaction. *Phytopathology*, *99*, 666–678.

Haikal, N. Z. (2008). Effect of filtrates of pathogenic fungi of soybean on seed germination and seedling parameters. *Journal of Applied Sciences Research*, *4*, 48–52.

Hampton, R., Ball, E., & De Boer, S. (1990). *Serological Methods for Detection and Identification of Viral and Bacterial Plant Pathogens: A Laboratory Manual*. APS Press.

Hawksworth, D. L., Kirk, P. M., Sutton, B. C., & Pegler, D. N. (1995). *Dictionary of the Fungi* (8th ed.). CAB International.

Hensen, J. M., & French, R. (1993). The polymerase chain reaction and plant disease diagnosis. *Annual Review of Phytopathology*, *31*, 81–109.

Higley, P., McGee, D., & Burris, J. (1993). Development of methodology for non-destructive assay of bacteria, fungi and viruses in seeds of large-seeded field crops. *Seed Science and Technology*, *21*, 399–409.

Horst, R. K. (2008). *Westcott's Plant Disease Handbook* (7th ed.). Springer Science.

Ishii, H., Tanoue, J., Oshima, M., Chung, W. H., Nishimura, K., Yamaguchi, J., Nemoto, F., So, K., Iwama, T., Yoshimatsu, H., & Shimizu, M. (2008). First application of PCR-Luminex system for molecular diagnosis of fungicide resistance and species identification of fungal pathogens. *Journal of General Plant Pathology*, *74*, 409–416.

Iwamoto, T., Sonobe, T., & Hayashi, K. (2003). Loop-mediated isothermal amplification for direct detection of *Mycobacterium tuberculosis* complex, *M. avium*, and *M. intracellulare* in sputum samples. *Journal of Clinical Microbiology*, *41*, 2616–2622.

Jacobsen, C. S. (1995). Microscale detection of specific bacterial DNA in soil with a magnetic capture-hybridization and PCR amplification assay. *Applied and Environmental Microbiology*, *61*, 3347–3352.

Jorgensen, J. (1977). Incidence of infections of barley seed by *Pyrenophora graminea* and *P. teres* as revealed by the freezing blotter method and disease counts in the field. *Seed Science and Technology*, *5*, 105–110.

Kaygusuz, T., & Coşkuntuna, A. (2022). Determination of seed-borne fungi in some medicinal and aromatic plants. *Applied Ecology and Environmental Research*, *20*(3), 2553–2564.

Kazinczi, G., & Horvath, J. (1998). Transmission of sowbane mosaic sobemovirus by seeds of *Chenopodium* species and viability of seeds. *Acta Phytopathologica et Entomologica Hungarica*, *33*, 21–26.

Kućmierz, J., & Gorajczyk, S. (1991). Mycoflora of perennial ryegrass (*Lolium perenne* L.) grains. *Phytopathology*, *1*, 58–61.

Kumar, P. L., Cuervo, M., Kreuze, J. F., Muller, G., Kulkarni, G., Kumari, S. G., Massart, S., Mezzalama, M., Alakonya, A., & Muchugi, A. (2021). Phytosanitary interventions for safe global germplasm exchange and the prevention of transboundary pest spread: The role of CGIAR germplasm health units. *Plants*, *10*, 328.

Kumar, R., Gupta, A., Srivastava, S., Devi, G., Singh, V. K., Goswami, S. K., Gurjar, M. S., & Aggarwal, R. (2020). Diagnosis and detection of seed-borne fungal phytopathogens. In *Seed-Borne Diseases of Agricultural Crops: Detection, Diagnosis & Management*. Springer.

Lange, L. (1986). The practical application of new developments in test procedures for the detection of viruses in seeds. *Development in Applied Biology*, *1*, 269–281.

Li, S. (2011). Phomopsis seed decay of soybean. In A. Sudaric (Ed.), *Soybean–Molecular Aspects of Breeding*.

Lievens, B., & Thomma, B. P. H. J. (2005). Recent developments in pathogen detection arrays: Implications for fungal plant pathogens and use in practice. *Phytopathology*, *95*, 1374–1380.

Limonard, T. (1966). A modified blotter test for seed health. *Netherlands Journal of Plant Pathology*, *72*, 319–321.

Lopez, M. M., Bertolini, E., Olmos, A., Caruso, P., Gorris, M. T., Llop, P., Penyalver, R., & Cambra, M. (2009). *Innovative Tools for Detection of Plant Pathogenic Viruses and Bacteria*. Springer.

Lyons, N. F., & Taylor, J. D. (1990). Serological detection and identification of bacteria from plants by the conjugated *Staphylococcus aureus* slide agglutination test. *Plant Pathology*, *39*, 584–590.

Magan, N., & Lacey, J. (1988). Ecological determinants of mold growth in stored grain. *International Journal of Food Microbiology*, *7*, 245–256.

Magan, N., Sanchis, V., & Aldred, D. (2004). Role of spoilage fungi in seed deterioration. In *Fungal Biotechnology in Agricultural, Food and Environmental Applications*. Marcel Dekker.

Maguire, J. D., Gabrielson, R. L., Mulanax, M. W., & Russell, T. S. (1978). Factors affecting the sensitivity of 2,4-D assays of crucifer seed for *Phoma lingam*. *Seed Science and Technology, 6*, 915–924.

Marcinkowska, J. (1998). Variability of dry seed mycobiota of *Pisum sativum*. *Acta Mycologica, 33*, 91–99.

Marcinkowska, J. Z. (2002). Methods of finding and identification of pathogens in seeds. *Plant Breeding and Seed Science, 46*, 31–48.

Marrou, J., & Messiaen, C. M. (1967). The *Chenopodium quinoa* test: A critical method for detecting seed transmission of lettuce mosaic virus. *Proceedings of the International Seed Testing Association, 32*, 49–57.

Mashilla, D. (2004). Grain storage methods and their effect on sorghum grain quality in Hararghe, Ethiopia (PhD thesis). Swedish University of Agricultural Science.

Maude, R. B. (1999). *Seedborne Diseases and Their Control: Principles and Practice*. CAB International.

Mbofung, G., & Pryor, B. (2010). A PCR-based assay for detection of *Fusarium oxysporum* f. sp. *lactucae* in lettuce seed. *Plant Disease, 94*, 860–866.

McCartney, H. A., Foster, S. J., Fraaije, B., & Ward, E. (2003). Molecular diagnostics for fungal plant pathogens. *Pest Management Science, 59*, 129–142.

McGee, D. C. (1995). Approach to disease management through seed technology. *Annual Review of Phytopathology, 33*, 445–466.

McLaughlin, R. J., & Chen, T. A. (1990). *ELISA Methods for Plant Pathogenic Prokaryotes*. APS Press.

Mehrotra, R. S., & Aggarwal, A. (2003). *Plant Pathology* (2nd ed.). Tata McGraw-Hill Publishing Company Limited.

Mguni, C. M., Mortensen, C. N., Keswani, C. L., & Hockenhull, J. (1999). Detection of the black rot pathogen (*Xanthomonas campestris* pv. *campestris*) and other xanthomonads in Zimbabwean and imported *Brassica* seed. *Seed Science and Technology, 27*, 447–454.

Misra, J. K., Mew, T., & Marca, S. D. (1994). Rice seed health and quarantine. In *A Manual of Rice Seed Health Testing*. IRRI.

Mori, Y., Kitao, M., Tomita, N., & Notomi, T. (2004). Real-time turbidimetry of LAMP reaction for quantifying template DNA. *Journal of Biochemical and Biophysical Methods, 59*, 145–157.

Mullis, K. B., & Faloona, F. A. (1987). Specific synthesis of DNA *in vitro* via a polymerase-catalyzed chain reaction. *Methods in Enzymology, 155*, 335–350.

Mumford, R., Boonham, N., Tomlinson, J., & Barker, I. (2006). Advances in molecular phytodiagnostics–New solutions for old problems. *European Journal of Plant Pathology, 116*, 1–19.

Munkvold, G. P. (2009). Seed pathology progress in academia and industry. *Annual Review of Phytopathology, 47*, 285–311.

Murakishi, H. H. (2002). Purple seed stain of soybean. *Phytopathology, 41*, 305–318.

Nameth, S. T. (1998). Priorities in seed pathology research. *Proceedings of the Seed Biology Symposium*, Piracicaba, Brazil.

Naqvi, S., Shiden, T., Merhawi, W., & Mehret, S. (2013). Identification of seed-borne fungi on farmer saved sorghum (*Sorghum bicolor* L.), pearl millet (*Pennisetum glaucum* L.) and groundnut (*Arachis hypogaea* L.) seeds. *Research Journal of Agricultural Science, 3*, 107–114.

Nasir, N. (2003). Detecting seed-borne fungi of soybean by different incubation methods. *Pakistan Plant Pathology Journal, 2*, 114–118.

Neergaard, P. (1997). *Seed Pathology* (Vols. 1–2). The MacMillan Press.

Niaz, I., & Dawar, S. (2009). Detection of seed-borne mycoflora in maize (*Zea mays* L.). *Pakistan Journal of Botany, 41*, 443–451.

Nie, X. Z. (2005). Reverse transcription loop-mediated isothermal amplification of DNA for detection of potato virus Y. *Plant Disease, 89*, 605–610.

Niessen, L., & Vogel, R. F. (2010). Detection of *Fusarium graminearum* DNA using a loop-mediated isothermal amplification (LAMP) assay. *International Journal of Food Microbiology, 140*, 183–191.

Notomi, T., Okayama, H., Masubuchi, H., Yonekawa, T., Watanabe, K., Amino, N., & Hase, T. (2000). Loop-mediated isothermal amplification of DNA. *Nucleic Acids Research, 28*, 63.

Ojeda, S., & Verdier, V. (2000). Detecting *Xanthomonas axonopodis* pv. *manihotis* in cassava true seeds by nested polymerase chain reaction assay. *Canadian Journal of Plant Pathology, 22*, 241–247.

Okuda, M., Matsumoto, M., Tanaka, Y., Subandiyah, S., & Iwanami, T. (2005). Characterization of the tufB-secE-nusG-rplKAJL-rpoB gene cluster of the citrus greening organism and detection by loop-mediated isothermal amplification. *Plant Disease, 89*, 705–711.

Old, R. W., & Primrose, S. B. (1985). *Principles for Gene Manipulation: An Introduction to Genetic Engineering* (3rd ed.). Blackwell Scientific Publications.

Olesen, M. H., Carstensen, J. M., & Boelt, B. (2011). Multispectral imaging as a potential tool for seed health testing of spinach (*Spinacia oleracea* L.). *Seed Science and Technology, 39*, 140–150.

Oliver, R. (1993). Nucleic acid-based methods for detection and identification. CAB International.

Olsvik, O., Popovic, T., Skjerve, E., Cudjoe, K. S., Hornes, E., Ugelstad, J., & Uhlen, M. (1994). Magnetic separation techniques in diagnostic microbiology. *Clinical Microbiology Reviews, 7*, 43–54.

Ora, N., Faruq, A. N., Islam, M. T., Akhtar, N., & Rahman, M. M. (2011). Detection and identification of seed-borne pathogens from some cultivated hybrid rice varieties in Bangladesh. *Middle-East Journal of Scientific Research, 10*, 482–488.

Osorio, J. A., & McGee, D. C. (1992). Effects of freeze damage on soybean seed mycoflora and germination. *Plant Disease, 76*, 879–882.

Paplomatas, E. J. (2006). Molecular diagnostics of fungal pathogens. *Arab Journal of Plant Protection, 24*, 147–158.

Pellegrino, C., Gilardi, G., Gullino, M. L., & Garibaldi, A. (2010). Detection of *Phoma valerianellae* in lamb's lettuce seeds. *Phytoparasitica, 38*, 159–165.

Peres, A. P., Nasser, L. C. B., & Machado, J. C. (2002). Use of semi-selective media for detection of *Sclerotinia sclerotiorum* on bean and soybean seeds. *Fitopatologia Brasileira, 27*, 123–127.

Peters, J., Sledz, W., Bergervoet, J. H. W., & van der Wolf, J. M. (2007). An enrichment microsphere immunoassay for the detection of *Pectobacterium atrosepticum* and *Dickeya dianthicola* in potato tuber extracts. *European Journal of Plant Pathology*, *117*, 97–107.

Phan, H., Ford, R., Bretag, T., & Taylor, P. (2002). A rapid and sensitive polymerase chain reaction (PCR) assay for detection of *Ascochyta rabiei*, the cause of ascochyta blight of chickpea. *Australasian Plant Pathology*, *31*, 31–39.

Plucknett, D. L. (1989). Quarantine and the exchange of crop genetic resources. *Bioscience*, *39*, 16–23.

Pooja, U., & Singh, S. P. (2019). Detection methods for seed-borne pathogens. *International Journal of Current Microbiology and Applied Sciences*, *8*(3), 318–323.

Prosen, D., Hatziloukas, E., Panopoulos, N. J., & Schaad, N. W. (1991). Direct detection of the halo blight pathogen *Pseudomonas syringae* pv. *phaseolicola* in bean seed by DNA amplification. *Phytopathology*, *81*, 1159.

Pryor, B. M., & Gilbertson, R. L. (2001). A PCR-based assay for detection of *Alternaria radicina* on carrot seed. *Plant Disease*, *85*, 18–23.

Pyndji, M. M., Sinclair, J. B., & Singh, T. (1987). Soybean seed thermotherapy with heated vegetable oils. *Plant Disease*, *71*, 213–216.

Rabelo, G. F., Enes, A. M., Braga, R., Dal, F., & Fabbro, I. M. (2011). Frequency response of biospeckle laser images of bean seeds contaminated by fungi. *Biosystems Engineering*, *110*, 297–301.

Rahman, M. M. E., Ali, M. E., Ali, M. S., Rahman, M. M., & Islam, M. N. (2008). Hot water thermal treatment for controlling seed-borne mycoflora of maize. *International Journal of Sustainable Crop Production*, *3*, 5–9.

Rajeshwari, N., Shylaja, M. D., Krishnappa, M., Shetty, H. S., Mortensen, C. N., & Mathur, S. B. (1998). Development of ELISA for the detection of *Ralstonia solanacearum* in tomato: Its application in seed health testing. *World Journal of Microbiology and Biotechnology*, *14*, 697–704.

Rajput, M. A., Pathan, M. A., Lodhi, A. M., Shah, G. S., & Khanzada, K. A. (2005). Studies on seed-borne fungi of wheat in Sindh Province and their effect on seed germination. *Pakistan Journal of Botany*, *37*, 181–185.

Randall-Schadel, B., Bailey, J., & Beute, M. (2001). Seed transmission of *Cylindrocladium parasiticum* in peanut. *Plant Disease*, *85*, 362–370.

Rao, N. K., Hanson, J., Dulloo, M. E., Ghosh, K., Nowell, D., & Larinde, M. (2006). *Manual of seed handling in genebanks*. Bioversity International.

Rasmussen, O. F., & Wulff, B. S. (1991). Detection of *Pseudomonas* pv. *pisi* using PCR. In *Proceedings of the 4th International Working Group on Pseudomonas syringae pathovars* (pp. 369–376).

Reeves, J. (1998). Molecular diagnostics for pathogen detection in seeds and planting materials. *Plant Cell, Tissue and Organ Culture*, *52*, 33–39.

Reeves, J. C. (1995). Nucleic acid techniques in testing for seedborne diseases. In *New Diagnostics in Crop Sciences*. CAB International.

Reeves, J. C., Rasmussen, O. F., & Simpkins, S. A. (1994). The use of DNA probe and PCR for the detection of *Pseudomonas syringae* pv. *pisi* in pea seed. In *Plant Pathogenic Bacteria: Proceedings of the 8th International Conference* (pp. 383–389).

Roberts, S. J., Brough, J., & Chakrabarty, S. (2002). Non-destructive seed testing for bacterial pathogens in germplasm material. *Seed Science and Technology*, *30*, 69–85.

Rodrigues, A. A. C., & Menezes, M. (2005). Identification and pathogenic characterization of endophytic *Fusarium* species from cowpea seeds. *Micropathological, 159*, 79–85.

Roth, D. A. (1989). *Review of Extraction and Isolation Methods*. APS Press.

Roy, K., Baird, R., & Abney, T. (2000). A review of soybean (*Glycine max*) seed, pod, and flower mycofloras in North America, with methods and a key for identification of selected fungi. *Mycopathologia, 150*, 15–27.

Russell, R. (1985). Longevity studies with wheat seed and certain seed-borne fungi. *Canadian Journal of Plant Science, 38*, 29–33.

Saiki, R. K., Gelfand, D. H., Stoffel, S., Scharf, S. J., Higuchi, R., Horn, G. T., Mullis, K. B., & Erlich, H. A. (1988). Primer-directed enzymatic amplification of DNA with a thermostable DNA polymerase. *Science, 239*, 487–491.

Schaad, N. W. (1989). *Detection and Identification of Bacteria*. APS Press.

Schaad, N. W., & Donaldson, R. C. (1980). Comparison of two methods for detection of *Xanthomonas campestris* in infected crucifer seeds. *Seed Science and Technology, 8*, 383–391.

Schaad, N. W., & Frederick, R. D. (2002). Real-time PCR and its application for rapid plant disease diagnostics. *Canadian Journal of Plant Pathology, 24*, 250–258.

Schaad, N. W., Bonde, M. R., & Hatziloukas, E. (1997). BIO-PCR: A highly sensitive technique for detecting seedborne fungi and bacteria. CAB International.

Schaad, N. W., Cheong, S. S., Tamaki, S., Hatziloukas, E., & Panopoulos, N. J. (1995). A combined biological and enzymatic amplification (BIO-PCR) technique to detect *Pseudomonas syringae* pv. *phaseolicola* in bean extracts. *Phytopathology, 85*, 243–248.

Scheel, C. (1997). Review on policy development with regard to seed health testing and seed treatment in the Nordic countries with special reference to Denmark. CAB International.

Schena, L., Nigro, F., Ippolito, A., & Gallitelli, D. (2004). Real-time quantitative PCR: A new technology to detect and study phytopathogenic and antagonistic fungi. *European Journal of Plant Pathology, 110*, 893–908.

Sharma, J. K., Ali, M. I. M., & Sutherland, J. R. (1997). Seed pathology of tropical hardwoods. In Z. Prochazkova (Ed.), *Proceedings of the ISTA Tree Seed Pathology Meeting, Opocno, Czech Republic, 9–11 October* (pp. 74–81).

Sheppard, J. W., Roth, D. A., & Saettler, A. W. (1989). *Detection of Xanthomonas campestris* pv. *phaseoli* in bean. APS Press.

Sheppard, J. W., Wright, P. F., & DeSavigny, D. H. (1986). Methods for the evaluation of ELISA tests for use in the detection of seed-borne diseases. *Seed Science and Technology, 14*, 49–59.

Shovan, L. R., Bhuiyan, M. K. A., Sultana, N., Begum, J. A., & Pervez, Z. (2008). Prevalence of fungi associated with soybean seeds and pathogenicity tests of the major seed-borne pathogens. *Journal of Sustainable Crop Production, 3*, 24–33.

Taylor, J. D. (1970). Bacteriophage and serological methods for the identification of *Pseudomonas phaseolicola* (Burkh.) Dowson. *Annals of Applied Biology, 66*, 387–395.

Taylor, J. D., Bevan, J. R., Crute, I. R., & Reader, S. L. (1989). Genetic relationship between races of *Pseudomonas syringae* pv. *pisi* and cultivars of *Pisum sativum*. *Plant Pathology, 38*, 364–375.

Taylor, J. L. (1993). A simple, sensitive, and rapid method for detecting seed contaminated with highly virulent *Leptosphaeria maculans*. *Applied and Environmental Microbiology, 59*, 3681–3685.

Thai, H. T. C., Le, M. Q., Vuong, C. D., Parida, M., Minekawa, H., Notomi, T., Hasebe, F., & Morita, K. (2004). Development and evaluation of a novel loop-mediated isothermal amplification method for rapid detection of severe acute respiratory syndrome coronavirus. *Journal of Clinical Microbiology, 42*, 1956–1961.

Thomson, J. R. (1970). *Proceedings of the International Seed Testing Association, 37*(1), 9–17.

Tomlinson, J. A., Boonham, N., Hughes, K. J. D., Griffin, R. L., & Barker, I. (2005). On-site DNA extraction and real-time PCR for detection of *Phytophthora ramorum* in the field. *Applied and Environmental Microbiology, 71*, 6702–6710.

Trojanowska, K. (1991). Evaluation of cereal grain quality using mycological methods. In C. Chelkowski (Ed.), *Cereal grain: Mycotoxins, Fungi and Quality in Drying and Storage*. Elsevier.

Trujillo, G. E., & Saettler, A. W. (1979). A combined semi-selective medium and serology test for the detection of *Xanthomonas* blight bacteria in bean seed. *Journal of Seed Science and Technology, 4*, 35–41.

Tseng, T. C., Tu, J. C., & Tzean, S. S. (1991). Mycoflora and mycotoxins in dry bean (*Phaseolus vulgaris*) produced in Taiwan and in Ontario, Canada. *Botanical Bulletin of Academia Sinica, 36*, 229–234.

Vallad, G., Bhat, R., Koike, S., Ryder, E., & Subbarao, K. (2005). Weedborne reservoirs and seed transmission of *Verticillium dahliae* in lettuce. *Plant Disease, 89*, 317–324.

Van Vuurde, J. W. L. (1997). Immunofluorescence colony-staining as a tool for sample indexing and for the determination of field thresholds for pathogenic seedborne bacteria. CAB International.

Van Vuurde, J. W. L., & Van den Bovenkamp, G. W. (1981). Routine methods for the detection of halo-blight (*Pseudomonas phaseolicola*) in beans. In *Report of the 17th International Workshop on Seed Pathology*. International Seed Testing Association.

Vivian, A. (1992). Identification of plant pathogenic bacteria using nucleic acid technology. Blackwell Scientific Publications.

Wakuliński, W., & Chelkowski, J. (1993). *Fusarium* species transmitted with seeds of wheat, rye, barley, oats and triticale. *Hodowla Roślin, Aklimatyzacja i Nasiennictwo, 37*, 131–136.

Walcott, R. R. (2003). Detection of seedborne pathogens. *HortTechnology, 13*, 40–47.

Walcott, R. R., & Gitaitis, R. D. (2000). Detection of *Acidovorax avenae* subsp. *citrulli* in watermelon seed using immunomagnetic separation and the polymerase chain reaction. *Plant Disease, 84*, 470–474.

Walcott, R. R., Gitaitis, R. D., & Langston, D. B. (2004). Detection of *Botrytis aclada* in onion seed using magnetic capture hybridization and the polymerase chain reaction. *Seed Science and Technology, 32*, 425–438.

Ward, E., Foster, S., Fraaije, B., & McCartney, H. A. (2004). Plant pathogen diagnostics: Immunological and nucleic acid-based approaches. *Annals of Applied Biology, 145*, 1–16.

Weller, S. A., Elphinstone, J. G., Smith, N. C., Boonham, N., & Stead, D. E. (2000). Detection of *Ralstonia solanacearum* strains with a quantitative, multiplex, real-time, fluorogenic PCR (TaqMan) assay. *Applied and Environmental Microbiology*, *66*, 2853–2858.

Wiewióra, B. (2009). Long-time storage effect on the seed health of spring barley grain. *Plant Breeding and Seed Science*, *59*, 3–12.

Wiwart, M., & Korona, A. (1997). The mycoflora of triticale grains in relation to cultivation system. *Biuletyn IHAR*, *201*, 263–268.

Yeh, C. C., & Sinclair, J. B. (1982). Effect of *Cercospora kikuchii* on soybean seed germination and its interaction with *Phomopsis* sp. *Phytopathology*, *105*, 265–270.

CHAPTER 17

Biotechnological Intervention for Sustainable Production of Quality Planting Material of Critically Endangered Medicinal Plant Species

AMAN DABRAL,[1,*] KAJAL MOG CHAUDHURI,[2] NARESH SINGH,[1] BIPLOV CH. SARKAR,[3] GOPAL SHUKLA,[4,*] C. P. SURESH,[5] and MANMOHAN J. DOBRIYAL[6]

[1]*Department of Forest Biology and Tree Improvement, College of Forestry, Ranichauri, VCSG Uttarakhand University of Horticulture and Forestry, Uttarakhand, India*

[2]*Department of Seed Science and Technology, Uttar Banga Krishi Viswavidyalaya, Pundibari, Cooch Behar, West Bengal, India*

[3]*Faculty of Agricultural Sciences, GLA University, Mathura, Chaumuhan, Uttar Pradesh, India*

[4]*Department of Forestry, North-Eastern Hill University, Tura Campus, Tura, Meghalaya, India*

[5]*Department of Forestry, North Eastern Hill University, Tura Campus, West Garo Hills District, Chasingre, Tura, Meghalaya, India*

[6]*College of Horticulture and Forestry, Rani Lakshmi Bai Central Agricultural University, Jhansi, Uttar Pradesh, India*

*Corresponding author

Seed Technology for Medicinal and Aromatic Plants: Production, Testing, Storage, and Marketing. Priyanka Sharma, Kishore Kumar Sharma, and Gopal Shukla (Eds.)

DOI: 10.1201/9781779641199-17

ABSTRACT

The conservation of medicinal plants is highly significant due to their crucial role in traditional and contemporary medicinal practices and their economic value. However, the increasing demand for plant-based medicines leads to the excessive utilization and the fragmentation of their natural habitats and poses a threat to numerous medicinal plant species. Notably, tools such as biotechnological methods offer viable approaches to tackling the issues connected with the long-term production of high-quality planting material for critically endangered medicinal plants. Procedures such as tissue culture, cryopreservation, and genetic engineering facilitate medicinal plants' reproduction, conservation, and genetic enhancement, surmounting obstacles arising from habitat destruction and climate shifts. Biotechnology also aids in investigating bioactive compounds, augmenting their medicinal attributes and therapeutic efficacy. Importantly, the application of biotechnology approaches such as genomics, metabolomics, and bioinformatics provides a summary of the genetic basis of medicinal properties and the metabolic pathways involved in the synthesis of bioactive compounds. Collective endeavors among scientists, policymakers, and local communities are vital for the successful implementation of biotechnology-driven conservation strategies.

17.1 INTRODUCTION

Historically, humankind has been using medicinal plants for the care and prevention of numerous ailments (Feyisa et al., 2024). These plants provide a variety of bioactive chemicals that provide health advantages and are widely utilized in pharmaceutical formulations (Yuan et al., 2016). Herbal treatments have been fundamental to human healthcare since antiquity, and their application continues to be widespread in both developed and developing countries for many medical diseases. The worldwide market for therapeutic plant-derived medicines is significant, with yearly sales projected to surpass $100 billion (Sofowora et al., 2013). Medicinal herbs are essential resources for basic healthcare in rural communities, underpinning both traditional and contemporary medical practices. Furthermore, they offer economic advantages, especially in areas where items sourced from wild-harvested medicinal plants enhance local lives (Hamilton, 2004).

17.1.1 SIGNIFICANCE OF MEDICINAL PLANTS

Approximately, the vascular plant species range from 350,000 to 500,000 (Joppa et al., 2011; Pimm et al., 2014); however, 10% of these species possess therapeutic properties (Grover et al., 2002). Their first discovery relied on trial and error, as early people utilized plants to treat diseases or enhance well-being (Kunle et al., 2014). Over years, the conventional therapeutic application of these plants has been refined, resulting in their extensive acknowledgment and acceptance. The World Health Organization (WHO) defines traditional medicine as "the total of the knowledge, skills, and practices based on the theories, beliefs, and experiences indigenous to different cultures, whether explicable or not, used in the prevention, diagnosis, improvement, or treatment of physical and mental illnesses" (WHO, 2000). It is widely acknowledged that all civilizations have used plants found in their native environments for medicinal purposes (Gurib, 2006; Houghton, 1995). Some researchers even suggest that pharmacy and medicine originated from this inherited knowledge. Today, many higher plants are still cultivated for the extraction of valuable compounds used in pharmaceutical production. The medicinal properties of plants have enabled the development of plant-derived pharmaceuticals (Jones et al., 2006). Until the 18th century, the medicinal effects and modes of administration for many plants were well-known, even though their active components had not yet been identified. For instance, a foundational work authored by the Persian physician and scientist Ibn Sina, that is Avicenna's Canon of Medicine, presented wide-ranging descriptions of medicinal plants and their use (Koh, 2009). Importantly, the discovery of the microscope and the development of methodology for the isolation of active compounds from plants leads to the extensive use of such important plants. These findings established the groundwork for the laboratory synthesis of active compounds, resulting in the development of contemporary pharmaceuticals. Notwithstanding the rising prevalence of synthetic medications, traditional medicinal plants persist in widespread usage, especially in regions where they are accessible and economically viable (Gertsch, 2009). The pharmaceutical industry, fundamental to contemporary medicine, remains significantly dependent on plant-derived active compounds as raw materials. In numerous impoverished nations, traditional medicine continues to be an essential resource owing to its cost-effectiveness and accessibility, particularly when juxtaposed with contemporary synthetic pharmaceuticals (Salmerón et al., 2020).

17.1.2 CONTEMPORARY STATUS OF CRITICALLY ENDANGERED MEDICINAL PLANT SPECIES

Conservation of biodiversity, especially the conservation of critically endangered medicinal plant species, is an increasingly complex endeavor. Conserving these species encounters obstacles at societal, political, and scientific dimensions (Pellettier et al., 2018). The extensive variety of plant species and constrained financial resources for conservation require a prioritization strategy, wherein species most at risk are discovered, evaluated, and selected for intervention. Medicinal plants, numerous of which are severely endangered, are essential to both traditional and contemporary medicine; yet, their existence is jeopardized by overharvesting, habitat destruction, climate change, and invasive species. The International Union for Conservation of Nature (IUCN) created the "Red List of Threatened Species" to tackle these difficulties, serving as an essential instrument for evaluating the global conservation status of plant species (Brooks et al., 2004; Pollock et al., 2003). Moreover, the Red List classifies species from least concern (LC) to critically endangered (CR) and extinct (EX); the criteria for categorizing the species was population size, geographic distribution, and decline rate (Collen et al., 2016; Khoshbakht and Hammer, 2005). Notably, many critically endangered medicinal plants, such as *Taxus wallichiana* (Himalayan yew), and *Saussurea costus* (Kuth), are being used for paclitaxel production, that is, a cancer therapy medication; overexploitation of these plants leading to extinction (Kala, 2005). *Panax ginseng*, a significant medicinal plant in traditional Asian medicine, is critically threatened in its natural habitat, mostly due to unsustainable harvesting practices and habitat destruction (Chen et al., 2008). The probable extinction of many therapeutic plant species could significantly impact global health, as numerous contemporary medications, like aspirin, morphine, and artemisinin, are derived from plants. The increasing demand for herbal-based treatments and medicines has exaggerated the negative impact on wild populations of medicinal plants (Hamilton, 2004). Conservation of critically endangered medicinal plants should be done through sustainable harvesting methods, more stringent enforcement of trade regulations, and cultivation of endangered species in restricted settings to reduce pressure on wild populations (Chen et al., 2016). Moreover, some crucial conservation strategies include *in situ* methods, that is, protection of natural habitats, and *ex situ* conservation strategies, such as botanical gardens, seed banks, and tissue culture technologies (Maunder et al., 2001). These approaches are crucial for restoring genetic diversity, and further help future research and restoration advantages. Notably, local communities also can play a major role in conservation as they will be provided with education and awareness (Kala, 2005). Some supportive roles by governments, conservation authorities, researchers, and local communities may be helpful for the conservation of critically endangered medicinal plants. Since the conservation of these flora is crucial for both healthcare and economic advancement for humankind. Therefore, sustainable conservation measures should be highlighted to settle growing needs with biodiversity conservation (Salmerón et al., 2020).

17.2 NEED FOR SUSTAINABLE PRODUCTION

17.2.1 ECONOMIC BENEFIT

Medicinal plants are experiencing a surge in global appeal owing to their distinctive therapeutic characteristics and advantages. Herbal medicine is gaining recognition as a prevalent alternative to traditional medical therapies, particularly owing to its historical cultural utilization, widespread acceptability, and supposed negligible adverse effects associated with these plants. In numerous underdeveloped nations, herbal remedies are advocated as an economical substitute for synthetic pharmaceuticals, hence enhancing accessibility for broader population segments. Moreover, initiatives are underway to identify, register, and patent derivatives of medicinal plants, hence advancing innovation in this domain. Nevertheless, numerous plant-derived medications are still in the requisite drug approval stages, with their efficacy and safety not yet fully substantiated. Pharmaceutical firms focused on natural products and herbal medicine can leverage the therapeutic benefits of these plants, fostering future market growth. The global market for medical plants and related products, valued at USD 60 billion in 2010, is anticipated to attain USD 5 trillion by 2050 (Zahra et al., 2020). Asian nations, especially those rich in medicinal plant resources, lead the export market for these items. By advocating for the sustainable utilization and production of medicinal plants, economically disadvantaged Asian nations may bolster their economies and generate employment prospects for local populations. Nonetheless, safeguarding these invaluable medicinal plant species is equally crucial, as many are threatened by overharvesting due to significant market demand. Sustainable harvesting procedures and measures to promote their growth are essential for guaranteeing the long-term availability of these species for future medical and economic uses (Zahra et al., 2020).

17.2.2 OVEREXPLOITATION AND HABITAT DESTRUCTION

The increasing worldwide population, along with the rising demand for medicinal plants, is exerting significant pressure on natural resources. This has resulted in the overexploitation of numerous wild plant species, pushing them towards depletion and, in certain instances, extinction. The swift deterioration of natural habitats caused by deforestation, urbanization, and agricultural development intensifies this issue. The reduction of wild habitats results in a corresponding decrease in the diversity of medicinal plants, causing a troubling drop in biodiversity (Mir et al., 2021). Medicinal plants, integral to both ancient and contemporary healthcare systems, are frequently plucked indiscriminately from their natural habitats. Unregulated and unsustainable harvesting practices have greatly contributed to the extinction of high-demand species such as *Taxus baccata* (European yew), *Nardostachys jatamansi* (Spikenard), and *Picrorhiza kurroa* (Kutki), which are now severely endangered (Hamilton, 2004; Kala, 2005). Overharvesting endangers the survival of these species and disrupts ecosystems by diminishing the availability of essential plant species that fulfil ecological roles such as pollination, seed dissemination, and habitat supply for other organisms. Habitat destruction significantly contributes to the decline of medicinal plant variety. The transformation of forests, grasslands, and wetlands into agricultural or urban areas results in the fragmentation and deterioration of ecosystems, hence restricting the natural habitats of medicinal plants. In tropical climates, where several valuable medicinal plants are located, deforestation for timber and land clearance is especially detrimental. Tropical forests are predicted to be disappearing at a rate of approximately 13 million hectares annually, severely affecting the richness of medicinal plants (Hamilton, 2004; Myers et al., 2000). In reaction to the escalating dangers of overexploitation and habitat degradation, various conservation programs have been planned and executed globally. *Ex situ* conservation techniques, including the creation of seed banks, botanical gardens, and tissue culture facilities, are essential for safeguarding the genetic diversity of medicinal plants. These initiatives seek to preserve endangered species for future generations and facilitate their reintroduction into natural habitats when conditions permit (Maunder et al., 2001). Moreover, *in situ* conservation, which emphasizes the preservation of plants within their native habitats, has received considerable advocacy. Protected places, including national parks and animal sanctuaries, are essential for preserving medicinal plant species and their habitats from further deterioration (Kala, 2005). Furthermore, incorporating local communities into conservation initiatives is increasingly seen as an essential approach for sustainable management. Communities dependent on medicinal plants for their livelihoods can serve as useful allies in conservation when equipped with adequate education and other income streams. Community-based conservation initiatives that priorities sustainable harvesting and agricultural practices can alleviate pressure on wild species while providing economic advantages for local communities (Hamilton, 2004; Mir et al., 2021). In the absence of decisive measures, the synergistic impacts of overexploitation and habitat destruction may result in the irreversible extinction of numerous medicinal plant species, thereby affecting both biodiversity and human health. Consequently, conservation methods must be enhanced to alleviate these concerns and foster the sustainable utilization of medicinal plant resources.

17.3 ROLE OF BIOTECHNOLOGY IN CONSERVATION AND PRODUCTION

Biotechnology is essential in improving plant conservation efforts by augmenting and elaborating on existing methods instead of entirely supplanting them. A key contribution of biotechnology is its emphasis on the sustainable management, characterization, and utilization of plant genetic resources (PGR). This comprehensive strategy seeks to guarantee the preservation and sustainable utilization of plant species, especially those that are rare, endangered, or possess significant therapeutic properties (Benson, 1999). Furthermore, biotechnology enables the swift multiplication of plants via micropropagation, guaranteeing the generation of genetically homogenous and disease-free specimens. This is essential for both conservation and the sustained cultivation of therapeutic plants, as well as economically and culturally significant crops. Through the integration of sophisticated molecular techniques, including DNA fingerprinting and marker-assisted selection, biotechnology facilitates the accurate characterization of PGR, hence ensuring the efficacy of conservation initiatives (FAO, 2010). Biotechnology employs *in vitro* procedures and cryopreservation to safeguard plant diversity and significantly contributes to breeding programs that enhance plant resilience and adaptability to environmental changes.

17.3.1 MICROPROPAGATION

Micropropagation, a technique within plant tissue culture, has been an essential instrument in crop development and enhancement for more than a century. This technology facilitates the generation of disease-free plant material via *in vitro* techniques, essential for the reliable and extensive growing of crops (Hussain et al., 2012). Micropropagation's primary application is the regeneration of transgenic plants, allowing researchers to incorporate desirable features like disease resistance and enhanced yields (Sedaghati et al., 2012). In recent years, micropropagation has progressed, enabling the alleviation of dormancy and the multiplication of superior plants with distinct chemotypes. In 2019, Mohan et al. effectively demonstrated the micropropagation of superior medicinal plant species with favorable chemical profiles, facilitating the mass production of plants with uniform bioactive components. This method also aids in the *in vitro* preservation of elite germplasm, guaranteeing the sustained availability of valuable genetic resources (Lata et al., 2016). Micropropagation is particularly advantageous for the conservation of rare and endangered plant species, as it offers a sustainable method for their multiplication without depleting wild populations. Moreover, it facilitates breeding initiatives for diverse crops and the encapsulation of seeds, providing a means to preserve genetic diversity and improve crop yield. Furthermore, investigations into plant biosynthesis through cell and root cultures, coupled with analyses of the symbiotic interactions between endophytes and their host plants, have unveiled novel pathways for comprehending and enhancing plant growth and secondary metabolite synthesis (Rosa et al., 2014). These medicinal plants are essential to the herbal economy, as they supply naturally occurring secondary active biomolecules that are crucial for human health and nourishment. Nevertheless, conventional plant breeding techniques frequently fall short in attaining both qualitative and quantitative enhancements in secondary metabolite levels. Due to the swift progress in plant genomics, researchers possess an enhanced comprehension of the intricate genetics and metabolic mechanisms that govern the manufacture of these vital chemicals. This understanding facilitates the advancement of more accurate genetic improvement methods, enabling plant breeders to enhance the synthesis of bioactive chemicals more efficiently (Chaturvedi et al., 2022).

17.3.2 SHORT- (IN VITRO PROPAGATION) AND MEDIUM-TERM PRESERVATION (SLOW GROWTH)

In vitro propagation is a valuable alternative to conventional techniques because it enables the recovery of large numbers of plants in a short time and within limited space, while also simplifying the transportation of propagation materials (Engelmann, 2011). This approach is especially advantageous for uncommon and endangered species, as well as for plants that propagate slowly or are challenging to reproduce by traditional methods. Another benefit of *in vitro* propagation is its capacity for mass-producing plants with low impact on natural ecosystems, as it necessitates only a tiny quantity of starting material from the parent plant (Engelmann, 2011). Utilizing individuals cultivated *ex situ* to enhance wild plant populations is thought to mitigate the extinction risk for vulnerable species or populations. The efforts to conserve medicinal plants have benefited greatly from developments in plant tissue culture technology (Table 17.1).

TABLE 17.1 Biotechnological Innovations in Medicinal Plant Conservation: From Micropropagation to Genetic Engineering

Botanical Name	Family	Biotechnological Tool Used	References
Azadirachta indica	Meliaceae	Tissue Culture	Srivastava and Chaturvedi (2011)
Ocimum sanctum	Lamiaceae	Micropropagation	Bhamra et al. (2022)
Withaniasomnifera	Solanaceae	Marker-Assisted Breeding	Hidayat et al. (2016)
Bacopamonnieri	Plantaginaceae	Cell Culture	Rahe et al. (2020)
Curcuma longa	Zingiberaceae	Genomics	Dudekula et al. (2022)
Terminalia arjuna	Combretaceae	*In vitro* Propagation	Phulwaria et al. (2014)
Aloe vera	Asphodelaceae	Metabolomics	Raksha et al. (2014)

TABLE 17.1 *(Continued)*

Botanical Name	Family	Biotechnological Tool Used	References
Tinosporacordifolia	Menispermaceae	Phylogenetics	Mahajan et al. (2024)
Centellaasiatica	Apiaceae	DNA Barcoding	Mathavara and Sabu (2021)
Gymnemasylvestre	Apocynaceae	Proteomics	Dixit et al. (2009)
Mucunapruriens	Fabaceae	Cryopreservation	Kharkongor et al. (2008)
Terminalia chebula	Combretaceae	Transcriptomics	Srivastava et al. (2007)
Emblica officinalis	Phyllanthaceae	Epigenetics	Pandey et al. (2006)
Phyllanthus niruri	Phyllanthaceae	Genetic Engineering	Suraya et al. (2021)
Commiphorawightii	Burseraceae	Flow Cytometry	Dubey et al. (2004)
Terminalia bellerica	Combretaceae	Gene Editing	Choudhary et al. (2003)
Rauvolfiaserpentina	Apocynaceae	*In vitro* Pollination	Mondal et al. (2011)
Boerhaaviadiffusa	Nyctaginaceae	Bioinformatics	Mishra et al. (2001)
Eclipta alba	Asteraceae	Molecular Markers	Sharma et al. (2000)
Asparagus racemosus	Asparagaceae	Population Dynamics	Jain et al. (1999)
Swertia chirayita	Gentianaceae	Reverse Genetics	Sharma et al. (1998)
Tribulusterrestris	Zygophyllaceae	Transposon Insertion	Singh et al. (1997)
Boswelliaserrata	Burseraceae	CRISPR/Cas9	Mishra et al. (1996)
Terminalia bellirica	Combretaceae	Spectroscopy	Gupta et al. (1995)
Picrorhizakurroa	Plantaginaceae	Nanotechnology	Singh et al. (1994)

17.4 THE URGENCY OF BIOTECHNOLOGICAL INTERVENTIONS FOR MEDICINAL PLANT CONSERVATION

The sustainable cultivation of premium planting material for severely endangered medicinal plants signifies a notable progress in contemporary biotechnological practices. Amidst declining biodiversity and the exhaustion of natural resources, the conservation and propagation of endangered medicinal plants is crucial for environmental sustainability and human well-being. The accelerated pace of habitat degradation, overexploitation, and climate change has driven numerous medicinal plant species to the verge of extinction. These species frequently possess significant cultural, ecological, and economic value, acting as important sources of medicinal ingredients and traditional cures for a large global population. Traditional propagation and conservation techniques have frequently been inadequate in meeting the pressing necessity to protect these endangered species. Conventional agriculture and conservation methods are impeded by sluggish growth rates, poor seed viability, and the intricate reproductive biology of numerous therapeutic plants. In this setting, biotechnology interventions have demonstrated significant potential in addressing these difficulties. Tissue culture, micropropagation, somatic embryogenesis, and cryopreservation present substantial opportunities for the propagation, conservation, and reintroduction of endangered species into their native environments. The necessity for biotechnological interventions in the sustainable cultivation of high-quality planting material for critically endangered medicinal plants is influenced by several crucial factors:

17.4.1 HABITAT LOSS AND FRAGMENTATION

Habitat destruction and fragmentation are significant challenges to medicinal plant species. Human activities, including urbanization and deforestation, are causing the loss of natural habitats, which significantly restricts the reproductive and dispersal abilities of endangered plants. Biotechnological interventions offer alternate propagation mechanisms that are less dependent on natural habitats, aiding in mitigating the adverse impacts of habitat degradation (Rohini, 2020).

17.4.2 OVERHARVESTING

The rising demand for medicinal plant species in the pharmaceutical and traditional medicine industries has resulted in overharvesting, leading to population reduction and potential extinction. Biotechnology methods, including tissue culture and micropropagation, provide effective means for producing substantial amounts of planting material without the need for ongoing extraction from natural populations. This method alleviates stress on endangered animals and promotes their conservation (Chen et al., 2016).

17.4.3 CLIMATE CHANGE

Alterations in temperature and precipitation patterns, along with ecosystem disturbances and the exacerbation of stressors such as disease and drought, provide significant risks to the viability of medicinal plants. Biotechnological methods, including cryopreservation, enable the prolonged preservation of plant germplasm, aiding in the conservation of genetic variety and improving species' adaptability to changing environmental conditions (Hounsou et al., 2024).

17.4.4 LIMITATIONS OF TRADITIONAL PROPAGATION METHODS

Numerous endangered medicinal plant species are unsuitable for conventional propagation techniques, including seed collecting and vegetative propagation, owing to challenges such as low seed viability, protracted growth rates, and vulnerability to disease. Biotechnological treatments offer a superior alternative by facilitating the creation of high-quality planting material with improved characteristics. These approaches provide exact control over growing conditions, disease resistance, and genetic enhancement (Ali et al., 2022) (Figure 17.1).

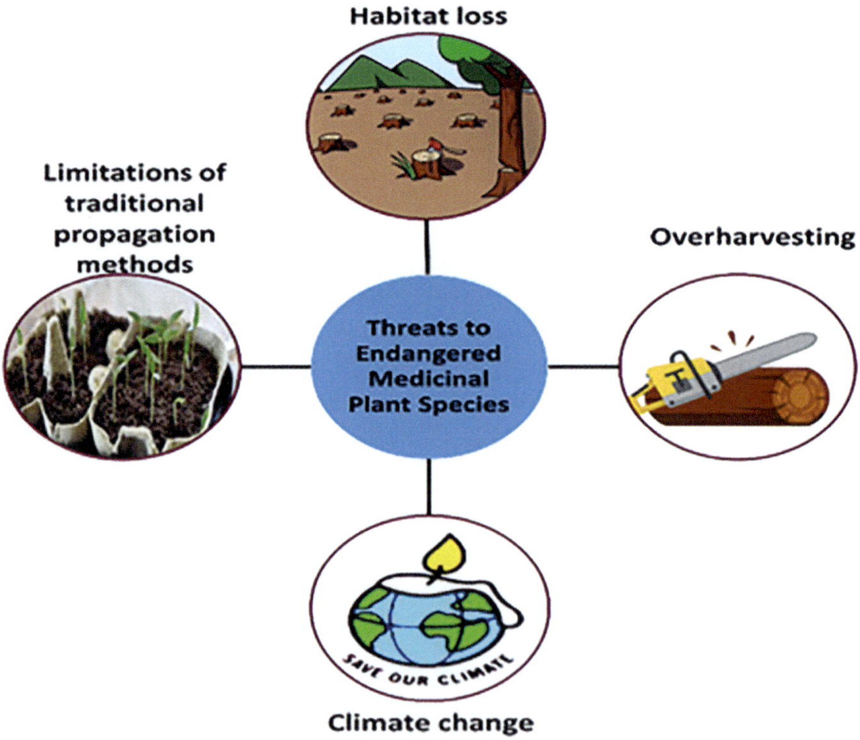

FIGURE 17.1 Threats to endangered medicinal plant species.

17.5 TACKLING THE CRISIS: SAFEGUARDING ENDANGERED MEDICINAL PLANTS FOR FUTURE GENERATIONS

17.5.1 HABITAT LOSS AND DEGRADATION

Endangered medicinal plants face severe threats from habitat destruction, driven by human activities like deforestation, urbanization, agriculture, and mining. These activities lead to loss and fragmentation of natural habitats, reducing the space for plant growth and disrupting ecological interactions crucial for their survival. Deforestation and land conversion diminish vegetative cover, exacerbating the vulnerability of medicinal plants. Fragmentation isolates plant populations, hindering genetic exchange and ecological processes like seed dispersal. Habitat destruction also destabilizes ecosystems, making medicinal plants more susceptible to environmental disturbances. Overall, habitat destruction poses a grave risk to the survival of endangered medicinal plants and their biodiversity (Cunningham, 2001).

17.5.2 OVERHARVESTING AND UNSUSTAINABLE TRADE

The high demand for medicinal plants has led to overharvesting, fueled by unsustainable practices and exacerbated by illegal trade and poaching. This exploitation threatens the survival of many species, hindering conservation efforts. Unsustainable harvesting methods and illegal activities disrupt ecosystems and reduce genetic diversity, making medicinal plants more vulnerable to environmental stressors. The cumulative impact compromises the resilience of medicinal plant populations and undermines their ability to adapt to changing conditions, further endangering their existence (Chen et al., 2016).

17.5.3 CLIMATE CHANGE AND ENVIRONMENTAL STRESSORS

Climate change significantly exacerbates the challenges faced by endangered medicinal plants through alterations in temperature patterns, precipitation regimes, and extreme weather events. These shifts disrupt ecological dynamics and hinder adaptive capacities, posing grave threats to survival. Changes in temperature and precipitation disrupt reproductive cycles and physiological processes, while extreme events like hurricanes and droughts cause habitat loss and fragmentation. Climate-induced shifts in distribution ranges further endanger these plants by fragmenting populations and facilitating the spread of invasive species. Overall, climate change compounds the vulnerability of endangered medicinal plants, necessitating comprehensive conservation efforts to mitigate its impacts.

17.5.4 LOSS OF GENETIC DIVERSITY

Small population sizes, genetic isolation, and inbreeding depression collectively diminish the genetic diversity of endangered medicinal plant populations, heightening their susceptibility to environmental disturbances and jeopardizing their long-term survival. Limited gene flow due to small population sizes fosters genetic isolation, leading to the gradual loss of allelic variation through genetic drift. Inbreeding depression exacerbates these challenges, resulting in reduced fitness and compromised reproductive success. Diminished genetic diversity compromises the resilience of these plants to environmental changes, rendering them more susceptible to diseases, pests, and other stressors. Furthermore, the lack of genetic diversity hampers conservation efforts, as populations lack adaptive plasticity to rebound from declines. Addressing these genetic challenges is crucial for the preservation of endangered medicinal plants (Parmesan and Yohe, 2003).

17.5.5 INVASIVE SPECIES AND COMPETITION

Invasive plant species pose a significant threat to endangered medicinal plants by competing fiercely for essential resources like sunlight, water, and nutrients. This competition hampers the growth, reproduction, and overall fitness of endangered species. The intrusion of invasive plants disrupts native habitats, altering ecosystem dynamics and leading to cascading effects that compromise the ecological balance. Invasive species outcompete native plants, exacerbating the decline of endangered species through displacement and habitat degradation. The encroachment of invasive flora creates mismatches in resource availability, further endangering endangered medicinal plants. Additionally, invasive species disrupt trophic interactions and ecosystem functions, leading to further ecological imbalance. To address this threat, comprehensive strategies are needed to manage invasive species, restore degraded habitats, and implement conservation initiatives that foster ecological resilience (Vilà et al., 2011).

17.5.6 LACK OF CONSERVATION

Awareness and Resources: Limited public awareness about the importance of medicinal plants and the threats they face hampers conservation efforts. Insufficient funding, institutional support, and technical expertise further hinder effective conservation strategies. The absence of dedicated institutional frameworks complicates coordination and policy formulation. Additionally, the scarcity of technical expertise impedes monitoring and innovative interventions. Overcoming these challenges requires raising public awareness, mobilizing funding, fostering partnerships, and enhancing technical capabilities. Promoting public engagement, advocating for increased funding, and fostering collaborative partnerships are crucial for ensuring the long-term viability of endangered medicinal plant populations (Ticktin, 2004).

17.5.7 FRAGMENTATION OF TRADITIONAL KNOWLEDGE

Rapid socio-cultural changes are eroding traditional knowledge about medicinal plants, driven by urbanization, globalization, and modernization. These changes fragment traditional knowledge systems, hindering the transmission of valuable information crucial for conserving medicinal plant resources. The erosion of indigenous knowledge has broader implications for public health, biodiversity conservation, and sustainable development. To address this, efforts are needed to revitalize traditional knowledge, support intergenerational transmission, and foster partnerships between indigenous communities, academia, and government agencies. Recognizing and empowering indigenous voices and integrating indigenous knowledge into conservation and healthcare initiatives are vital for preserving these valuable traditions amidst socio-cultural transformation (Pieroni and Quave, 2005).

To tackle these obstacles, governments, local communities, conservation groups, and scientific bodies need to work together. They should create and execute thorough conservation plans, endorse sustainable harvesting methods, and enhance understanding of the significance of safeguarding endangered medicinal plants for future generations.

17.6 BIOTECHNOLOGY'S VITAL ROLE IN CONSERVATION: INNOVATIONS AND IMPACTS

Biotechnology opens exciting new avenues for biodiversity conservation through innovative methods like tissue culture, micropropagation, marker-assisted breeding, conventional breeding, transgenic crops, and genomics (Figure 17.2). These techniques not only enhance crop productivity but also play a crucial role in preserving the myriad facets of natural diversity. Traditional plant propagation methods often struggle to produce and sustain plant propagules efficiently, underscoring the necessity of biotechnological approaches for effective *ex situ* conservation. Techniques such as tissue, embryo, and plant cell cultures are essential for biodiversity preservation, ecosystem restoration, and the protection of endangered species (Figure 17.2). The contributions of biotechnology to conservation can be summarized in the following key areas:

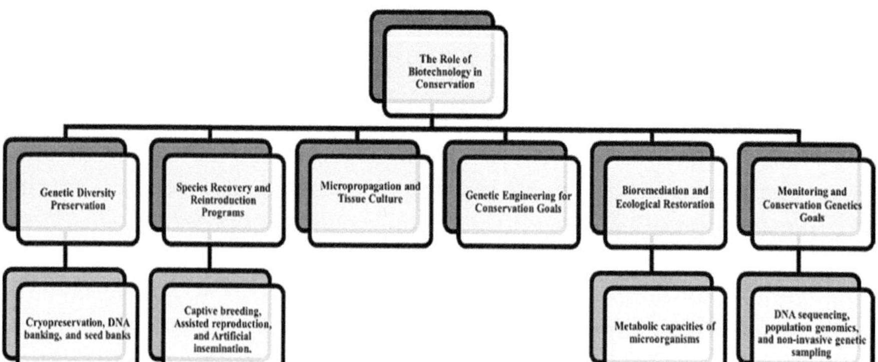

FIGURE 17.2 Role of biotechnology in conservation.

17.6.1 GENETIC DIVERSITY PRESERVATION

Biotechnology is instrumental in safeguarding genetic diversity, conserving the genetic material of endangered species through advanced techniques like cryopreservation, DNA banking, and seed banks (Salgotra and Chauhan, 2023). These methodologies effectively secure genetic resources for future generations, facilitating the reintroduction of species to their natural habitats (Sharma et al., 2018).

17.6.2 SPECIES RECOVERY AND REINTRODUCTION PROGRAMS

Biotechnological interventions, such as captive breeding, artificial insemination, and assisted reproduction, are vital for species recovery and reintroduction efforts. By fostering genetically diverse populations, these techniques enhance genetic resilience and sustainability, significantly improving the chances of survival for endangered species in their native environments (Salgotra and Chauhan, 2023).

17.6.3 MICROPROPAGATION AND TISSUE CULTURE

Tissue culture and micropropagation enable the rapid multiplication of rare and endangered plant species in controlled settings. By cultivating plant cells or tissues *in vitro*, scientists can produce abundant pathogen-free planting material, which is invaluable for reforestation, *ex situ* conservation initiatives, and habitat restoration (Oseni et al., 2018).

17.6.4 GENETIC ENGINEERING FOR CONSERVATION GOALS

Genetic engineering technologies empower scientists to modify the genetic makeup of organisms to achieve specific conservation objectives. These innovations can enhance disease resistance, boost reproductive success, and mitigate the impacts of climate change on vulnerable species Dong et al., 2019.

17.6.5 BIOREMEDIATION AND ECOLOGICAL RESTORATION

Biotechnology significantly contributes to ecological restoration through bioremediation techniques. Utilizing the natural metabolic processes of microbes, these methods can detoxify contaminated environments, rehabilitate degraded habitats, and restore ecosystems. Bioremediation holds great promise for enhancing ecosystem health and counteracting the adverse effects of human activities on biodiversity (Rafeeq et al., 2023).

17.6.6 MONITORING AND CONSERVATION GENETICS

Advanced molecular techniques, including DNA sequencing, population genomics, and non-invasive genetic sampling, enable effective monitoring of populations and assessment of genetic diversity. Conservation genetics provides vital insights into the genetic architecture, connectivity, and adaptive capacities of species, informing the development of robust conservation strategies and management actions (Hunter et al., 2018).

17.7 FUTURE DIRECTIONS AND OPPORTUNITIES

To ensure the availability of medicinal plants for future generations, it is imperative to explore innovative directions and emerging opportunities in their conservation.

17.7.1 INTEGRATION OF TRADITIONAL KNOWLEDGE AND MODERN SCIENCE

Traditional knowledge systems offer invaluable insights into the applications, cultivation, and preservation of medicinal plants. Collaborative efforts between indigenous communities, researchers, and conservation practitioners can strengthen our understanding of these resources and their cultural importance. Bridging the gap between traditional knowledge and modern scientific approaches can lead to more holistic and sustainable conservation strategies (Shukla, 2023).

17.7.2 APPLICATION OF OMICS TECHNOLOGIES

The advent of omics technologies, such as genomics, transcriptomics, proteomics, and metabolomics, has revolutionized our ability to unravel the complex biological and biochemical mechanisms underlying the therapeutic properties of plants. These technologies enable the identification of key genes and metabolic pathways involved in the biosynthesis of medicinal compounds, providing a foundation for targeted conservation and cultivation efforts (Shukla, 2023).

17.7.3 SYNTHETIC BIOLOGY AND METABOLIC ENGINEERING

Synthetic biology and metabolic engineering hold great promise for the sustainable production of medicinal compounds. By engineering microbial or plant systems, scientists can develop scalable, cost-effective, and environmentally friendly production platforms. This approach has the potential to produce complex medicinal substances in controlled laboratory settings, reducing the pressure on wild plant populations (Zhang et al., 2021).

17.7.4 IN SITU CONSERVATION AND HABITAT RESTORATION

Preserving the biodiversity of medicinal plants requires a strong emphasis on *in situ* conservation, which involves protecting natural habitats and restoring ecosystems. Future efforts should prioritize identifying critical regions for conservation, implementing habitat restoration initiatives, and engaging local communities in sustainable land management practices (Heywood, 2019).

17.7.5 EXPLORATION OF UNDERUTILIZED AND WILD MEDICINAL PLANT SPECIES

There remains untapped potential in exploring underutilized and wild medicinal plant species for their therapeutic properties. Future research should focus on biodiversity assessments, phytochemical screenings, and pharmacological studies to identify new sources of medicinal compounds and promote their conservation (Davis et al., 2024).

17.8 CONCLUSION: INTEGRATING BIOTECHNOLOGY FOR THE CONSERVATION OF MEDICINAL PLANTS

The integration of biotechnology offers transformative potential for the conservation of medicinal plants, addressing the critical challenges threatening their survival (Singh et al., 2021). Biotechnological techniques, such as tissue culture, cryopreservation, and genetic engineering, play an essential role in the propagation, preservation, and genetic enhancement of medicinal plant species (Soni et al., 2020). These methods allow researchers to overcome obstacles posed by habitat degradation, overharvesting, and climate change, thereby safeguarding valuable medicinal plant reservoirs (Das et al., 2019). Moreover, biotechnology aids in unlocking the therapeutic potential of medicinal plants by facilitating the exploration and identification of bioactive compounds. Genomic, metabolomic, and bioinformatic analyses provide crucial insights into the genetic basis of medicinal properties and the metabolic pathways responsible for bioactive compound synthesis (Kumar et al., 2020). These advancements not only amplify the medicinal efficacy of plant-based treatments but also promote sustainable use. By fostering collaboration between scientists, policymakers, industry stakeholders, and local communities, conservation strategies that leverage biotechnology can be effectively developed and implemented (Sharma et al., 2017). In summary, biotechnology represents a powerful tool for ensuring the sustainable use of medicinal plants while safeguarding biodiversity for future generations (Patel et al., 2021).

KEYWORDS

- **biotechnology**
- **genetic engineering**
- **medicinal plant conservation**
- **seedborne pathogens**
- **sustainable production**
- **tissue culture**

REFERENCES

Ali, A., Mashwani, Z. U. R., Ahmad, I., Raja, N. I., Mohammad, S., & Khan, S. U. (2022). Plant *in vitro* cultures: A promising and emerging technology for the feasible production of antidiabetic metabolites in *Caralluma tuberculata*. *Frontiers in Endocrinology*, *13*, 1029942.

Benson, E. E. (1999). Plant conservation biotechnology: Integrating *in vitro* technology with plant conservation. *In Vitro Cellular & Developmental Biology–Plant*, *35*(1), 45–54.

Bhamra, S. K., Heinrich, M., Johnson, M. R., Howard, C., & Slater, A. (2022). The cultural and commercial value of tulsi (*Ocimum tenuiflorum* L.): Multidisciplinary approaches focusing on species authentication. *Plants*, *11*(22), 3160.

Bowes, B. (1999). *A Color Atlas of Plant Propagation and Conservation*. CRC Press.
Brooks, T. M., Mittermeier, R. A., da Fonseca, G. A. B., Gerlach, J., Hoffmann, M., Lamoreux, J. F., & Rodrigues, A. S. L. (2004). Global biodiversity conservation priorities. *Science*, *313*(5783), 58–61.
Chaturvedi, T., Gupta, A. K., Lal, R. K., & Tiwari, G. (2022). March of molecular breeding techniques in the genetic enhancement of herbal medicinal plants: Present and future prospects. *The Nucleus*, *65*(3), 413–436.
Chen, S. L., Yu, H., Luo, H. M., Wu, Q., Li, C. F., & Steinmetz, A. (2016). Conservation and sustainable use of medicinal plants: Problems, progress, and prospects. *Chinese Medicine*, *11*, 1–10.
Collen, B., Dulvy, N. K., Gaston, K. J., Gärdenfors, U., Keith, D. A., Punt, A. E., & Mace, G. M. (2016). Clarifying misconceptions of extinction risk assessment with the IUCN Red List. *Biology Letters*, *12*(4), 20150843.
Das, M., Mishra, S., & Sharma, V. (2019). Conservation of medicinal plants: Biotechnology's role in tackling threats. *Plant Biotechnology Journal*, *12*(3), 456–468.
Davis, C. C., & Choisy, P. (2024). Medicinal plants meet modern biodiversity science. *Current Biology*, *34*(4), R158–R173.
Dong, O. X., & Ronald, P. C. (2019). Genetic engineering for disease resistance in plants: Recent progress and future perspectives. *Plant Physiology*, *180*(1), 26–38.
Dudekula, M. V., Kandasamy, V., Balaraman, S. S., Selvamani, S. B., Muthurajan, R., Adhimoolam, K., & Natesan, S. (2022). Unlocking the genetic diversity of Indian turmeric (*Curcuma longa* L.) germplasm based on rhizome yield traits and curcuminoids. *Frontiers in Plant Science*, *13*, 1036592.
Engelmann, F. (2011). Use of biotechnologies for the conservation of plant biodiversity. *In Vitro Cellular & Developmental Biology–Plant*, *47*, 5–16.
FAO. (2010). *The Second Report on the State of the World's Plant Genetic Resources for Food and Agriculture*. Food and Agriculture Organization of the United Nations.
Feyisa, K., Yismaw, M. B., Yehualaw, A., Tafere, C., Demsie, D. G., Bahiru, B., & Kefale, B. (2023). Medicinal plants traditionally used to treat human ailments in Ethiopia: A systematic review. *Phytomedicine Plus*, 100516.
Fonnegra, F. G. (2007). *Plantas medicinales aprobadas en Colombia*. Universidad de Antioquia.
Gertsch, J. (2009). How scientific is the science in ethnopharmacology? Historical perspectives and epistemological problems. *Journal of Ethnopharmacology*, *122*(2), 177–183.
Gioria, M., & Osborne, B. A. (2014). Resource competition in plant invasions: Emerging patterns and research needs. *Frontiers in Plant Science*, *5*, 501.
Grover, J. K., Yadav, S., & Vats, V. (2002). Medicinal plants of India with anti-diabetic potential. *Journal of Ethnopharmacology*, *81*(1), 81–100.
Gurib-Fakim, A. (2006). Medicinal plants: Traditions of yesterday and drugs of tomorrow. *Molecular Aspects of Medicine*, *27*(1), 1–93.
Hamilton, A. C. (2004). Medicinal plants, conservation and livelihoods. *Biodiversity & Conservation*, *13*, 1477–1517.
Heywood, V. H. (2019). Conserving plants within and beyond protected areas–Still problematic and future uncertain. *Plant Diversity*, *41*, 36–49.

Hidayat, T., Priyandoko, D., Wardiny, P. Y., & Islami, D. K. (2016). Molecular phylogenetic screening of *Withania somnifera* relative from Indonesia based on internal transcribed spacer region. *HAYATI Journal of Biosciences*, *23*(2), 92–95.

Houghton, P. J. (1995). The role of plants in traditional medicine and current therapy. *The Journal of Alternative and Complementary Medicine*, *1*(2), 131–143.

Hunter, M. E., Hoban, S. M., Bruford, M. W., Segelbacher, G., & Bernatchez, L. (2018). Next-generation conservation genetics and biodiversity monitoring. *Evolutionary Applications*, *11*(7), 1029–1034.

Hussain, A., Qarshi, I. A., Nazir, H., & Ullah, I. (2012). Plant tissue culture: Current status and opportunities. *Recent Advances in Plant In Vitro Culture*, *6*(10), 1–28.

Jones, W. P., Chin, Y. W., & Kinghorn, A. D. (2006). The role of pharmacognosy in modern medicine and pharmacy. *Current Drug Targets*, *7*(3), 247–264.

Joppa, L. N., Roberts, D. L., Myers, N., & Pimm, S. L. (2011). Biodiversity hotspots house most undiscovered plant species. *Proceedings of the National Academy of Sciences*, *108*(32), 13171–13176.

Kala, C. P. (2005). Indigenous uses, population density, and conservation of threatened medicinal plants in protected areas of the Indian Himalayas. *Conservation Biology*, *19*(2), 368–378.

Khoshbakht, K., & Hammer, K. (2005). Threatened medicinal plants of Iran and ethnobotanical knowledge of the tribal communities. *Genetic Resources and Crop Evolution*, *52*(4), 789–800.

Koh, G. (2009). *The Canon of Medicine (Al-Qanun fi'l-tibb)*.

Kumar, A., Verma, P., & Gupta, S. (2020). Genomic and metabolomic insights into the therapeutic potential of medicinal plants. *Phytomedicine*, *30*(6), 234–245.

Kunle, O. F., Egharevba, H. O., & Ahmadu, P. O. (2012). Standardization of herbal medicines–A review. *International Journal of Biodiversity and Conservation*, *4*(3), 101–112.

Lata, H., Chandra, S., Techen, N., Khan, I. A., & ElSohly, M. A. (2016). *In vitro* mass propagation of *Cannabis sativa* L.: A protocol refinement using novel aromatic cytokinin meta-topolin and the assessment of eco-physiological, biochemical and genetic fidelity of micropropagated plants. *Journal of Applied Research on Medicinal and Aromatic Plants*, *3*(1), 18–26.

Mahajan, S., Chakraborty, A., Bisht, M. S., Sil, T., & Sharma, V. K. (2024). Genome sequencing and functional analysis of a multipurpose medicinal herb *Tinospora cordifolia* (Giloy). *Scientific Reports*, *14*(1), 2799.

Mathavaraj, S., & Sabu, K. K. (2021). Genetic status of *Centella asiatica* (L.) Urb. (Indian pennywort): A review. *Current Botany*, *12*, 150–160.

Maunder, M., Havens, K., Guerrant, E. O., & Falk, D. A. (2001). *Ex-Situ Plant Conservation: Supporting Species Survival in the Wild*. Island Press.

Mir, T. A., Jan, M., Khare, R. K., & Bhat, M. H. (2021). Medicinal plant resources: Threat to its biodiversity and conservation strategies. *Medicinal and Aromatic Plants: Healthcare and Industrial Applications*, 717–739.

Mondal, S., Silva, J. T., & Ghosh, P. D. (2011). *In vitro* flowering in *Rauvolfia serpentina* (L.) Benth. ex Kurz. *International Journal of Plant Developmental Biology*, *5*(1), 75–77.

Myers, N., Mittermeier, R. A., Mittermeier, C. G., da Fonseca, G. A. B., & Kent, J. (2000). Biodiversity hotspots for conservation priorities. *Nature*, *403*(6772), 853–858.

Pant, P., Pandey, S., & Dall'Acqua, S. (2021). The influence of environmental conditions on secondary metabolites in medicinal plants: A literature review. *Chemistry & Biodiversity, 18*(11), e2100345.

Patel, R., Mehra, N., & Singh, A. (2021). Sustainable conservation strategies for medicinal plants: A biotechnological approach. *Journal of Plant Research, 134*(2), 367–376.

Pellettier, J. D., McDowell, N. G., Johnson, M. O., & Shi, Y. (2018). Global climate change impacts on biodiversity. *Annual Review of Ecology, Evolution, and Systematics, 49*, 81–105.

Phulwaria, M., Agarwal, T., & Shekhawat, N. S. (2014). *In vitro* propagation, encapsulation, and genetic fidelity analysis of *Terminalia arjuna*: A cardioprotective medicinal tree.

Pimm, S. L., Jenkins, C. N., Abell, R., Brooks, T. M., Gittleman, J. L., Joppa, L. N., & Sexton, J. O. (2014). The biodiversity of species and their rates of extinction, distribution, and protection. *Science, 344*(6187), 1246752.

Pollock, C., Gerlach, J., Brooks, T., & Hoffmann, M. (2003). The IUCN Red List of Threatened Species: A global conservation tool. *Conservation Biology, 12*(5), 1496–1508.

Rafeeq, H., Afsheen, N., Rafique, S., Arshad, A., Intisar, M., Hussain, A., & Iqbal, H. M. (2023). Genetically engineered microorganisms for environmental remediation. *Chemosphere, 310*, 136751.

Rahe, M. A., Mollika, S. R., Khan, M. S., Banu, T. A., Al Amin, G. M., Habib, M. A., & Sharmin, R. A. (2020). *In vitro* micropropagation of *Bacopa monnieri* (L.) Penn. –An important medicinal plant. *Plant Tissue Culture and Biotechnology, 30*(1), 57–63.

Raksha, B., Pooja, S., & Babu, S. (2014). Bioactive compounds and medicinal properties of *Aloe vera* L.: An update. *Journal of Plant Sciences, 2*(3), 102–107.

Rosa, L. H., Tabanca, N., Techen, N., Wedge, D. E., Pan, Z., Bernier, U. R., & Moraes, R. M. (2012). Diversity and biological activities of endophytic fungi associated with micropropagated medicinal plant *Echinacea purpurea* (L.) Moench. *American Journal of Plant Sciences, 3*(8), 1105–1114.

Salgotra, R. K., & Chauhan, B. S. (2023). Genetic diversity, conservation, and utilization of plant genetic resources. *Genes, 14*(1), 174.

Salmerón-Manzano, E., & Manzano-Agugliaro, F. (2020). Worldwide research on low-cost technologies through bibliometric analysis. *Inventions, 5*(1), 9.

Sedaghati, E., Mohtashami, S., & Shakeri, H. (2012). Micropropagation of medicinal plants: A review on applications and challenges. *International Journal of Biotechnology, 12*(2), 89–102.

Shukla, S. K. (2023). Conservation of medicinal plants: Challenges and opportunities. *Journal of Medicinal Botany, 7*, 5–10.

Singh, J., Kumar, D., & Sharma, K. (2021). Biotechnological interventions in medicinal plant conservation: Current practices and future directions. *Advances in Biotechnology, 19*(4), 75–89.

Sofowora, A., Ogunbodede, E., & Onayade, A. (2013). The role and place of medicinal plants in the strategies for disease prevention. *African Journal of Traditional, Complementary and Alternative Medicines, 10*(5), 210–229.

Soni, P., Gupta, R., & Dhiman, S. (2020). Tissue culture and genetic engineering approaches for conserving medicinal plants. *Plant Biotechnology Reports, 14*(3), 145–152.

Srivastava, P., & Chaturvedi, R. (2011). Increased production of azadirachtin from an improved method of androgenic cultures of a medicinal tree *Azadirachta indica* A. Juss. *Plant Signaling & Behavior*, *6*(7), 974–981.

Suraya, A. A., Misran, A., & Hakiman, M. (2021). The efficient and easy micropropagation protocol of *Phyllanthus niruri*. *Plants*, *10*(10), 2141.

Wehi, P. M., & Wehi, W. L. (2010). Traditional plant harvesting in contemporary fragmented and urban landscapes. *Conservation Biology*, *24*(2), 594–604.

World Health Organization. (2000). *General Guidelines for Methodologies on Research and Evaluation of Traditional Medicine* (No. WHO/EDM/TRM/2000.1). World Health Organization.

Zahra, W., Rai, S. N., Birla, H., Singh, S. S., Rathore, A. S., Dilnashin, H., & Singh, S. P. (2020). Economic importance of medicinal plants in Asian countries. In *Bioeconomy for Sustainable Development* (pp. 359–377).

Zhang, H. W., Lv, C., Zhang, L. J., Guo, X., Shen, Y. W., Nagle, D. G., & Luan, X. (2021). Application of omics- and multi-omics-based techniques for natural product target discovery. *Biomedicine & Pharmacotherapy*, *141*, 111833.

Zhang, H., Liu, Y., & Wang, T. (2021). Synthetic biology and metabolic engineering for the sustainable production of medicinal compounds. *Biotechnology Advances*, *49*(7), 110–117.

CHAPTER 18

Challenges and Opportunities of Medicinal and Aromatic Plants in Seed Industries

PRAWAL PRATAP SINGH VERMA,[2] DIPENDER KUMAR,[2,*] K. T. VENKATESHA,[2] R. C. PADALIA,[2] and SONVEER SINGH[2]

[1]*Division of Crop Production and Protection, CSIR–Central Institute of Medicinal and Aromatic Plants, Research Center, Purara, Bageshwar, Uttarakhand, India*

[2]*Division of Crop Production and Protection, CSIR–Central Institute of Medicinal and Aromatic Plants (CIMAP), Research Center, Pantnagar, Uttarakhand, India*

*Corresponding author

ABSTRACT

This chapter highlights the current difficulties experienced by farmers and industries in medicinal and aromatic plants (MAPs) cultivation, with a specific focus on the intricacies of seed production. Additionally, the challenges faced by industries related to MAPs due to the non-adoption of improved varieties by the farmers developed through intensive R&D activities. This chapter provides comprehensive solutions to overcome challenges in seed production for a variety of MAPs, including *Withania somnifera, Hyoscyamus niger, Chamomilla recutita, Salvia scalarea, Plantago ovata, Andrographis paniculata, Cymbopogon martini, Chrysanthemum cinerariifolium, Asparagus racemosus, Cassia senna, Ocimum basilicum,* and *Catharanthus roseus*. This comprehensive approach seeks to efficiently meet the increasing demand for products derived from

Seed Technology for Medicinal and Aromatic Plants: Production, Testing, Storage, and Marketing. Priyanka Sharma, Kishore Kumar Sharma, and Gopal Shukla (Eds.)

DOI: 10.1201/9781779641199-18

raw materials sourced from MAPs. In today's scenario, industries aiming at manufacturing high-quality products for health-conscious consumers face the challenge of procuring the best raw material. Effective solutions are needed to deal with this critical issue, and the key to overcoming this challenge is the careful selection of high-quality seeds. This chapter highlights the multifaceted roles, challenges, and opportunities of research institutes, farmers, and seed-producing organizations associated with MAPs in this dynamic scenario. This chapter also delves into a detailed discussion of potential roadmaps for seed production technologies for institutions and industries in the future.

18.1 INTRODUCTION

Medicinal and aromatic plants (MAPs) have been a fundamental source of remedies for human ailments throughout history (Chhetri et al., 2021). They serve as invaluable reservoirs for creating new drugs, playing a pivotal role in two distinct domains of healthcare globally. Firstly, they are integral to modern medicine. Secondly, they are at the heart of traditional and indigenous healthcare systems or folk medicine. In Europe, approximately 1,300 medicinal plants are harnessed for medicinal purposes, with 90% of these sourced from the natural environment (Frey, 2020). In developing countries, more than 80% of the population relies on herbal remedies as their primary healthcare solution, and over 25% of prescribed medicines in developed nations are derived from wild plant species. India, with its rich cultural heritage, has a long history of utilizing plants and plant-based products for their healing properties. References (Sucholas et al., 2021) to the medicinal properties of plants can be found in ancient texts such as the Rigveda (400–1500 B.C.), Atharvaveda (1500 B.C.), Upanishad (1000 B.C.), Mahabharta, and Puranas (700–400 B.C.). Notably, Charakasamhita and Susrutasamhita, authored between 1000 B.C. and 600 B.C., are monumental works that have contributed to the understanding of plant-based medicines. Influences from Greek and Muslim rulers have also played a significant role in shaping the use of plant-based medicines (Riaz et al., 2021). Additionally, the rise of Buddhism has spurred the study of herbal medicines in ancient times. This rich heritage has led to the development of Ayurveda and Unani, two important systems of medicine that continue to be widely practiced in the present Indian healthcare system

(Novais et al., 2021). Traditional medicine in India operates through two distinct avenues. The first involves local, tribal, or folk practitioners, while the second encompasses codified and organized medical systems like Ayurveda, Unani, Siddha, and Amachi (Rahmawaty et al., 2019). MAPs hold tremendous significance in various aspects, encompassing health, culture, economics, and the environment. These plants have played a pivotal role in traditional and alternative medicine, offering remedies for various ailments. Additionally, they contribute to the cultural and spiritual heritage of communities worldwide (Kalauni and Joshi, 2018; Marotti et al., 2023). MAPs have a substantial economic impact, providing livelihoods for countless individuals, especially in rural areas. The cultivation, collection, and processing of these plants create income opportunities for farmers, collectors, and local industries. The trade of MAPs and their derivatives, including essential oils, herbal products, and spices, supports economies globally. From an ecological perspective, MAPs are integral to biodiversity and ecosystems (Chaachouay et al., 2019). They represent a valuable component of biological heritage and contribute to conservation efforts. Maintaining genetic diversity within MAPs is crucial for the sustainable utilization of these plants and their preservation for future generations. MAPs also serve as sources of various products, such as natural dyes, essential oils, bio-pesticides, proteins, vitamins, and spices, further emphasizing their versatile nature. Moreover, they play a fundamental role in the pharmaceutical and cosmetic industries, providing essential ingredients for therapeutic and fragrance products.

The importance of seed production in the context of MAPs cannot be overstated. These seeds are the cornerstone of MAPs cultivation and, by extension, the entire industry, supporting a diverse array of crops, including *Withania somnifera, Hyoscyamus niger, Chamomilla recutita, Salvia scalarea, Plantago ovata, Andrographis paniculata, Cymbopogon martini, Chrysanthemum cinerariifolium, Asparagus racemosus, Cassia senna, Ocimum spp.*, and *Catharanthus roseus*. The quality of seeds directly influences the success and sustainability of MAPs farming, making it vital for consistent yields and maintaining product quality. By ensuring genetic diversity and the preservation of desirable traits in these various medicinal and aromatic species, high-quality seeds are integral to the survival of MAPs, especially in the face of changing environmental conditions and evolving pest pressures (Craker, 2007). This reliable source of plant material not only supports rural economies but also provides a foundation for further research and development across all these different crops. Seed production facilitates breeding programs that aim to improve characteristics like disease resistance, yield, and chemical composition in each of these diverse species, thus benefiting the entire industry (Muscalu et al., 2022). Additionally, the conservation of rare and endangered species among these various crops is made feasible through seed banks and controlled propagation, making it a crucial component of preservation efforts across the board. In sum, the role of seed production in MAPs goes beyond mere propagation; it is the linchpin that sustains the industry, supports livelihoods, fosters research, and aids in conservation across a wide spectrum of MAP crops.

In summary, this chapter delves into the complex landscape of seed production in the context of MAPs. It seeks to address a wide array of challenges and opportunities within the MAPs seed industry, with the goal of establishing sustainable seed production systems for various valuable MAPs. Emphasizing innovation, research, and collaborative efforts, this industry has the potential to play a vital role in meeting the ever-growing global demand for MAPs-derived products. Subsequent sections will provide a more in-depth exploration of these challenges and opportunities, offering valuable insights and strategies to enhance seed production in MAPs.

18.2 CURRENT CHALLENGES IN MAPS CULTIVATION AND OPPORTUNITIES

Cultivating MAPs presents a multifaceted set of challenges that extend across various dimensions of agriculture, from environmental considerations to economic viability (Rana et al., 2020). Successfully addressing the intricacies of MAPs cultivation necessitates a comprehensive grasp of the obstacles faced by both growers and the industry. One of the foremost challenges emanates from environmental factors, most notably climate change, which is actively reshaping growing seasons and mandating adjustments in traditional agricultural practices (Spinozzi et al., 2021). Moreover, the vitality and productivity of MAPs hinge on soil quality and land utilization. Sustainable cultivation practices demand immediate attention to counteract soil degradation and implement effective land management methods (Solomou et al., 2019). Pests and diseases pose yet another formidable challenge. MAPs are susceptible to a gamut of potential threats that can adversely affect crop yields and undermine the overall quality of the produce. It is imperative to institute judicious strategies for pest and disease management. Sustainability serves as a cornerstone for the long-term viability of MAPs cultivation (Rai, 2012). Embracing environmentally responsible practices, such as organic and agroecological methods, not only mitigates the industry's ecological footprint but also elevates the quality of the end products. Beyond the sphere of environmental concerns, the economic sustainability of MAPs cultivation is intrinsically tied to market dynamics, fluctuations in prices, and shifts in consumer demand. Skillfully navigating these economic challenges is paramount for ensuring the well-being of MAPs farmers and the industry (Oztruk and Hakeem, 2018). These multifaceted challenges are profoundly interconnected and represent linchpins for the successful cultivation of MAPs. As we delve deeper into these complexities, our objective is to provide a comprehensive understanding of the multifarious obstacles faced by growers and the broader industry. A pivotal challenge within the realm of MAPs cultivation is the constrained acceptance of improved varieties and cultivars (Schippmann et al., 2005). These advanced plant varieties, meticulously developed through extensive research and development endeavors, offer the promise of elevated yields, superior quality, and heightened resilience against environmental adversities. Yet, their underutilization not only curtails these advantages but also hinders the progress of the entire MAPs industry. It is imperative to magnify our focus on this issue to fully comprehend its implications and explore potential remedies for the betterment of MAPs cultivation.

The cultivation of MAPs presents a promising landscape with several opportunities that can be harnessed for sustainable growth in the industry. These opportunities hold the potential to address the existing challenges and pave the way for a more vibrant and robust MAPs seed production sector. Firstly, the growing global demand for MAPs presents a significant opportunity. The world's increasing recognition of the therapeutic and aromatic properties of these plants has led to a rising global demand. This presents a significant opportunity for seed producers to cater to the expanding market, both domestically and internationally. By focusing on producing quality seeds, the industry can capitalize on this heightened demand and contribute to the export market. Secondly, the adaptability of MAPs to local climates offers a unique advantage (Kalamartzis et al., 2023).

These plants often thrive in diverse climatic conditions. Seed producers have the opportunity to adapt their offerings to suit local climates and specific growing regions. By developing climate-resilient seed varieties, they can enhance crop productivity and support local farmers in optimizing their yields. Investment in R&D activities forms a critical avenue for seed production. Advancements in breeding, genetics, and cultivation techniques can lead to the development of superior MAPs seed varieties. Collaborative efforts between research institutions, government agencies, and the private sector can further innovation and enhance seed quality. Furthermore, promoting sustainable farming practices not only benefits the environment but also offers a competitive advantage. Seed producers can support and advocate for sustainable cultivation methods among farmers. By incorporating organic and agroecological approaches, the industry can ensure the long-term viability of seed production and meet the increasing demand for eco-friendly products. Diversification of MAPs also stands as an opportunity. The wide variety of MAPs available provides an opportunity for diversification in seed production. By expanding the range of MAPs cultivated and offered as seeds, producers can tap into niche markets and cater to specific industries, including pharmaceuticals, cosmetics, and herbal remedies. Government initiatives and support provide another avenue for growth. Indian government recognizes the economic potential of MAPs and offers support in the form of subsidies, incentives, and favorable policies. Seed producers can leverage these government initiatives to expand their operations, enhance seed quality, and explore new markets. Increasing awareness about the value of MAPs in various industries can boost demand. Seed producers can play a crucial role in educating the market about the benefits of MAPs and their applications, thereby creating a more receptive environment for their products. Lastly, access to high-value markets is a noteworthy opportunity. Certain MAPs, such as saffron, truffles, and vanilla, command high prices in international markets. By focusing on these high-value crops and ensuring quality seed production, the industry can tap into lucrative markets and enhance profitability.

18.3 SPECIFIC PROBLEM AND SOLUTIONS FOR DIFFERENT MAPS

The cultivation of MAPs poses unique challenges, as these plants are not as commonly grown among farmers, high-quality seeds are often scarce, and specific seed preparation methods are required (Figure 18.1).

Additionally, the lack of knowledge regarding nursery management after seed germination exacerbates the issues faced by farmers in the production of MAPs. Following are some comprehensive solutions to address these challenges.

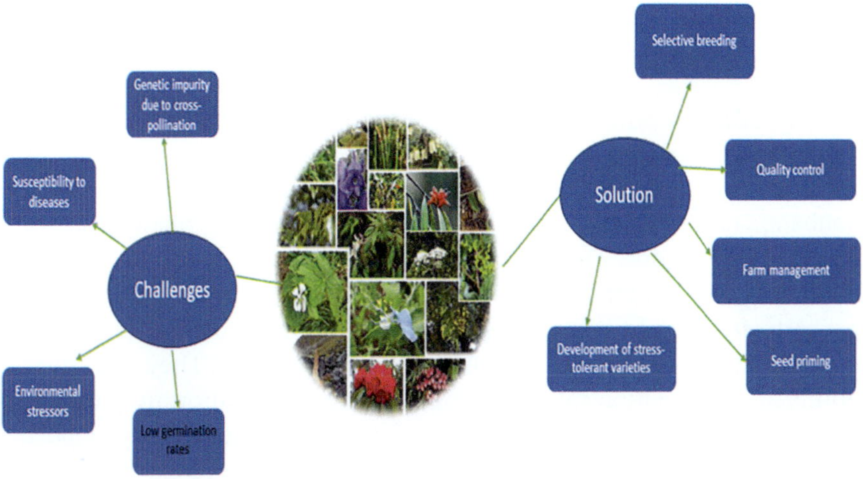

FIGURE 18.1 Challenges and solutions in seed production of various medicinal and aromatic plants.

18.3.1 INDIAN GINSENG (WITHANIA SOMNIFERA)

Withania somnifera, commonly known as Ashwagandha or Indian Ginseng, has long been cherished for its therapeutic properties rooted in the Ayurvedic tradition. However, its propagation through seed-based methods faces distinctive challenges. One challenge is the inconsistent germination of *Withania somnifera* seeds. These seeds exhibit irregular germination patterns, and the seeds of different accessions display varying germination behaviors. Another challenge is the suboptimal field conditions. Under natural field conditions, the germination of *Withania somnifera* seeds is often poor, and seedling mortality is high. In the quest for addressing the challenges associated with *Withania somnifera* seed germination, several solutions have been explored (Kalamartzis et al., 2023). Notably, the application of growth regulators, specifically gibberellic acid (GA_3), has proven to be highly effective. An optimal concentration of 150 μg/ml of GA_3, combined with a temperature of 25°C, significantly promotes germination. Temperature indeed plays a pivotal role in the germination of *Withania somnifera*. Research has established that an ideal temperature for germination is 25°C, which provides a substantial boost to germination rates. Additionally, the role of light conditions cannot be underestimated in this process. Continuous light has been found to foster germination, underscoring the significance of photoperiod as a determinant in the germination process.

18.3.2 BLACK HENBANE (HYOSCYAMUS NIGER L.)

Hyoscyamus niger L., commonly known as Khurasani-ajvayan or black henbane, holds a significant place among medicinal plants due to its diverse applications. The leaves of this plant serve as a valuable source of medicinal drugs, particularly for their sedative, narcotic, and therapeutic properties in conditions like asthma and whooping cough. *Hyoscyamus niger* L. seeds can be quite challenging to germinate. They exhibit variability in germination rates, and sometimes, germination may not occur efficiently. These seeds have a hard coat, making them a bit stubborn in their germination behavior. To overcome the problem of low germination rates in seeds, the researchers conducted experiments involving various treatments and environmental conditions (Verma et al., 2014). These treatments included soaking the seeds in water for different durations (24 and 48 hours), using gibberellic acid (GA_3) at various concentrations (200 and 500 ppm), applying potassium nitrate (KNO_3) at 0.2%, and briefly exposing the seeds to concentrated sulfuric acid (H_2SO_4) for 15 seconds. The study revealed that the most effective treatment for enhancing germination was soaking the seeds in GA_3 (200 ppm). This treatment resulted in the highest percentage of normal seedlings (73.0%) under specific conditions, such as substrate and temperature, after 12 days.

18.3.3 CHAMOMILE (CHAMOMILLA RECUTITA)

In a study on chamomile (*Matricaria chamomilla* L.) seed quality conducted in Ethiopia (Garkebo and Melese, 2023), several significant data values were obtained. The study found that the germination percentage of chamomile seeds declined as storage periods increased. Initially, the germination percentage was 78.2%, but after 12 months of storage, it dropped to 42%. This represents a substantial reduction in seed quality over time. The research suggested that the choice of storage method significantly impacted seed quality. For instance, seeds stored in polypropylene bags recorded the lowest germination percentages after 12 months, while those stored in bottles maintained higher germination percentages. The data showed that bottle storage resulted in a 60% germination rate after 12 months, which was better compared to polypropylene bags. Furthermore, the moisture content of seeds stored in polypropylene bags was significantly higher after 12 months, reaching 55%. In contrast, the bottles maintained lower moisture content (47%) over the same period.

18.3.4 CLARY SAGE (SALVIA SCALAREA L.)

This study investigates the effect of temperature on seed germination and emergence of *Salvia sclarea* L. in sub-tropical climatic conditions. *Salvia sclarea* L. has been introduced as a winter season crop in sub-tropical regions, but the problem identified is the variation in seed germination and emergence under different temperature conditions in these areas. The study highlights (Javaid et al., 2018) the crucial role of temperature in regulating seed germination and emergence of *Salvia sclarea* L., emphasizing that understanding the optimal temperature conditions is essential for successful cultivation in these regions. The research revealed that temperature significantly influenced the germination of *Salvia sclarea* L. seeds. The maximum seed germination (78.50%) occurred at 25°C, followed by 20°C (74.44%) and 15°C (65.80%), while the lowest germination was observed at 5°C (2.33%). Similarly, the first seed emergence occurred at four days at 25°C. In conclusion, the study recommends a temperature range of 15°C to 25°C as suitable for seed germination of *Salvia sclarea* L. in sub-tropical conditions. To ensure successful seed germination and emergence in sub-tropical climates, it is advisable to sow the seeds during the favorable temperature conditions, typically in the months of October and November, when the temperature falls within the optimal range.

18.3.5 ISABGOL (PLANTAGO OVATA)

Seed shattering poses a significant challenge in the cultivation of *Plantago ovata*. This natural phenomenon refers to the plant's tendency to release mature seeds from the seed heads, making it a complex issue for farmers.

To combat this problem, precise timing is critical. Harvesting Isabgol seeds must be done when they are just mature but before they fully shatter. This requires close monitoring and a deep understanding of the plant's growth stages. Gentle harvesting techniques should be employed to minimize seed loss. Whether through hand harvesting or specialized equipment designed to reduce seed shattering, the objective is to handle the plants with care and minimize mechanical stress, which can trigger seed release. Additionally, selecting non-shattering varieties can be a worthwhile consideration (Patel and Amin, 1970). Breeding and choosing Isabgol cultivars with reduced seed shattering tendencies can offer a long-term solution to this issue. Some growers may explore the use of growth regulators or chemicals that inhibit seed abscission to minimize seed shattering, although the safety and regulations surrounding their use should be thoroughly evaluated. Furthermore, after the harvest, it is crucial to ensure that the seeds are properly dried and stored to prevent further shattering. Excess moisture can lead to premature seed release, affecting the overall yield and quality. Continuous research and development efforts are also essential in identifying and developing solutions to mitigate seed shattering in Isabgol cultivation. This persistent problem can have a significant impact on crop yield and quality, making it vital for agricultural researchers and practitioners to work together to find effective strategies for addressing this challenge. By implementing these strategies and remaining dedicated to ongoing improvement, farmers can reduce the impact of seed shattering in Isabgol cultivation, leading to higher yields and improved seed quality.

18.3.6 KALMEGH [ANDROGRAPHIS PANICULATA (BURM. F.) WALL EX., NEES]

Andrographis paniculata (Burm. F.) Wall ex. Nees, a valuable medicinal herb, faces a challenge of low seed germination rates due to its hard seed coat. To overcome this issue, researchers tested various treatments (Zuffo et al., 2017). They found that soaking the seeds in 50°C hot water for 5 minutes significantly improved germination, with a maximum germination percentage of 93%. This treatment helped soften the hard seed coat, facilitating germination. In contrast, hormonal treatments with Gibberellic acid and other growth regulators did not yield significant improvements. The study provides a practical solution for enhancing seed germination in

Andrographis paniculata (Burm. F.) Wall ex. Nees, which is essential for its propagation and medicinal use.

18.3.7 PALMAROSA (*CYMBOPOGON MARTINI*)

Cymbopogon martini is a fragrant grass species of economic and medicinal significance due to its essential oil. However, seed production in *Cymbopogon martini* faces several notable challenges. One primary concern is the low seed set, where the plants produce a limited number of seeds, hampering the availability of seeds for cultivation and essential oil extraction. Low seed set can be attributed to factors like inadequate pollination and genetic variations. Another challenge is the species' dependence on cross-pollination, making it essential to maintain a diverse plant population to facilitate effective pollination and mitigate reduced seed yields. Furthermore, variability in seed viability can lead to inconsistent germination rates, affecting crop establishment. The susceptibility of Palmarosa seeds to viability loss during storage complicates the preservation of seed quality over time. Additionally, seed dormancy can hinder germination under suitable conditions, necessitating methods to break dormancy for successful crop establishment. To tackle these issues in *Cymbopogon martini* seed production, various strategies can be employed. These include improved pollination management, such as maintaining diverse plant populations and ensuring the presence of pollinators. Regular seed viability testing helps select high-quality seeds for sowing and crop establishment (Pinzón-Torres et al., 2014). Effective seed storage protocols should be developed to prolong seed viability and maintain a consistent seed supply. Additionally, exploring methods to break seed dormancy can improve germination rates. Genetic improvement through breeding programs can address specific challenges, such as increasing seed set and reducing seed dormancy. Addressing these challenges is essential to ensure a stable and reliable seed supply for Palmarosa cultivation and essential oil extraction, thereby supporting the economic and medicinal value of this plant species.

18.3.8 PYRETHRUM (*CHRYSANTHEMUM CINERARIIFOLIUM*)

Chrysanthemum cinerariifolium seed production encounters various challenges that hinder the efficient and effective generation of quality seeds.

Key issues include low germination rates, which often lead to poor stand establishment and reduced crop yields, impacting overall farm productivity. Additionally, maintaining genetic purity is crucial, but cross-pollination with other Tanacetum species can introduce genetic impurity, resulting in inconsistent *C. cinerariifolium* plants. Susceptibility to diseases, such as powdery mildew and foliar pathogens, further affects seed quality and germination rates. Growing pyrethrum in regions susceptible to environmental stressors, including drought, salinity, and temperature fluctuations, compounds these problems. To address these challenges and improve pyrethrum seed production, several solutions can be considered (Jawhari et al., 2022; Nikolova, 2016). Selective breeding programs can be implemented to develop pyrethrum varieties with higher germination rates and genetic purity, necessitating stringent monitoring and crop isolation to prevent cross-pollination. Effective disease management strategies, such as using disease-resistant varieties, proper crop rotation, and suitable fungicides, can prevent and control diseases that affect seed quality. The development and promotion of stress-tolerant pyrethrum varieties can ensure more consistent seed production, even in challenging environmental conditions (Li et al., 2011). Utilizing seed priming techniques like hydropriming can enhance seed quality and germination rates, improving seedling establishment. Implementing best practices in farm management, including soil preparation, irrigation, and pest control, creates optimal conditions for pyrethrum seed production. Rigorous quality control measures at every stage of seed production, from planting to harvesting and processing, help ensure that only high-quality seeds are available to farmers.

18.3.9 SHATAVARI (ASPARAGUS RACEMOSUS)

The production of seeds for *Asparagus racemosus* presents various challenges and opportunities that are crucial for the cultivation and conservation of this medicinal plant. One of the primary challenges is the low and erratic germination rate of *A. racemosus* seeds. Achieving synchronized and early germination is vital for successful cultivation. To address this challenge, researchers have explored pre-sowing seed treatments, such as soaking in hot water and dilute sulfuric acid. These treatments have shown promise in overcoming seed dormancy and improving germination rates (Parveen et al., 2020). Another significant problem associated with *A. racemosus* is the high demand in the herbal medicine market, leading to overharvesting and habitat destruction. To meet market demand while conserving the plant, mass cultivation is proposed as a solution. Improving seed germination is a key step in enabling large-scale cultivation efforts. Physiological dormancy in *A. racemosus* seeds has been identified as a barrier to achieving high germination rates. Studies suggest the use of hot water and dilute sulfuric acid treatments to break this dormancy effectively. These treatments are not only practical but also cost-effective and can be implemented under field conditions, making them valuable tools for seed production.

18.3.10 SENNA (CASSIA AURICULATA)

Breaking seed dormancy in *Cassia auriculata* presents a critical challenge for successful cultivation. To address this issue, a comprehensive study explored various treatments, including sulfuric acid scarification, hot water soaking, and gibberellic acid (GA_3) application (Khan, 2013). Among these treatments, soaking the seeds in hot water (100°C) for 15 minutes, followed by a 3-hour immersion in 100 ppm GA_3, proved to be the most effective method, resulting in a remarkable increase in germination rates to 98%, significant growth in root and shoot lengths, and improved dry matter production and vigor index. This approach was particularly successful in weakening the seed coat, a primary dormancy factor, and the application of GA_3 further promoted germination by stimulating crucial metabolic processes. The findings of this study offer valuable insights into addressing seed dormancy in *Cassia auriculata*, contributing to enhanced crop establishment and productivity in the cultivation of this medicinal plant.

18.3.11 BASIL (OCIMUM BASILICUM)

To overcome problems in *Ocimum basilicum* seed production, it's essential to isolate varieties to prevent cross-pollination, monitor for and address pests and diseases, space *O. basilicum* densely for self-pollination, attract pollinators through companion planting or hand-pollination, control temperature and start seeds indoors in unfavorable climates, wait for plant maturity before seed collection, ensure seeds are fully dried before storage, store them in a cool, dry place in airtight containers, and select healthy, robust parent plants for the best seed quality, thus promoting successful basil seed production. To tackle problems in *O. basilicum* seed production, consider these solutions: First, address cross-pollination issues by isolating basil varieties or resorting to hand-pollination techniques. Second, actively monitor plants for pests and diseases, promptly implementing organic or chemical remedies when necessary, and explore companion planting to deter common pests. Third, ensure sufficient plant density by properly spacing basil plants for enhanced self-pollination. By implementing these solutions, you can improve your chances of successful basil seed production (Thongtip et al., 2022).

18.3.12 PERIWINKLE (CATHARANTHUS ROSEUS)

The seed production of *Catharanthus roseus* may encounter several challenges that can be effectively addressed with appropriate solutions. One common issue is the low germination rate of the seeds, which can be improved by using fresh seeds and subjecting them to a period of cold stratification. Cross-pollination is another concern, and to ensure true-to-type seeds, isolation of different varieties and hand-pollination can be employed. Additionally, managing seed quality and viability by harvesting from healthy plants and storing seeds in optimal conditions is crucial. Proper seed collection, sowing, and growth conditions are essential for maximizing germination, while regular disease control and crop rotation help maintain seed health (Nabaei and Amooaghaie, 2019). These measures collectively contribute to the successful seed production of *C. roseus*.

18.4 ROLE OF HIGH-QUALITY SEEDS IN MEETING INDUSTRY DEMANDS

High-quality MAPs seeds are of paramount importance in fulfilling the demands of multiple industries, including pharmaceuticals, herbal products, cosmetics, and traditional medicine. These seeds serve as the cornerstone for a consistent and dependable supply of MAPs with specific properties. First and foremost, high-quality MAPs seeds are the backbone of the pharmaceutical and herbal medicine sectors. They are essential for cultivating plants with distinct medicinal attributes, such as analgesic, anti-inflammatory, or antimicrobial properties. These cultivated plants serve as the primary source of raw materials to produce drugs and herbal remedies that address the healthcare needs of the population. Crucially, these seeds contribute to consistency in active compounds found in medicinal plants (Medar, 2018; Nagaraju et al., 2018). These active compounds have therapeutic effects, and high-quality seeds ensure that the plants consistently contain the desired levels of these compounds. This consistency is vital for the pharmaceutical industry to manufacture standardized medicines with predictable efficacy and safety. Moreover, high-quality MAPs seeds play a pivotal role in ensuring adherence to quality control standards and regulatory compliance. Regulatory authorities mandate that pharmaceutical and herbal product manufacturers meet stringent quality requirements. High-quality seeds are fundamental in achieving the standards necessary to comply with regulations related to product efficacy and safety. The cosmetic and fragrance industry also relies heavily on aromatic plants grown from high-quality seeds. These plants provide the essential oils and fragrances used in cosmetics, perfumes, and aromatherapy products. The fragrance industry depends on the consistency of fragrance profiles and essential oil content, which can only be achieved using high-quality seeds. Traditional and alternative medicine systems, such as Ayurveda and Traditional Chinese Medicine, depend on specific medicinal plants. High-quality seeds are instrumental in the cultivation of plants with the right properties for traditional remedies, ensuring the continued effectiveness of these treatments. Furthermore, the herbal products and dietary supplements industry requires a reliable supply of medicinal plants (Sinjushin, 2015). High-quality seeds are vital for meeting the demand for herbal teas, dietary supplements, and other wellness products that incorporate plant-based ingredients. Sustainability is a growing concern in these industries. The pharmaceutical and herbal sectors are increasingly focused on sustainable sourcing practices. High-quality seeds enable the cultivation of medicinal plants in a sustainable manner, reducing the impact on natural plant populations and ecosystems.

Research and development in botany, pharmacology, and traditional medicine rely on high-quality MAPs seeds. These seeds are essential for conducting studies on plant properties, active compounds, and their effects on health and well-being. They underpin innovation and product development, driving progress in these fields. Moreover, high-quality MAPs seeds also play a role in the conservation of threatened or endangered medicinal plant species. By cultivating these species, industries can contribute to their preservation while sustainably meeting demand. Finally, the MAPs seed industry contributes to economic growth by providing a sustainable source of raw materials for various sectors. It supports job creation, value addition, and export opportunities, particularly in regions with rich biodiversity. High-quality MAPs seeds are indispensable for fulfilling the diverse demands of industries relying on MAP. These seeds ensure the production of consistent, effective, and safe products in the pharmaceutical, herbal, cosmetic, and traditional medicine sectors. They also contribute to sustainability, research, and economic development while promoting the responsible use and conservation of medicinal plant resources.

18.5 COLLABORATIVE EFFORTS FOR MAPS SEED PRODUCTION

18.5.1 INVOLVEMENT OF RESEARCH INSTITUTES

Research institutes play a pivotal role in advancing the quality and quantity of MAPs seed production. Their involvement is multifaceted, encompassing various critical aspects. Firstly, research institutes are instrumental in conducting extensive studies on MAPs, including their genetic characteristics, growth patterns, and optimal cultivation techniques. Through research, they identify the most suitable species and varieties for cultivation and develop innovative strategies for enhancing seed quality and yield (Inoue et al., 2019). These institutes are also at the forefront of germplasm conservation, safeguarding the genetic diversity of MAPs. By maintaining seed banks and genetic repositories, they ensure that valuable genetic resources are preserved for future use, preventing the loss of potentially important plant varieties. Additionally, research institutes engage in breeding programs aimed at developing high-yielding and disease-resistant MAPs varieties. These efforts contribute to increased seed production while reducing the vulnerability of crops to pests and diseases. Through selective breeding, they can also enhance the content of specific bioactive compounds in MAPs, making them more valuable for various industries (Amin et al., 2022; Inoue and Craker, 2014a). Research institutes facilitate knowledge dissemination by conducting training programs and providing technical assistance to farmers and seed producers. They share their findings on best practices, seed treatment methods, and sustainable cultivation techniques (Tables 18.1 and 18.2; Figure 18.2), empowering stakeholders in the MAPs seed production chain.

TABLE 18.1 Detailed Information on Cultivation Parameters for Seed Production of *Withania somnifera, Hyoscyamus niger, Chamomilla recutita, Salvia scalarea, Plantago ovata, Andrographis paniculata, Cymbopogon martini, Chrysanthemum cinerariifolium, Asparagus racemosus, Cassia senna, Ocimum basilicum.*, and *Catharanthus roseus.*

S. No.	Name of Crop	Plant Spacing	Seed Requirement/ ha	Fertilizers Requirement/ ha	Irrigation During Growing Season	Isolation Distance for Seed ProductIon	Rouging	Seed Yield/ ha (kg)
1.	*Withania somnifera*	50X50 cm	5–7 kg	60–80:40–60:40–60 kg/ha	2–4	400–800 meters	Frequent inspection to remove weeds, off-types, and diseased plants throughout the growing season.	100–300
2.	*Hyoscyamus niger L.*	30X30 cm	2–3 kg	60–80:40–60:40–60 kg/ha	1–2	400–800 meters	Early rouging to eliminate unwanted plants and maintain purity.	50–200
3.	*Chamomilla recutita*	30X40 cm	1–2 kg	30–40:20–30:20–30 kg/ha	3–5	200–400 meters	Start early and maintain uniformity by removing unwanted plants.	150–200
4.	*Salvia scalarea L.*	50X50 cm	4–6 kg	30–40:20–30:20–30 kg/ha	2–4	400–800 meters	Begin rouging as plants emerge to remove weeds and off-types.	150–200
5.	*Plantago ovata*	40X40 cm	3–5 kg	40–50:20–30:20–30 kg/ha	4–6	200–400 meters	Early rouging to ensure purity and eliminate undesired plants.	400–500
6.	*Andrographis paniculata (Burm. F.) Wall ex., Nees*	30X30 cm	1–2 kg	40–60:20–30:20–30 kg/ha	2–4	400–800 meters	Continuous rouging to eliminate weeds and maintain purity.	200–250
7.	*Cymbopogon martini*	100X100 cm	4–6 kg	40–60:30–40:30–40 kg/ha	3–5	200–400 meters	Early rouging to maintain a pure stand by removing unwanted plants.	100–150
8.	*Chrysanthemum cinerariifolium*	50X50 cm	500–600 gm	50–60:40–50:40–50 kg/ha	3–4	400–800 meters	Early rouging to maintain the purity of the Pyrethrum variety.	180–200
9.	*Asparagus racemosus*	30X30 cm	1–2 kg	80–100:40–60:40–60 kg/ha	4–6	400–800 meters	Ongoing rouging throughout the season to remove weeds and off-types.	200–250
10.	*Cassia auriculata*	60X60 cm	15–20 kg	40–60:30–40:30–40 kg/ha	2–4	400–800 meters	Early rouging to eliminate undesired plants and maintain purity.	200–300
11.	*Ocimum basilicum*	40X40 cm	1–2 kg	60–80:30–40:30–40 kg/ha	4–6	200–400 meters	Early and continuous rouging to remove weeds and off-types.	200–250
12.	*Catharanthus roseus*	50X50 cm	200–400 gm	50–70:30–40:30–40 kg/ha	3–5	200–400 meters	Early rouging to maintain variety purity.	250–300

TABLE 18.2 Roadmaps for Future MAPs Seed Production Technologies

Opportunity	Key actions and considerations
1. Crop diversity	1. Explore the demand for various crop varieties, both traditional and novel.
	2. Assess regional adaptability of different crops.
2. Genetic research	1. Invest in genomic research to develop high-yield and resilient varieties.
	2. Collaborate with research institutions for genetic mapping.
3. Sustainability	1. Promote sustainable agriculture practices, including organic and regenerative farming.
	2. Implement eco-friendly seed production techniques.
4.Technological advancements	1. Embrace modern technologies such as precision farming, IoT, and blockchain for traceability.
	2. Train farmers in tech adoption.
5. Market research	1. Conduct market surveys to identify emerging trends and consumer preferences.
	2. Analyze competitors and assess market potential.

18.5.2 ROLE OF FARMERS IN SEED PRODUCTION

Farmers are the backbone of MAPs seed production. Their active involvement in cultivation is vital for ensuring a consistent and high-quality seed supply (Figures 18.3 and 18.4). They are responsible for the hands-on work of sowing, nurturing, and harvesting MAPs plants to obtain seeds (Shukla et al., 2022). To fulfil this role effectively, farmers must receive adequate training and support. Collaborative efforts with farmers involve providing them with access to quality seeds, which are the starting point of the entire process. Equipped with high-quality seeds, farmers can establish healthy MAPs crops that serve as the source of seeds for subsequent generations. They should also be educated on sustainable farming practices that consider soil health, water conservation, and pest management to ensure the longevity of seed production (Inoue and Craker, 2014b). To promote the cultivation of specific MAPs varieties that align with industry demands, collaboration with farmers can include incentive programs and contracts that guarantee the purchase of seeds at fair prices. Such initiatives encourage farmers to focus on the production of sought-after MAPs varieties, fostering a more responsive seed supply chain.

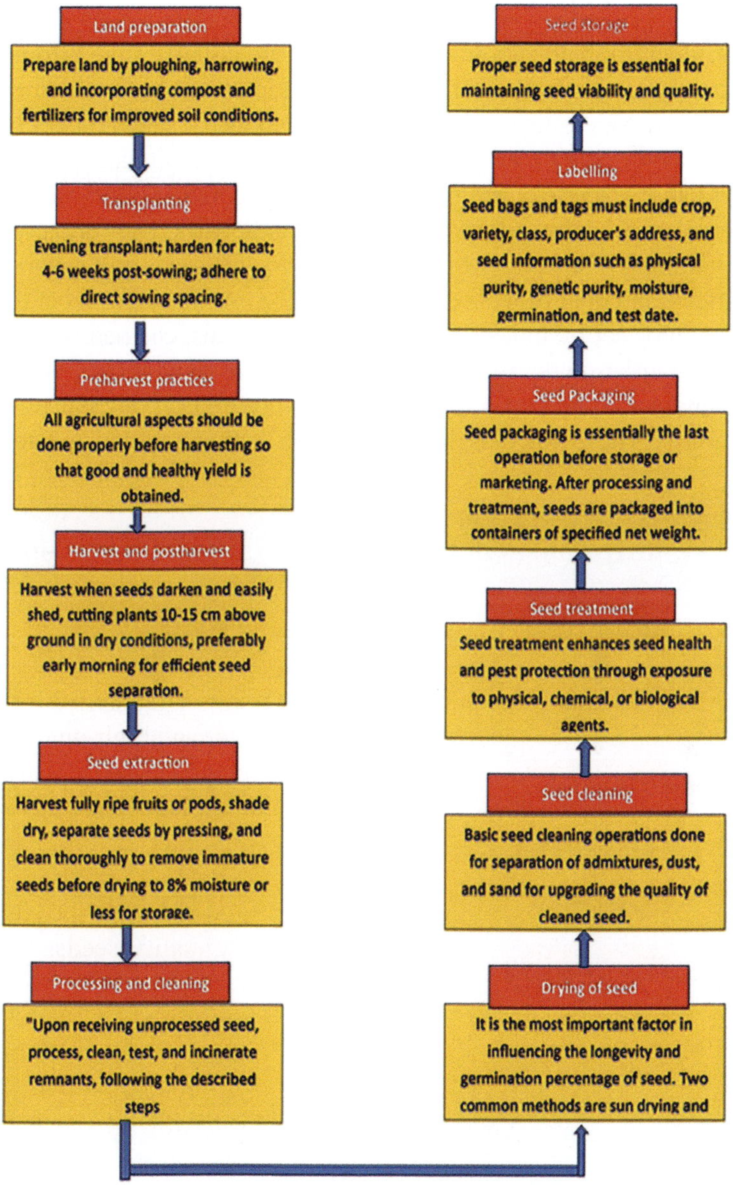

FIGURE 18.2 Comprehensive process for achieving successful seed production of *Withania somnifera, Hyoscyamus niger, Chamomilla recutita, Salvia scalarea, Plantago ovata, Andrographis paniculata, Cymbopogon martini, Chrysanthemum cinerariifolium, Asparagus racemosus, Cassia senna, Ocimum basilicum.*, and *Catharanthus roseus*.

18.5.3 SEED-PRODUCING ORGANIZATIONS

Seed-producing organizations are central to the collaborative efforts aimed at MAPs seed production. These organizations serve as intermediaries, connecting research institutes, farmers, and industry stakeholders (Figures 18.3 and 18.4). They play a coordinating role in facilitating the exchange of knowledge, expertise, and resources. Seed-producing organizations often function as seed banks and seed certification authorities. They are responsible for maintaining a repository of high-quality MAPs seeds and ensuring their authenticity and quality. This quality assurance is essential to meet the stringent standards set by industries like pharmaceuticals, cosmetics, and herbal products. Furthermore, these organizations may engage in research and development activities to improve seed quality and production techniques continuously. They work closely with research institutes to stay updated on the

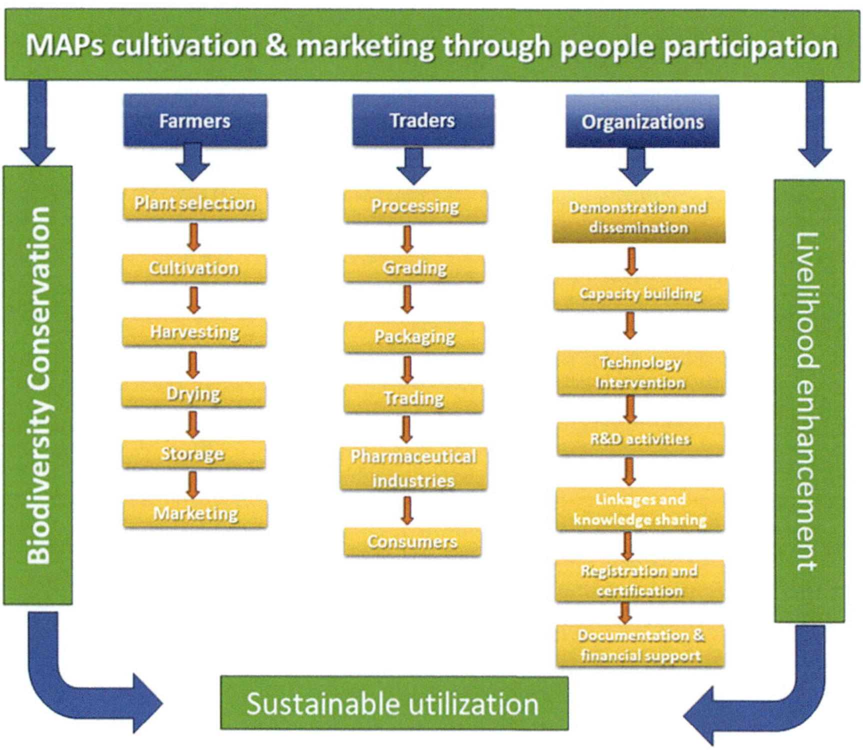

FIGURE 18.3 Cultivation, conservation, and marketing strategies involving farmers, traders, and organizations for medicinal and aromatic plants.

latest advancements in MAPs cultivation and incorporate these innovations into seed production processes. In summary, collaborative efforts for MAPs seed production require a tripartite approach involving research institutes, farmers, and seed-producing organizations. Research institutes contribute expertise, knowledge, and genetic resources, while farmers provide the essential cultivation and harvesting work. Seed-producing organizations act as intermediaries, ensuring quality control and facilitating the flow of information and resources. This synergy is essential to meet the growing demand for high-quality MAPs seeds across various industries.

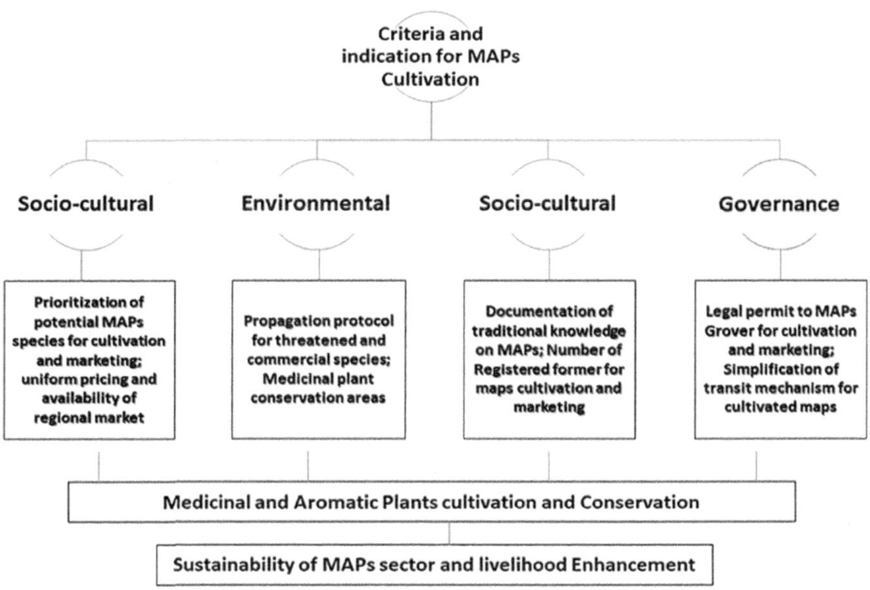

FIGURE 18.4 Criteria and indicators for medicinal and aromatic plants cultivation.

18.6 ENHANCING SEED QUALITY FOR MEDICINAL AND AROMATIC CROPS

Enhancing seed quality is crucial for the cultivation of medicinal and aromatic crops in India, given their economic and cultural importance. These crops often face challenges like dormancy and germination inhibitors, making seed enhancement techniques essential (Nagaraju et al., 2018). These techniques involve treating seeds with specific substances under precise conditions to improve germination (Table 18.3). Seed quality is vital for successful cultivation, especially for crops like MAPs with diverse applications. A significant challenge is their often-poor seed germination, which can be attributed to various factors. Accelerating the germination process through techniques like seed priming can help overcome these issues. Seed priming offers advantages such as improved germination rates and enzyme activity stimulation, enhancing seedling development. Various seed priming methods are available to meet MAPs' specific germination requirements.

TABLE 18.3 Seed Treatment Methods and Their Impact on Seed Quality in Various Crops

Treatment	Enhancing Seed Quality Through Treatments	Crop	References
Seed dormancy breaking treatments	Immerse seeds in a 1% sodium hypochlorite solution to overcome seed dormancy issues.	*Chrysanthemum cinerariifolium*	Somrug et al. (2021)
GA_3 treatment	Soak seeds in GA_3 (Gibberellic Acid) to enhance germination rates.	*Saussurea lappa*	Li et al. (2022)
Seed pelleting	Apply hydro priming for 12 hours, resulting in the highest seedling vigor, germination percentage, and seedling dry weight.	*Ocimum basilicum*	Nawrot-Chorabik et al. (2021)
Seed soaking in cold water	Submerge seeds in cold water to achieve the highest germination percentage.	*Catharanthus roseus*	Odoi et al. (2021)
Water soaking	Soak seeds in water for 18–24 hours to enhance the quality of aged seed lots.	*Tagetes erecta*	Asropi et al. (2019)
Hydro priming	Implement hydro priming for 12 hours, enhancing germination and promoting dry matter production.	*Trigonella Foenum-Graecum*	Kalita et al. (2021)
Acid scarification	Fresh seeds undergo scarification by immersing them in concentrated sulfuric acid (H_2SO_4).	*Rheum webbianum*	Šoch et al. (2023)
Seed priming with growth regulators	Utilize seed priming with growth regulators by soaking seeds in 100 ppm GA_3 for 6 hours to enhance germination.	*Andrographis paniculata*	Qamar et al. (2022)
	Soaking seeds in 50 ppm GA_3 for 3 hours maximizes seedling vigor.	*Asparagus sprengeri*	Elhindi et al. (2016)
	Soak seeds in 200 ppm IAA for 6 hours to achieve higher germination rates.	*Cassia obtusifolia*	Rout et al. (2017)

TABLE 18.3 *(Continued)*

Treatment	Enhancing Seed Quality Through Treatments	Crop	References
	1. Soak seeds in 200 ppm Ethrel for 6 hours to maximize seedling growth. 2. Soaking seeds in a 1% KH_2P0_4 solution for 1 hour is beneficial for *Withania somnifera.*	*Gymnema sylvistre (Gymnema)*	Nagaraju et al. (2018)
Priming with inorganic salts	Apply priming with inorganic salts, such as soaking seeds in 50 to 100 mM KNO_3, to achieve maximum germination compared to untreated seeds.	*Aconitum balfourii (Indian Atees)*	Dhillon et al. (2021)
	Soaking seeds in 130 mm thiourea leads to early germination.	*Aconitum heterophyllum*	Pandey et al. (2000)
	Soaking seeds in a 1% solution of sodium nitrate enhances plant height, root length, and other yield parameters.	*Withania somnifera*	Suryawanshi et al. (2013)
Seed priming with organic products	Apply seed priming with organic products, such as soaking seeds in 25% coconut water for 6 hours to enhance nursery emergence.	*Bixa orellana (Achiote)*	Lakmali and Seran (2022)
Hydration Treatments	1. Implement hydration treatments, including soaking in pre-chilling conditions at 4°C for 10 days. This approach stimulates germination and reduces the time required for germination. 2. Similarly, pre-chilling at 4°C for 10 days significantly improves germination in Discerning Sophia (Flixweed).	*Descurainia sophia*	Asgharpour et al. (2022)

18.6.1 HYDRO-PRIMING

In hydro-priming, seeds are soaked in water for a specified duration. This method allows seeds to imbibe water and initiate metabolic processes, preparing them for germination. Hydro-priming is often used for seeds that require hydration to break dormancy.

18.6.2 HALO-PRIMING

Halo-priming involves the hydration of seeds using inorganic salt solutions. This technique can be particularly effective for MAPs with hard seed coats or those that are sensitive to salinity.

18.6.3 OSMO-PRIMING

Osmo-priming utilizes solutions with substances like polyethylene glycol (PEG) to facilitate seed hydration. It is a valuable method for improving germination in seeds that require osmotic adjustments to break dormancy or enhance germination rates.

18.6.4 THERMO-PRIMING

Thermo-priming is based on subjecting seeds to specific temperature regimes, either high or low, before sowing. This technique can stimulate germination in some MAPs by breaking dormancy or promoting quicker and more uniform germination.

18.6.5 MATRI-PRIMING

In matri-priming, seeds are treated with solid matrices like hydrated sand, peat, or vermiculite. This method ensures uniform hydration of seeds and can be especially useful for small or irregularly shaped seeds.

18.6.6 BIO-PRIMING

Bio-priming involves coating seeds with beneficial microorganisms such as *Trichoderma spp.* or *Pseudomonas aureofaciens*. These microorganisms can assist in enhancing seed germination and protecting seeds from soilborne pathogens.

18.6.7 CHEMICAL PRIMING

Some MAPs may benefit from chemical priming using substances like gibberellic acid (GA_3) or other plant growth regulators. These chemicals can help break dormancy and improve germination rates.

18.7 CONCLUSION

The climate in our country is well-suited for the cultivation of MAPs. Currently, the cultivation of these plants is limited due to low demand compared to traditional crops like grains and vegetables. However, looking ahead, it is essential to consider expanding the cultivation of MAP. This is because health-conscious individuals are increasingly turning to naturopathy, and meeting this demand requires an expansion of the cultivation areas for these plants. By increasing the cultivation of these plants, our country can become a leading provider of raw materials, while also generating revenue through value addition and product manufacturing. Realizing this potential necessitates collaboration among farmers, traders, product manufacturers, and governing bodies. It is only through coordinated efforts, addressing current challenges, and planning for the future that we can overcome the issues related to seed production and cultivation of MAPs.

ACKNOWLEDGMENT

The authors extend their heartfelt gratitude to the authors and editors of the numerous articles, journals, and books that have been referenced and discussed throughout the creation of this chapter.

KEYWORDS

- **challenges**
- **market access**
- **opportunities**
- **seed germination**
- **seed production**
- **sustainability**

REFERENCES

Abbasi Khalaki, M., Moameri, M., Asgari Lajayer, B., & Astatkie, T. (2021). Influence of nano-priming on seed germination and plant growth of forage and medicinal plants. *Plant Growth Regulation, 93*(1). https://doi.org/10.1007/s10725-020-00670-9 (accessed on 12th August 2025).

Amin, M., Mahmood, K., Bodlah, I., & Qasim, M. (2022). Assessing diversity of aphids (*Hemiptera*: Aphididae) along elevation gradient in Azad Jammu and Kashmir–Pakistan. *International Journal of Tropical Insect Science, 42*(1). https://doi.org/10.1007/s42690-021-00573-w (accessed on 12th August 2025).

Asgharpour, M., Enayati, N., Rezaei Sadrabadi, M., Mohamadi Afrakati, M., Khavandegar, A., Mardi, P., Alirezaei, A., Taherinia, A., & Bakhtiyari, M. (2022). Effects of *Descurainia sophia* on oxidative stress markers and thirst alleviation in hemodialysis patients: A randomized double-blinded placebo-controlled cross-over clinical trial. *Evidence-Based Complementary and Alternative Medicine, 2022.* https://doi.org/10.1155/2022/2001441 (accessed on 12th August 2025).

Asropi, A., Bintoro, N., Karyadi, J. N. W., Rahayoe, S., & Saputro, A. D. (2019). Kinetika perubahan sifat fisik dan kadar tanin biji sorgum (*Sorghum bicolor* L.) selama perendaman. *AgriTECH, 39*(3). https://doi.org/10.22146/agritech.42610 (accessed on 12th August 2025).

Chaachouay, N., Benkhnigue, O., Fadli, M., El Ibaoui, H., & Zidane, L. (2019). Ethnobotanical and ethnopharmacological studies of medicinal and aromatic plants used in the treatment of metabolic diseases in the Moroccan Rif. *Heliyon, 5*(10). https://doi.org/10.1016/j.heliyon.2019.e02191 (accessed on 12th August 2025).

Chhetri, V. T., Shrestha, S., Thapa, S., & Timilsina, S. (2021). Status and role of medicinal and aromatic plants (MAPs) in Nepalese livelihood. *International Journal of Environment, 10*(1). https://doi.org/10.3126/ije.v10i1.38405 (accessed on 12th August 2025).

Craker, L. E. (2007). Medicinal and aromatic plants—Future opportunities. *Nutrition Business Journal*, Flexner 1910.

Dhillon, B. S., Kumar, V., Sagwal, P., Kaur, N., Singh Mangat, G., & Singh, S. (2021). Seed priming with potassium nitrate and gibberellic acid enhances the performance of dry direct seeded rice (*Oryza sativa* L.) in north-western India. *Agronomy, 11*(5). https://doi.org/10.3390/agronomy11050849 (accessed on 12th August 2025).

Elhindi, K. M., Dewir, Y. H., Asrar, A. W., Abdel-Salam, E., El-Din, A. S., & Ali, M. (2016). Improvement of seed germination in three medicinal plant species by plant growth regulators. *HortScience, 51*(7). https://doi.org/10.21273/hortsci.51.7.887 (accessed on 12th August 2025).

Frey, K. M. (2020). Structure activity relationship (SAR) maps: A student-friendly tool to teach medicinal chemistry in integrated pharmacotherapy courses. *Currents in Pharmacy Teaching and Learning, 12*(3). https://doi.org/10.1016/j.cptl.2019.12.014 (accessed on 12th August 2025).

Gantait, S., Kundu, S., Ali, N., & Sahu, N. C. (2015). Synthetic seed production of medicinal plants: A review on influence of explants, encapsulation agent, and matrix. *Acta Physiologiae Plantarum, 37*(5). https://doi.org/10.1007/s11738-015-1847-2 (accessed on 12th August 2025).

Garkebo, H., & Melese, B. (2023). Effects of storage periods and methods on seed quality of chamomile (*Matricaria chamomilla* L.) in Wondogenet, Ethiopia. *Global Academic Journal of Agriculture and Biosciences, 5*(03). https://doi.org/10.36348/gajab.2023.v05i03.001 (accessed on 12th August 2025).

H Patel, A. U., Amin, D. G. P. H. B. P. P. T. P. (1970). Evaluation of stable and non-shattering isabgol cultivar–Gujarat isabgol 4. *Journal of Spices and Aromatic Crops*. https://doi.org/10.25081/josac.2018.v27.i1.1022 (accessed on 12th August 2025).

Inoue, M., & Craker, L. E. (2014a). Medicinal and aromatic plants—Uses and functions. In *Horticulture: Plants for People and Places* (Vol. 2). https://doi.org/10.1007/978-94-017-8581-5_3 (accessed on 12th August 2025).

Inoue, M., & Craker, L. E. (2014b). Medicinal and aromatic plants—Uses and functions. In *Horticulture: Plants for People and Places, Volume 2: Environmental Horticulture*. https://doi.org/10.1007/10.1007/978-94-017-8581-5_3 (accessed on 12th August 2025).

Inoue, M., Hayashi, S., & Craker, L. E. (2019). Role of medicinal and aromatic plants: Past, present, and future. In *Pharmacognosy–Medicinal Plants*. https://doi.org/10.5772/intechopen.82497 (accessed on 12th August 2025).

Islam, M. S., & Bari, M. A. (2012). *In vitro* regeneration protocol for artificial seed production in an important medicinal plant *Mentha arvensis* L. *Journal of Bio-Science, 20*(0). https://doi.org/10.3329/jbs.v20i0.17722 (accessed on 12th August 2025).

Javaid, M. M., Florentine, S., Ali, H. H., & Weller, S. (2018). Effect of environmental factors on the germination and emergence of *Salvia verbenaca* L. cultivars (verbenaca and vernalis): An invasive species in semi-arid and arid rangeland regions. *PLoS ONE, 13*(3). https://doi.org/10.1371/journal.pone.0194319 (accessed on 12th August 2025).

Jawhari, F. Z., Imtara, H., El Moussaoui, A., Khalis, H., Es-Safi, I., Al Kamaly, O., Saleh, A., Parvez, M. K., Guemmouh, R., & Bari, A. (2022). Reproductive biology of the two varieties of *Anacyclus pyrethrum* L.—*Anacyclus pyrethrum* var. *pyrethrum* (L.) Link and *Anacyclus pyrethrum* var. *depressus* (Ball.) Maire—An endemic endangered species. *Plants, 11*(17). https://doi.org/10.3390/plants11172299 (accessed on 12th August 2025).

Kalamartzis, I., Menexes, G., & Dordas, C. (2023). Effect of low water availability on seed yield and seed quality of basil (*Ocimum basilicum*). *Plants, 12*(5). https://doi.org/10.3390/plants12051094 (accessed on 12th August 2025).

Kalauni, D., & Joshi, A. (2018). Status of medicinal and aromatic plants (MAPs) and socio-economic influence in Nepalese livelihood—A review research. *Acta Scientific Agriculture, 2*(9).

Kalita, D., Jain, S., Srivastava, B., & Goud, V. V. (2021). Sono-hydro priming process (ultrasound modulated hydration): Modelling hydration kinetics during paddy germination. *Ultrasonics Sonochemistry, 70*. https://doi.org/10.1016/j.ultsonch.2020.105321 (accessed on 12th August 2025).

Kavusi, E., Shahi Khalaf Ansar, B., Dehghanian, Z., Asgari Lajayer, B., Nobaharan, K., Ma, Y., & Glick, B. R. (2023). Delivery of beneficial microbes via seed coating for medicinal and aromatic plant production: A critical review. *Journal of Plant Growth Regulation, 42*(2). https://doi.org/10.1007/s00344-022-10597-2 (accessed on 12th August 2025).

Khan, M. R. (2013). The effect of pretreatments on seed germination of the *Cassia auriculata* Linn. and *Cassia tora* Linn. *Advances in Plant Sciences, 26*(1).

Lakmali, K. N. D., & Seran, T. H. (2022). Impact of seed priming with king coconut water on growth and yield of okra (*Abelmoschus esculentus* L.). *Journal of Food and*

Agriculture, 15(2). https://doi.org/10.4038/jfa.v15i2.5271 (accessed on 12th August 2025).

Li, J., Bai, Y., Xie, Y., & Gao, J. (2022). Ultrastructure change and transcriptome analysis of GA_3 treatment on seed germination of moso bamboo (*Phyllostachys edulis*). *Plant Signaling & Behavior, 17*(1). https://doi.org/10.1080/15592324.2022.2091305 (accessed on 12th August 2025).

Li, J., Yin, L. Y., Jongsma, M. A., & Wang, C. Y. (2011). Effects of light, hydropriming, and abiotic stress on seed germination, and shoot and root growth of pyrethrum (*Tanacetum cinerariifolium*). *Industrial Crops and Products, 34*(3). https://doi.org/10.1016/j.indcrop.2011.05.012 (accessed on 12th August 2025).

Marotti, I., Whittaker, A., Bağdat, R. B., Akin, P. A., Ergün, N., & Dinelli, G. (2023). Intercropping perennial fruit trees and annual field crops with aromatic and medicinal plants (MAPs) in the Mediterranean Basin. *Sustainability (Switzerland), 15*(15). https://doi.org/10.3390/su151512054 (accessed on 12th August 2025).

Muscalu, A., Tudora, C., Sorică, C., Mîrzan, O., Naie, M., & Barcanu-Tudor, E. (2022). Mechanised harvesting of inflorescences of medicinal and aromatic plant species cultivated on a small scale. *INMATEH–Agricultural Engineering, 67*(2). https://doi.org/10.35633/inmateh-67-42 (accessed on 12th August 2025).

Nabaei, M., & Amooaghaie, R. (2019). Interactive effect of melatonin and sodium nitroprusside on seed germination and seedling growth of *Catharanthus roseus* under cadmium stress. *Russian Journal of Plant Physiology, 66*(1). https://doi.org/10.1134/S1021443719010126 (accessed on 12th August 2025).

Nagaraju, K., Medar, V. S., & K, A. (2018). Seed quality enhancement techniques in medicinal and aromatic crops. *Journal of Pharmacognosy and Phytochemistry, 3.*

Nawrot-Chorabik, K., Osmenda, M., Słowiński, K., Latowski, D., Tabor, S., & Woodward, S. (2021). Stratification, scarification, and application of phytohormones promote dormancy breaking and germination of pelleted Scots pine (*Pinus sylvestris* L.) seeds. *Forests, 12*(5). https://doi.org/10.3390/f12050621 (accessed on 12th August 2025).

Nikolova, I. M. (2016). Effects of organic products with insecticidal action on key insect pests in alfalfa seed production. *Emirates Journal of Food and Agriculture, 28*(9). https://doi.org/10.9755/ejfa.2016-04-381 (accessed on 12th August 2025).

Novais, C., Pereira, C., Molina, A. K., Liberal, Â., Dias, M. I., Añibarro-Ortega, M., Alves, M. J., Calhelha, R. C., Ferreira, I. C. F. R., & Barros, L. (2021). Bioactive and nutritional potential of medicinal and aromatic plant (MAP) seasoning mixtures. *Molecules, 26*(6). https://doi.org/10.3390/molecules26061587 (accessed on 12th August 2025).

Odoi, J. B., Mugeni, D., Kiiza, R., Apolot, B., & Gwali, S. (2021). Effect of soaking treatment on germination of hard-coated tropical forest tree seeds. *Uganda Journal of Agricultural Sciences, 19*(2). https://doi.org/10.4314/ujas.v19i2.1 (accessed on 12th August 2025).

Ozturk, M., & Hakeem, K. R. (2018). *Plant and Human Health, Volume 1: Ethnobotany and Physiology*. Springer International Publishing.

Pandey, H., Nandi, S. K., Nadeem, M., & Palni, L. M. S. (2000). Chemical stimulation of seed germination in *Aconitum heterophyllum* Wall. and *A. balfourii* Stapf.: Important Himalayan species of medicinal value. *Seed Science and Technology, 28*(1).

Parveen, Kumar, A., Barthwal, S., & Kant, R. (2020). Identification of seed sources of *Asparagus racemosus* Willd. for higher productivity. *Acta Horticulturae, 1301*. https://doi.org/10.17660/ActaHortic.2020.1301.12 (accessed on 12th August 2025).

Ph D Scholar, A. K., & Medar, V. S. (2018). National conference on "Conservation, cultivation and utilization of medicinal and aromatic plants" seed quality enhancement techniques in medicinal and aromatic crops. *Journal of Pharmacognosy and Phytochemistry, 3*.

Pinzón-Torres, J. A., Porras, N. J. C., Garcia, D. C. D., Morales, J. R. M., & Stashenko, E. (2014). Green biomass production and quality of essential oils of palmarosa (*Cymbopogon martini*) with application of synthetic fertilizers and organic fertilizers. *Acta Agronomica, 63*(4). https://doi.org/10.15446/acag.v63n4.42840 (accessed on 12th August 2025).

Prasuna, V. S., Reddy, C. N., Rajagopal, M., Purna, G. S., & Srinivas, B. (2022). *In vitro* flowering and production of seeds in *Oxalis corniculata* L.—An important medicinal plant. *Plant Tissue Culture and Biotechnology, 32*(2). https://doi.org/10.3329/ptcb.v32i2.63553 (accessed on 12th August 2025).

Qamar, R., Khan, S., Safdar, M. E., ur-Rehman, A., Rehman, A., Javeed, H. M. R., Nadeem, M. A., Al-Yahyai, R., & Alkahtani, J. (2022). Seed priming with growth regulators modulates production, physiology, and antioxidant defense of Indian squash (*Praecitrullus fistulosus*) under semi-arid conditions. *PLoS ONE, 17*(4). https://doi.org/10.1371/journal.pone.0265694 (accessed on 12th August 2025).

Rahmawaty, Samosir, J. B., Batubara, R., & Rauf, A. (2019). Diversity and distribution of medicinal plants in the Universitas Sumatera Utara arboretum of Deli Serdang, North Sumatra, Indonesia. *Biodiversitas, 20*(5). https://doi.org/10.13057/biodiv/d200539 (accessed on 12th August 2025).

Rai, P. P. (2012). Status of medicinal and aromatic plants in Papua New Guinea: A review. *Pacific Journal of Medical Sciences, 9*(2).

Rana, S. K., Rana, H. K., Ranjitkar, S., Ghimire, S. K., Gurmachhan, C. M., O'Neill, A. R., & Sun, H. (2020). Climate-change threats to distribution, habitats, sustainability, and conservation of highly traded medicinal and aromatic plants in Nepal. *Ecological Indicators, 115*. https://doi.org/10.1016/j.ecolind.2020.106435 (accessed on 12th August 2025).

Riaz, U., Iqbal, S., Sohail, M. I., Samreen, T., Ashraf, M., Akmal, F., Siddiqui, A., Ahmad, I., Naveed, M., Khan, N. I., & Akhter, R. M. (2021). A comprehensive review on emerging importance and economical potential of medicinal and aromatic plants (MAPs) in current scenario. *Pakistan Journal of Agricultural Research, 34*(2). https://doi.org/10.17582/journal.pjar/2021/34.2.381.392 (accessed on 12th August 2025).

Rout, S., Khare, N., Patra, S. S., Beura, S., & Nayak, S. (2017). Effect of seed pre-treatment with different concentrations of gibberellic acid (GA_3) on seed germination and seedling growth of *Cassia fistula* L. *Journal of Medicinal Plants Studies, 5*(6).

Schippmann, U., Cunningham, A. B., Leaman, D. J., & Walter, S. (2005). Impact of cultivation and collection on the conservation of medicinal plants: Global trends and issues. *Acta Horticulture, 676*. https://doi.org/10.17660/actahortic.2005.676.3 (accessed on 12th August 2025).

Shukla, S., Singh, R., Bhutani, R., Tripathi, A., & Shukla, S. K. (2022). Restoration and conservation of plant genetic resources via molecular techniques: An important measure for sustainable agriculture. In *Agro-Biodiversity and Agri-Ecosystem Management*. https://doi.org/10.1007/978-981-19-0928-3_12 (accessed on 12th August 2025).

Sinjushin, A. (2015). Cytogenomics and mutagenomics in plant functional biology and breeding. In *PlantOmics: The Omics of Plant Science*. https://doi.org/10.1007/978-81-322-2172-2_5 (accessed on 12th August 2025).

Šoch, J., Šonka, J., & Ponert, J. (2023). Acid scarification as a potent treatment for an *in vitro* germination of mature endozoochorous *Vanilla planifolia* seeds. *Botanical Studies, 64*(1). https://doi.org/10.1186/s40529-023-00374-z (accessed on 12th August 2025).

Solomou, A. D., Skoufogianni, E., Mylonas, C., Germani, R., & Danalatos, N. G. (2019). Cultivation and utilization of "Greek mountain tea" (*Sideritis* spp.): Current knowledge and future challenges. *Asian Journal of Agriculture and Biology, 7*(2).

Somrug, K., Hemwong, S., & Jorjong, S. (2021). Effect of seed dormancy breaking treatment and storability on seed quality of indigo (*Indigofera tinctoria* L.). *Asian Journal of Plant Sciences, 20*(3). https://doi.org/10.3923/ajps.2021.389.395 (accessed on 12th August 2025).

Spinozzi, E., Maggi, F., Bonacucina, G., Pavela, R., Boukouvala, M. C., Kavallieratos, N. G., Canale, A., Romano, D., Desneux, N., Wilke, A. B. B., Beier, J. C., & Benelli, G. (2021). Apiaceae essential oils and their constituents as insecticides against mosquitoes—A review. *Industrial Crops and Products, 171*. https://doi.org/10.1016/j.indcrop.2021.113892 (accessed on 12th August 2025).

Sucholas, J., Ukhanova, M., Greinwald, A., & Luick, R. (2021). Wild collection of medicinal and aromatic plants (MAPs) for commercial purposes in Poland: A system's analysis. *Herba Polonica, 67*(3). https://doi.org/10.2478/hepo-2021-0014 (accessed on 12th August 2025).

Suryawanshi, J. B., Bhingarde, M. T., Waghmode, S. U., & Deshmukh, D. V. (2013). Effect of pre-soaking seed treatments on seed quality in ashwagandha (*Withania somnifera* Dunal.). *Ecology, Environment and Conservation, 19*(3).

Thongtip, A., Mosaleeyanon, K., Korinsak, S., Toojinda, T., Darwell, C. T., Chutimanukul, P., & Chutimanukul, P. (2022). Promotion of seed germination and early plant growth by KNO_3 and light spectra in *Ocimum tenuiflorum* using a plant factory. *Scientific Reports, 12*(1). https://doi.org/10.1038/s41598-022-11001-5 (accessed on 12th August 2025).

Verma, S. S., Verma, P. K., & Verma, U. (2014). Standardization of seed germination technique in khurasani-ajavayan (*Hyoscyamus niger*) seeds. *Indian Journal of Agricultural Sciences, 84*(11). https://doi.org/10.56093/ijas.v84i11.44643 (accessed on 12th August 2025).

Zuffo, A. M., Steiner, F., Zuffo Júnior, J. M., Bush, A., da Silva, J. R. M., Limede, A. C., & Oliveira, C. E. da S. (2017). Non-chemical methods to break seed dormancy of canafistula [*Peltophorum dubium* (Sprengel) Taubert (Fabaceae)]. *Australian Journal of Crop Science, 11*(12). https://doi.org/10.21475/ajcs.17.11.12.pne730 (accessed on 12th August 2025).

Index

D

E

N

O

P

Q

R

S

Z